M000201495

"*The Age of Sustainability* is an outstanding book. Int into the author's own years of personal experienc involvement. Swilling argues deliberative democracy won't do. We will need to tap into thymotic rage aligning the feminine principle of care with energy to really change our deeply unsustainable practices. A must-read."

Professor Maarten Hajer, Urban Futures Studio, Utrecht University, The Netherlands

"Mark Swilling asks deep questions of all of us: what does it mean to be human in the Anthropocene? What are the contours of a just 21st-century transition that can eradicate poverty and diminish inequality, without destroying natural systems across the world? How can we muster the positive rage necessary to fuel a passion for change and collective action to make this possible? He uses a mix of deep personal narrative, the practice of radical incrementalism in post-transition South Africa and meta-theoretical explorations to present responses to these existential challenges. A provocative read, especially for leaders from the Global South, who struggle with going beyond conventional binaries."

Aromar Revi, Indian Institute of Housing Studies and
United Nations Sustainable Development Solutions Network

"It is hard to imagine the fate of humanity unless our actions match our sustainability discourses. Just talking of sustainability in an unfair world of imbalanced wealth will lead us nowhere. Besides the time wasted while unfairly accumulated wealth grows, we would remain in the same world. This book presents a dimension that we have to apply and follow if humanity has to co-exist and survive on this one earth of ours."

Gete Zeleke, Water and Land Resource Centre, Addis Ababa University, Ethiopia

"*The Age of Sustainability* sketches a way out of the ecological apocalypse, both ambitious and realistic. Ambitious because it is a question of completely reshaping our institutions. Realistic because the author has no taste for revolutionary eschatologies but prefers precise roadmaps, rooted in the experience of grassroots communities whose lifestyle foreshadows our next world. At the crossroads of political science, economics, sociology and ethics, his book will be an indispensable reference for all those who want to build a just world in common."

Gaël Giraud, Agence Française de Développement and
Ecole Nationale des Ponts et Chaussées, France

"Global changes in the health of our climate, biodiversity and resources now threaten the very existence of life on Earth. These cannot be reversed by the generations of people and habits that caused them. It is now the next generation that must, for its own sake and for all of civilization, craft a better, more livable future. For this, all of us need a deeper and clearer understanding of the issues and possible solutions. Mark Swilling's book has done a wonderful job of providing this. It is now for all of us to do our job to bring about the transformation needed."

Ashok Khosla, Development Alternatives, India, and former President of IUCN and
former Co-Chair of the International Resource Panel

"This new book *The Age of Sustainability* by Professor Mark Swilling is a welcome addition to increasing our scientific knowledge on Just Transitions. The book provides useful insights in understanding the dynamics of transitions at a global scale. Professor Swilling makes compelling arguments to demonstrate that understanding sustainability and achieving the sustainable development goals require two necessary conditions: a theory of change and the passion for such changes."

Elias T. Ayuk, Former Director, United Nations
University Institute for Natural Resources in Africa, Accra, Ghana

"Swilling draws you into a breath-taking journey. Just transitions are to happen: eradicating poverty in our lifetime without destroying the planet's natural systems. The book provides all it takes to make this conceivable: a solid metatheoretical footing, a plausible theory of change that escapes the dualism of 'state or market' by radical incrementalism focused on developing the commons, and a driver: rage that overcomes fear. A rage that aligns with the feminine principle of care rather than the male principle of control ... Here speaks the author as a person who made change happen, collaborates with others and learns from experiences that supply the narratives in this book."

Marina Fischer-Kowalski, Institute of Social Ecology, University of Klagenfurt,
Austria and Chair of the Scientific Advisory Board of the Potsdam Institute on Climate
Impact Research, Germany, and Vice President of the European Society of Ecological Economics

"This is a brave and ambitious book from a pioneering activist scholar. Mark Swilling offers a persuasive account of our contradictory times—dangerous and hopeful—while pointing to clear steps to become part of a just transition to a more sustainable civilisation. This masterstroke deserves the widest possible audience within the academy and far beyond. This is what praxis-driven, decolonial and free thought sounds like in its purest expression."

Edgar Pieterse, African Centre for Cities, University of Cape Town, South Africa

"Mark Swilling takes us back to an era of optimism and reassuringly posits that the advent of the SDGs is the beginning of a positive thinking era in his tour de force book, *The Age of Sustainability: Just Transitions in a Complex World*. Swilling brings to the fore that niggling thought that has been at the edge of your conscience all along but you could not put it into words. That the efforts to create a better world and a better future through 'fixes', whether they be policies, regulations, technology, better government, efficient use of resources etc., miss the point. That without understanding global (increasingly complex and diverse) dynamics within which power, entrenched inequality and lack of social justice looms large and are commonalities, and that without understanding the nature and catalytic impacts of transitions for good or otherwise, we will not get a better world or future for all, and continue at a 'frenetic snails' pace'."

Tanya Abrahamse, former Chief Executive Officer of the South African
National Biodiversity Institute

"In this transformational moment, we witness numerous futile attempts that try to resolve the confluence of challenges through reductionist over-simplification. There are also efforts that are based on inter-disciplinary interventions which are at best incremental or at worst lead to more detailed complexities. This book provides a sound basis for understanding the dynamic complexity of the challenges we are facing through a transdisciplinary lens."

Desta Mebratu, Stellenbosch University, South Africa, and former Deputy-Director of the Africa
Regional Office of the United Nations Environment Programme

"A timely and magnificent contribution on the dynamics of transition. A powerful analysis of sustainable pathways towards just transitions, a subject neglected by economic orthodoxy for decades. Swilling's book makes a strong case in setting out the contested futures envisaged by the Sustainable Development Goals and offers a viable collective future for us all."

A. Erinç Yeldan, Bilkent University, Turkey

"In this magisterial book, Professor Swilling builds on his earlier work on sustainable transitions and strives to address the question of how we might galvanize the 'passion' in facilitating positive change for the planet and its inhabitants. In this quest for action, that brings forth our best instincts of head and heart, he presents rigorous empirical analysis conducted throughout his academic career and as an advocate for social and environmental justice. He also presents a range of case analysis and pedagogic concepts for just transitions from across the world, but particularly from his home continent of Africa. This book is essential reading for anyone who wants a nuanced view of the global sustainable development agenda and how to maximize its impact across all strata of society."

Saleem H. Ali, University of Delaware, USA

"New thinking beyond all variation of modernity is needed to govern the Anthropocene meeting the challenges of the 21st century. Mark Swilling taps into contemporary theories like system thinking, integral theory and theories of resonance to reflect the African experience, pre-colonial, colonial, and post-colonial, to contribute to, if not to lead, the relevant global discourses of sustainability and resilience looking for the more than needed new narrative."

Louis Klein, European School of Governance, Germany

THE AGE OF SUSTAINABILITY

With transitions to more sustainable ways of living already underway, this book examines how we understand the underlying dynamics of the transitions that are unfolding. Without this understanding, we enter the future in a state of informed bewilderment.

Every day we are bombarded by reports about ecosystem breakdown, social conflict, economic stagnation and a crisis of identity. There is mounting evidence that deeper transitions are underway that suggest we may be entering another period of great transformation equal in significance to the agricultural revolution some 13,000 years ago or the Industrial Revolution 250 years ago. This book helps readers make sense of our global crisis and the dynamics of transition that could result in a shift from the industrial epoch that we live in now to a more sustainable and equitable age. The global renewable energy transition that is already underway holds the key to the wider just transition. However, the evolutionary potential of the present also manifests in the mushrooming of ecocultures, new urban visions, sustainability-oriented developmental states and new ways of learning and researching.

Shedding light on the highly complex challenge of a sustainable and just transition, this book is essential reading for anyone concerned with establishing a more sustainable and equitable world. Ultimately, this is a book about hope but without easy answers.

Mark Swilling is Distinguished Professor of Sustainable Development in the School of Public Leadership, University of Stellenbosch, South Africa, where he is the Co-Director of the Stellenbosch Centre for Complex Systems in Transition.

Routledge Studies in Sustainable Development

This series uniquely brings together original and cutting-edge research on sustainable development. The books in this series tackle difficult and important issues in sustainable development including: values and ethics; sustainability in higher education; climate compatible development; resilience; capitalism and de-growth; sustainable urban development; gender and participation; and well-being.

Drawing on a wide range of disciplines, the series promotes interdisciplinary research for an international readership. The series was recommended in the Guardian's suggested reads on development and the environment.

Metagovernance for Sustainability
A Framework for Implementing the Sustainable Development Goals
Louis Meuleman

Survival: One Health, One Planet, One Future
George R. Lueddeke

Poverty and Climate Change
Restoring a Global Biogeochemical Equilibrium
Fitzroy B. Beckford

Achieving the Sustainable Development Goals
Global Governance Challenges
Edited by Simon Dalby, Susan Horton and Rianne Mahon, with Diana Thomaz

The Age of Sustainability
Just Transitions in a Complex World
Mark Swilling

For a full list of titles in this series, please visit www.routledge.com

THE AGE OF SUSTAINABILITY

Just Transitions in a Complex World

Mark Swilling

LONDON AND NEW YORK

from Routledge

First published 2020
by Routledge
2 Park Square, Milton Park, Abingdon, Oxon OX14 4RN

and by Routledge
52 Vanderbilt Avenue, New York, NY 10017

Routledge is an imprint of the Taylor & Francis Group, an informa business

British Library Cataloguing-in-Publication Data
A catalogue record for this book is available from the British Library

Library of Congress Cataloging-in-Publication Data
A catalog record for this book has been requested

ISBN: 978-0-367-17815-4 (hbk)
ISBN: 978-0-367-17816-1 (pbk)
ISBN: 978-0-429-05782-3 (ebk)

Typeset in Bembo
By Apex CoVantage, LLC

*To my sons Michael and Ranen, who have already
stepped into the future*

CONTENTS

ACKNOWLEDGEMENTS

A vast array of amazing people have contributed to this book in various ways over the years, in particular the students who come to study at the Sustainability Institute. However, specific mention must be made of my colleagues at Stellenbosch University located in the Sustainability Institute and the Centre for Complex Systems in Transition. These include Jannie Hofmeyr, Rika Preiser, John van Breda, Josephine Musango, Jess Schulschenk, Scott Drimie, Beatrix Steenkamp, Holle Wlokas, Monique Beukes and Cornelia Jacobs. A special thanks to two people who read early drafts, listened patiently to my verbal meanderings over the years and kept me going: Amanda Gcanga and Megan Davies. Thank you to Tom Graedel and all his colleagues at the Yale School of Forestry and Environmental Studies for hosting me at Yale in 2018 and to Amanda Mondesir for her support and insights during my time in the United States. I must also acknowledge a core group of my peers who have helped me shape my ideas over the years: Edgar Pieterse, Maarten Hajer, Eve Annecke, Desta Mebratu, John Benington and John van Breda. Others whom I have engaged with and who influenced my thinking include Gael Giraud, Johan Hattingh, Marina Fischer-Kowalski, Bagele Chilisa, Kevin Urama, Andrew Boraine, Thuli Madonsela, John Spiropolous, Mmatshilo Motsei, Mamphela Ramphele, Gary Jacobs, Adriana Allen, Serge Salat, Mila Popovich, Aromar Revi, Janez Patocnik, Ernst von Weizsacher, Ashok Khosla, Simon Marvin, Haroon Bhorat, Phumlani Nkontwana, Nontsikelelo Mngqibisa, the late Paul Cilliers and the many participants from African Universities in our Transdisciplinary Summer and Winter Schools over the years. The consistent collegial, institutional and moral support provided by the School of Public Leadership of Stellenbosch University and the Sustainability Institute, in general, is also acknowledged. Thanks to the Board of the Development Bank of Southern Africa for preventing me from detaching from the real world. Direct and indirect financial support for this project was received from the National Research Foundation and the Yale Institute of Biospheric Studies (which sponsored my eight-month sabbatical at Yale University as the Edward P. Bass Environmental Scholar for 2018) and the Open Society Foundation. The three reviewers are warmly acknowledged for providing insights that much improved the final result. Finally, without the personal inspirations catalysed by Ray Swilling and my students, none of this would have come to fruition.

PART I
Points of departure

1

INTRODUCTION

Change in the age of sustainability

Transforming our world

The adoption of the Sustainable Development Goals (SDGs) in 2015 by the United Nations (UN) was a turning point in our understanding of the challenges we face. It marks the start of the sustainability age – a time of crisis and transition when contested interpretations of sustainability provide the coordinates for future imaginaries. This does not refer to an age when sustainable modes of existence have been achieved in practice at the national and global levels. Building sustainable national and global systems may be the ultimate outcome if certain conditions materialize, in particular with respect to energy. For now, the SDGs – with all their imperfections – provide a shared language for engaging contested futures shaped by the language of the SDGs. This, read together with the real-world conditions discussed in this book that make the prolongation of industrial modernity in its current form unviable, makes it possible to argue that the sustainability age has begun. To be sure, the significance of the SDGs is questioned in the literature on post-developmentalism and degrowth (Escobar, 2015). Nevertheless, the Preamble to the document approved by the UN does conclude with the following profound words:

> If we realize our ambitions across the full extent of the Agenda, the lives of all will be profoundly improved and our world will be transformed for the better.

What a "transformed" world looks like is slightly elaborated earlier in the Preamble by the statement that "[w]e are resolved to free the human race from the tyranny of poverty and want to heal and secure our planet". Arguably, by adopting this commitment, the global community of nations has committed to eradicating poverty in our lifetime without destroying the planet's natural systems. It is this commitment that animates the sustainability age. Realizing this commitment is what this book will refer to as a *just transition*. But as Bruno Latour observed, such a commitment needs to be underpinned by a deep passion for change:

> [W]e are trapped in a dual excess: we have an excessive fascination for the inertia of the existing socio-technical systems and an excessive fascination for the total, global and radical nature of the changes that need to be made. The result is a frenetic snails' pace. An apocalypse in slow motion . . . Changing trajectories means more than a mere apocalypse, and is more demanding than a mere revolution.
>
> *(Latour, 2010)*

In practice, an unjust transition is highly likely: this will happen if the planetary systems we depend on are saved on terms that serve the elites, while poverty is allowed to persist. After reviewing the way in which 'securocratic' post–Cold War thinking and doomsday predictions of climate science have converged within military intelligence circles in the global North, Christian Parenti concluded that a just "political adaptation" is definitely not what is being contemplated by these strategists:

> [T]he military-led strategy for dealing with climate change suggests another type of political adaptation is already under way, which might be called the "politics of the armed lifeboat": responding to climate change by arming, excluding, forgetting, repressing, policing and killing. One can imagine a green authoritarianism emerging in rich countries, while the climate crisis pushes the Third World into chaos. . . . The Pentagon and its European allies are actively planning a militarized adaption, which emphasizes the long-term, open-ended containment of failed or failing states – a counterinsurgency forever.
>
> *(Parenti, 2016:35–36)*

This outcome of an unjust transition – the 'armed lifeboat' – will emerge from one particular conception of sustainability, namely a conception that is focused on technocratic solutions aimed at fixing the planetary systems to retain (what will become tightly micromanaged/monitored via the new 5G infrastructures) ecosystems without in any way reducing the powers and wealth of the rich and super-rich. As will be argued in Chapter 3, this would amount to a (militarized) 'deep' transition but definitely not a 'just' transition. A 'just transition' draws on traditions that are now particularly alive in the global South, namely commitments to social justice premised on the assumption that without social justice the planetary systems can never be restored to support the web of life.

Whereas the adoption of the SDGs can be interpreted as (a highly contested) recognition at the global policy level that fundamental change of some sort is required, a similar trend is evident in the academy, where interest in 'transition studies' has emerged as a major and productive new field of interdisciplinary research. As Escobar argues,

> The forceful emergence of transition discourses in multiple sites of academic and activist life over the past decade is one of the most anticipatory signs of our times. This emergence is a reflection of both the steady worsening of planetary ecological, social, and cultural conditions and of the inability of established policy and knowledge institutions to imagine ways out of such crises.
>
> *(Escobar, 2015:452)*

This rapidly expanding body of work generally referred to as 'transition studies' includes two quite different schools of thought. As discussed in detail in Chapter 3, there are those interested in the different dimensions of sustainability transitions (STs), including socio-technical transitions (Grin, Rotmans, Schot, Geels, and Loorbach, 2010), socio-metabolic transitions (Haberl, Fischer-Kowalski, Krausmann, Martinez-Alier, and Winiwarter, 2011), techno-industrial transition (Perez, 2016), long-term development cycles (Gore, 2010) and now most recently 'deep transitions' (Schot and Kanger, 2018). In general, these authors are interested in the dynamics of structural change in light of socio-technical advances and ecological limits. Most do not necessarily subscribe to a post-capitalist alternative, nor do they pay much attention to the commons – the shared resources needed for all species to flower and prosper (Bauwens and Ramos, 2018).

The second group of authors "posit[s] a profound cultural, economic, and political transformation of dominant institutions and practices" (Escobar, 2015:454). Following Escobar's excellent overview, this group envisages a post-development, non-neoliberal, post/non-capitalist, biocentric and post-extractivist future and includes those who write from the global North about the commons (Bollier and Helfich, 1978), transition towns (Hopkins, 2018), degrowth (D'Alisa, Demaria and Kallis, 2015), the "Great Transition"[1], the "Great Turning" (Macy and Brown, 1998), the "Great Work" (Berry, 1999), "Enlivenment" (Weber, 2013), transition from an Age of Separation to an Age of Reunion (Eisenstein, 2013) and the transitions from "Enlightenment to Sustainability" (Fry, 2012). In the global South, the reference points for this perspective are post-developmentalism (Escobar, 1995), "crisis of civilisation" (Ahmed, 2017), the Latin American narrative of *Buen Vivir* and the rights of nature, the commons and communal initiatives (see certain case studies in Bollier and Helfich, 1978) and transitions to post-extractivism (Lang and Mokrani, 2013).

The problem with the second group of transition studies referred to here by Escobar as the "post-development"/"post-capitalist" perspective is that it may be strong when it comes to critique and alternative visions, but it is relatively weak when it comes to understanding *how* fundamental change will actually happen in practice. Who, in other words, can best shape the directionality of the 'great transformation'?

The sustainability transitions literature and the post-development literature referred to in the previous two paragraphs need not be regarded as antithetical. They frame what this book is about – a search for a way of understanding the global dynamics of transition and what this means for actions aimed at bringing about a transition that envisages radical alternatives to the dominant configurations of power and mainstream practices.[2]

I will argue in this book that we need a better understanding of a just transition. A just transition is a process of increasingly radical incremental changes that accumulate over time in the actually emergent transformed world envisaged by the SDGs and sustainability. The outcome is a state of well-being (Fioramonti, 2015) founded on greater environmental sustainability and social justice (including the eradication of poverty). These changes arise from a vast multiplicity of struggles, each with their own context-specific temporal and spatial dimensions.

What really matters is the evolutionary potential of the present (a phrase borrowed from Snowden, 2015) and the incrementalist actions that are required to instigate the changes that are needed. The question then becomes: what is emerging now that is suggestive of the kind of future embodied in the vision of a transformed world? And what are the conditions that make these changes possible, and cumulatively do they add up to more than the sum of the parts? In short, what is our understanding of change?

Drawing on transdisciplinary studies, I will persistently explore throughout this book the kind of knowledge that equips us to act in an increasingly complex world. Transdisciplinary research makes a useful distinction between three types of knowledge (Regeer and Bunders, 2009): systems knowledge, which is knowledge about current social-ecological systems in order to arrive at conclusions about whether they need to be changed or not; target knowledge, which is knowledge about desired endstates or futures; and transformation knowledge, which is knowledge about change, that is, how to get from where we are now to where we want to be. This book has been written as a contribution to the kind of transformation knowledge that I believe is required in the world today. Obviously, this cannot be done without systems knowledge, which will of course be discussed in all the chapters. Target knowledge is also present in many chapters but not in the form of prescriptive policy solutions. Systems and target knowledge without transformation knowledge is what causes the consternation expressed in Latour's quote mentioned earlier.

Just as a revolution in the classical sense (as in storming the Bastille or seizing the Winter Palace[3]) is unlikely, so too is it unlikely that the transition will be triggered by a sudden cataclysmic apocalypse. We need a theory of change that is less obsessed with uni-centric structural change or faith in sudden ecologically driven system collapses on the other (or some mix of both). Instead, such a theory should be more focused on the efficacy of a multiplicity of incremental changes that emerge from socio-political struggle and provide glimpses of certain kinds of desired futures, whether or not there are an increasing number of catastrophes along the way. We need to avoid both faith in techno-fixes and the faithlessness of doom and gloom.

As will be argued in the various chapters of this book, this middle way – often referred to as post-capitalism – lies between market fundamentalism and statism (both the socialist and social democratic varieties) (see Chapter 2). As discussed further in Chapter 2, the most coherent perspective within the broader post-capitalist spectrum of thinking is associated with the burgeoning literature on the 'commons' and the practices of 'commoning'. The commons literature reveals the practical governance arrangements that are neither statist nor market oriented, without excluding aspects of both. Commoning – the practices involved in building the commons – gives radical incrementalism a specific normative direction (see the end of Chapter 2 for further discussion of post-capitalism and the commons).

When read together, the strategy for building the post-capitalist commons is what will be referred to in this book as 'radical incrementalism'. In many ways, this book has been written in defence of radical incrementalism (Chapters 5 and 6) during a time of deep transition (Chapter 4) – an alternative to a belief in revolutionary ruptures, systemic collapses or over-optimistic modernizing techno-fixes to 'green' the status quo.

Notwithstanding the dreams of many revolutionaries who have called for world revolution time and again, the classical conception of revolution (from approximately the eighteenth through to the twentieth centuries) is inescapably constrained within the boundaries of the nation-state and is focused on the seizure of state power (i.e. the 'nation-state'). From the French and Russian revolutions against corrupt aristocracies to revolutions in the Americas waged against imperial control during the 1700s and 1800s by occupying settler communities (who, of course, oppressed local indigenous populations), the anti-colonial revolutions in Africa in the 1950s/1960s, the 'velvet revolutions' in Eastern Europe in the late 1980s/early 1990s and the 'second wave' of African revolutions against tyrannical regimes (e.g. Ethiopia) during the 1990s (including, of course, the democratic transition in South Africa in 1994), the focus has been the seizure of state power within the boundaries of a nation-state despite the mounting evidence that the nation-state may have become a highly problematic reference point for progressive politics (see editorial to a special edition of *Territory, Politics Governance* by Agnew, 2017). The assumption was that state power would provide revolutionary elites with the means to transform society more or less in a top-down way. While this may be partially true in some instances, my experience with the South African democratic project is this: on the morrow of the revolution the challenges remain – how to transform institutions, create working experiments and demonstrate that alternatives are possible without succumbing to the temptations of certainty, whether of the statist or the market fundamentalist varieties. After all, it's the quest for certainty that is the greatest threat to democracy.

Most of those who worry about the state of the world's environment share the broad scientific consensus that has evolved over the past three decades (Barnosky, Brown, Daily, Dirzo, Ehrlich, et al., 2014). In essence, this book fully accepts the opening sentence of the authoritative summary of this consensus:

> Earth is rapidly approaching a tipping point. . . . Human impacts are causing alarming levels of harm to our planet. As scientists who study the interaction

of people with the rest of the biosphere using a wide range of approaches, we agree that the evidence that humans are damaging their ecological life support systems is overwhelming.

(Barnosky, Brown, Daily, Dirzo,
Ehrlich, et al., 2014:79)

Barnosky et al. identify five primary drivers of global change that all interact with each other: climate disruption, extinction of biodiversity, the wholesale loss of vast ecosystems, pollution and ever-increasing consumption of resources. They conclude as follows:

The vast majority of scientists who study the interactions between people and the rest of the biosphere agree on a key conclusion: that the five interconnected dangerous trends listed above are having detrimental effects and, if continued, the already-apparent negative impacts on human quality of life will become much worse within a few decades.

(Barnosky, Brown, Daily, Dirzo,
Ehrlich, et al., 2014:81)

This book also accepts the widely held scientific view that the "great acceleration" since the 1950s has been the primary driver of the processes that have resulted in this approaching tipping point (Steffen, Broadgate, Deutsch, Gaffney and Ludwig, 2015). This publication updates the well-known hockey stick graphs that reveal how rapidly consumption has increased since the 1950s with respect to, for example, population, GDP, foreign direct investment, energy use, fertilizer consumption, water use and transportation. These increases in consumption more or less track the upward trends in environmental indicators with respect to carbon dioxide, methane, ocean acidification, nitrogen, domestic land use and so forth.

If the tipping point is approaching rapidly and the unabated great acceleration remains a key driver, the question this book asks is simple: what is our theory of change? This is the question posed by Latour in the quotation above.

There is also a widely held scientific view that we now live in the "Anthropocene" – an era when humans – and by implication, *all* humans in equal measure – have become a geophysical force of nature (Crutzen, 2002). Many who share this somewhat problematic view (for the Anthropocene debate, see Malm and Hornborg, 2014; Maslin and Lewis, 2015) seem to think that a cataclysmic catastrophe is inevitable, with a climate-related catastrophe probably the most likely catalyst of radical change. Inspired by writers like James Lovelock (Lovelock, 2006), this 'doom and gloom' brigade have largely given up on the potential for socially induced changes commensurate with the challenges we face. In other words, these people have given up on what Paul Erhlich called "conscious evolution" (Ehrlich, 2002): a broad process of socio-cultural evolution induced by the conscious construction of alternative narratives about the future of humankind on the planet. Their alternative is a grand ecological reset that could, in turn, wipe out more than half the global population. One can only wonder what would happen if they are right: will runaway climate change and its consequent destructive natural disasters for humans (in particular) usher in the post-Anthropocene – a time when humans lose control?

A similar pessimism pervades those who prefer to refer to the 'capitalocene' – an era when capital seems so arrogantly hegemonic and all powerful (Moore, 2016). No matter what happens, it is always, ultimately, about the reassertion of capitalist power. This gives rise to the oft-quoted phrase: "It is easier to imagine the end of the world than the end of capitalism".

The alternative to ecological catastrophe or an anti-capitalist revolution is what one could call 'radical incrementalism'. Most people who share the view that radical transformation is needed without assuming that a classical revolution is necessary (or that we should wait for catastrophe) tend to share an assumption that the best way to bring about change is via dialogue of various kinds (Hajer, Nilsson, Raworth, Bakker, Berkhout, et al., 2015). However, it seems to me that there is a certain incommensurability about – as Latour put it in the quote above – our "excessive fascination for the total, global and radical nature of the changes that need to be made" (Latour, 2010) and our obsession with dialogue – the latter seems so feeble compared to the great heroic field and street battles of the old-fashioned revolutions of the eighteenth, nineteenth and early twentieth centuries. Dialogue is so often presented as a kind of panacea – as long as we are having dialogues at global round tables, national consultative meetings, city-wide gatherings and town hall meetings and in our local projects, all will be well. There is now an entire global industry of dialogue facilitators employing thousands of people using a countless number of formalized 'futuring methodologies' and almost everything that needs to be implemented now must be preceded by dialogue.

It is this obsession with dialogue that has emerged in response to the shift from a structure-centred notion of 'government' to the relational notion of 'governance' over the past few decades (Jessop, 2016). If governance is about collaboration between diverse stakeholders, dialogue becomes a key capability, often requiring the skills of a trained facilitator. While there is nothing wrong with most kinds of dialogues per se, what seems to be lacking is a theory of change that can be used to assess the efficacy of any given set of dialogues as they ideally express themselves in action. For me, this is what radical incrementalism can offer – a way of thinking about actions that change things rather than dialogues that are held in the hope that somehow they will bring about change.

This introductory chapter should, ideally, be read together with Chapter 2. Chapter 2 provides a conceptually dense elaboration of the meta-theoretical synthesis that informs the meso-level analyses of various dimensions of the global crisis and potential solutions that are discussed from Chapter 3 onwards. However, readers unfamiliar with social theory can skip Chapter 2. Subsequent chapters are written in ways that do not presuppose an understanding of Chapter 2. What follows in this chapter after a personal account of my journey of discovery as an activist academic is a succinct summation of my axiological point of departure, followed by a summary overview of the argument elaborated across the chapters of the book, including a simplified description of the concepts in Chapter 2 that get implicitly deployed in the subsequent chapters.

Reflections of an enraged incrementalist: a personal journey

The arguments presented in the rest of this book are derived in part from who I am, my experiences and my identity. Thus, before proceeding to elaborate a conceptual framework that justifies the use of the notion of radical incrementalism and why I believe rage is so significant in today's discussion about change, it is necessary to relate the personal journey that brings me to this particular vantage point on the world we live in and why I think an alternative is already emerging. By doing so, I am declaring what is referred to in the social sciences as my 'positionality', while remaining faithful to my commitment to a balance between epistemology and ontology (explained further in Chapter 2).

For the past 17 years, I have devoted a large part of my energy to the building of South Africa's first intentional socially mixed ecologically designed community (see Swilling and Annecke, 2012: Chapter 10). I did this as part of my involvement in the wider process of democratization that led to South Africa's first democratic non-racial elections in 1994 (see my interview in Callinicos, 1992; Swilling, 1999) and the struggle against the tyrannical rule of President Jacob Zuma that ended in 2018 (see the foreword and prologue to Chipkin and Swilling, 2018). At the centre of the Lynedoch EcoVillage was the Sustainability Institute, which I co-founded in 1999 with my former wife, Eve Annecke. This experience has fundamentally reshaped my research and teaching because of the way it allowed me to connect a broader vision of the future (that inspired my involvement in the struggle for democracy) with what it means to build something in practice through experimentation and innovation. At the centre of this EcoVillage is the Sustainability Institute (SI)[4] – an extraordinary space for learning and activism. Edgar Pieterse, Naledi Mabeba, Roshieda Shabodien and Adrian Enthoven were the founding board members and Eve Annecke was the founding director.[5]

The Lynedoch EcoVillage has matured and is nearly 20 years old now.[6] Most of the core founding group are still involved in one way or another, and the community-based collaborative governance system has remained largely intact and viable. The socio-technical systems to sustainably manage water, sewage, solid waste and energy that were designed in the late 1990s have proven to be viable (except for the verti-cally integrated wetland [VIW] that was replaced in 2016 with a horizontal wetland which, to date, has not performed much better than the VIW). In 2017, ESKOM – South Africa's state-owned electricity utility – installed a state-of-the-art renew-able energy smart grid, which includes 1,500 watts of photovoltaic (PV) panels, an inverter, batteries and smart meter for every house at no cost to the EcoVillage. For ESKOM, the EcoVillage is an experimental laboratory in grid-tied sustainable and renewable energy smart grid systems. (They needed a socially mixed community to set up their lab and sadly could find only one.)

The trees and gardens have matured, and, as originally anticipated (following Christopher Alexander's 'pattern language'), it is the 'spaces-in-between' that nature makes possible rather than the buildings themselves that make it such a beautifully textured holding space (Alexander, 1977). It is not the product of the imagination of a single architect or designer, and as a result a wide range of aesthetic preferences are expressed across income groups; as a consequence, many different materials have been used. Quite often, though, the selection of materials is determined by what banks or government funding agencies will accept, and they have preconceived ideas about what these should be (Swilling, 2015). The original idea of having small sites with 80% coverage by buildings succeeded in creating a large amount of shared space. This is particularly important for families who moved there because they wanted children to feel safe and have large open spaces to roam.

Before reflecting on the dynamics of experimentation during the design and construction of the EcoVillage within the wider context of national democratization and development in post-1994 South Africa, more needs to be said about the story of the SI. The inspiration for the vision of the SI came from experiencing two places – the Schumacher College in Devon, UK, and the Goree Institute on Goree Island, Senegal. The former inspired the possibility of a pedagogical approach that combines what Schumacher College calls "soil, soul and society" – or alternatively "ecology, spirit and community". The latter demonstrated what can be done to create an authentic African aesthetic and working space. The result was a versatile space that is absolutely ideal for interactive, immersive, discussion-based experiential learning. There are flat open classrooms that can be constantly rearranged, outside spaces for group discussions and land art experiences, a large hall for gatherings from yoga stretches to communal meals, a café with coffee and organic food, small meeting spaces, a beautiful sense of embracement by nature's greenery and organic food gardens where students work before class – as I tell my students, "I judge how much you know about sustainability by how much dirt there is under your fingernails".

Over the past 17 years, I have coordinated an academic programme that has expressed in practice a desire to synthesize discussion-based learning about an African interpretation of the sustainability challenges underpinned by complexity theory (see chapter 10). However, instead of being satisfied with critique, I have focused on what I have often referred to as 'phronesis' – the capacity for sound judgement appropriate to the context (Flyvbjerg, 1998). This is the third of Aristotle's three conceptions of knowledge – the other two being techné (technical knowledge) and episteme (general wisdom). In short, going against the fashionable postmodernist predilection for critique and 'deconstruction', our emphasis has been on action appropriate to the context within an actually existing experimental community. The programme has been delivered by a unique partnership between the SI, the School of Public Leadership and the Centre for Complex Systems in Transition (CST) at Stellenbosch University. The academic programme comprises a Diploma in Sustainable Development (for school leavers), a Postgraduate Diploma in Sustainable Development (honours-level equivalent), an MPhil in Sustainable

Development, a Transdisciplinary PhD in Complex Systems in Transition and post-doctoral research. Connecting complexity theory, sustainability science, transdisciplinary research and social innovation, this combination of degrees delivered within an actual living and learning context with this specific pedagogical orientation is unique in the world.

Unsurprisingly, the SI is now led and managed by graduates from the programme, and an increasing number of modules are taught by former graduates. They bring with them their own orientations. For most of the first eight to ten years, the focus was on sustainability – reconciling development with ecological sustainability expressed in the book *Just Transitions* (Swilling and Annecke, 2012). Since about 2000, there has been a tendency to emphasize transitions to a more sustainable world, reflected in the interest in global transitions (Swilling, Musango and Wakeford, 2015, 2016), urban transitions (Robinson et al., 2013; Hodson, Marvin, Robinson and Swilling, 2012; Swilling, Robinson, Marvin and Hodson, 2013), resilience (Preiser, Biggs, de Vos and Folke, 2018) and transdisciplinarity (Muhar, Visser and van Breda, 2013; van Breda and Swilling, 2018). From about 2015–2016 onwards, the new generation has started to shift the orientation towards social entrepreneurship and social innovation. Phronesis and a sense of the spiritual have remained consistently significant (Annecke, 2013). My intuition tells me that the next wave waiting to break will be about the commons and commoning.

I come from a generation that believed it was possible to imagine and build a progressive democratic state (Swilling, 2008) – the new generation are disillusioned with both the state and free market thinking. Social entrepreneurship and innovation attract them as a kind of 'third way', but they realize this still means building capable states and viable socially embedded markets. However, unlike others in this burgeoning field, repairing the future becomes the new *raison d'etre* for the endeavours of this particular group.

Starting in 1999, I spent a decade pretty obsessed with the idea of imagining, designing and constructing a socially mixed community that could also live in ecologically sustainable ways. Without being able to draw on a precedent-setting model in the South African context, we had to muddle our way through, working closely with Gita Goven and Alastair Rendall, both architects (now with their own Cape Town practice called ARG Design). We visited other places, such as Crystal Waters near Brisbane, Australia, Vauban in Germany and Tlolego in the North West Province; I built for my family a sustainable log house in Kuthumba, a largely white, middle-class and socially unsuccessful ecovillage near Plettenbergbay.[7]

I played a leading role in the process of stitching together the key components of the Lynedoch project: the urban and architectural design, an appropriate mode of community-based governance, a viable financial model (including raising the loan finance and obtaining housing subsidies from the government), convincing the provincial government to fund a new intersection (costing R6 million) and an effective project management system to ensure infrastructures and houses were built within the confines of a very tight budget (which was only partially successful). It was a decade of continuous experimentation that ran in parallel to the building of the academic programme and ongoing involvement in the wider dynamics of a young democracy. It was a decade that also, effectively, destroyed my health.

What is not recognized, even by those who know a lot about the history of the development since 1999, is the critical role played by various state institutions in making possible the Lynedoch EcoVillage and the SI. This is probably best revealed by stating the extent of funding from various state institutions: housing subsidies for 11 houses from the Western Cape Government's Department of Local Government and Housing (R35,000 each); R6 million from the Western Cape Government's Department of Transport and Public Works to restructure the intersection between Baden Powel Drive and Annandale Road (which was a development approval condition imposed by Stellenbosch Municipality); subsidies for the labour costs of constructing the 11 houses built for people who qualified for housing subsidies, which came from the Construction Sector Education and Training Authority, which, in turn, lowered the costs of the houses for the buyers; help from the Western Cape Government's Department of Education, which has consistently from the start contributed to the operating costs and rental of the primary school, even after it was converted from a government school into the Spark School (i.e. a private school); the continuous annual flow of capital and operating funds into the SI from Stellenbosch University (a publicly funded university); the off-and-on support for the preschool from the Department of Welfare; the recently completed baffle-reactor-connected horizontal wetland funded by the Western Cape Government's Department of Transport and Public Works as a component of the new road that went through the previous dysfunctional horizontal wetland; the urban design and infrastructure design costs funded by the World Bank–linked International Finance Corporation (with support from Stellenbosch Municipality); the recent R2.5 million renewable energy smart-grid system installed by ESKOM; and last, but not least, the R3 million loan from the state-owned Development Bank of Southern Africa (DBSA) obtained in 1999 for the infrastructure (water, energy, sanitation, roads). Every single one of these interventions entailed dozens of conversations and meetings. The SI provided the organizational space and capacity for making all this happen. In reality, the SI animated, integrated and coordinated a multi-pronged state investment strategy in innovations that resulted in what exists now, namely a community-based socially mixed ecologically sustainable urban settlement that expresses in practice the values encapsulated within the South African Constitution. The likelihood of these state institutions collaborating and doing this without an external animating agency like the SI is as likely as a sandstorm transforming itself into an adobe brick house.

The state's role in the making of the Lynedoch EcoVillage provides some confirmation that progressive, innovative, state-supported development is possible.[8] However, the vision and coordination capacity came from *outside the state*, that is, it was provided by an NGO – the SI – working with a university partner (Stellenbosch University) that actively engaged and transacted with a multiplicity of state institutions. Some of these state institutions were strong institutions but unable to act strategically (e.g. Western Cape Department of Local Government and Housing), while others were weak institutions and reasonably well intentioned (e.g. Stellenbosch Municipality). Without this capacity to organize the state from the outside to mount and execute a multi-year strategy, the Lynedoch EcoVillage, the Spark School and the SI in its current form would not have existed. Factor in the long-term support and commitment provided by the DBSA and the nearby Spier Wine Estate (owned by the Endhoven family) from day one (including approximately R3 million in grants and buildings) and the full dimensions of the partnership approach animated by the SI start to emerge. This holds many lessons for those who mistakenly expect these kinds of innovations to emerge exclusively from *within the state* or from NGOs funded by the corporate sector.

It is this experience that has made me particularly sympathetic to Mazzucato's conceptions of the role of the state in the innovation cycle (discussed later in this chapter and in Chapters 2 and 6, even though she is not sensitive to this inside-outside dynamic) and Jessop's notion of 'collibration' which refers to the 'governance of governance' by a new generation of institutions that facilitate partnering, stakeholder engagement and strategic alignment across institutional boundaries (explained further in Chapter 2). I also recognize that they both place insufficient emphasis on the role of the exogenous animator operating outside the formal state sector. However, reflecting back on nearly two decades of working with state institutions to build a community-controlled urban space without any personal financial reward (as a shareholder or property developer) has led me to realize that we as a group were (largely unwittingly) part of a much wider global movement to rebuild a shared 'commons'. We were – in the language of this movement – 'commoners' collaborating to create a shared space that could be the home for families, schools, social enterprises and university programmes. As discussed further at the end of Chapter 2, institutionalizing the governance of the commons is an authentic 'third way' between statist and market-driven solutions. The way we worked to harness the resources of several different state institutions was how we incrementally built an ecocultural commons that has a very distinct sense of a social commonwealth reflected in the 'buzz' and energy that infuses the entire space every day.

The Lynedoch EcoVillage experience was the start of my relationship with the DBSA. Established during the Apartheid era to fund infrastructures that supported 'bantustanization', the DBSA, after 1994, became a key thought leader and animator of developmentally oriented infrastructure, albeit largely within a neoliberal framework. In 2013, I was appointed by the minister of finance to the Board of the DBSA, and, in December 2018, I was appointed chairperson of the Board with effect from 1 January 2019. This experience reinforced my conviction that Mazzucato is spot on when she argues that state-owned development finance institutions (DFIs) have a key role to play in reducing risk during the early phases of the innovation cycle (Mazzucato and Penna, 2015). The DBSA played a key role in establishing and funding South Africa's rapidly expanding renewable energy sector. During the course of 2018, it adopted a progressive development vision that positioned it to intervene more meaningfully in actual development processes, with the Lynedoch EcoVillage acting as a key role model.

In 2017, I led a team of academics that published a report entitled "Betrayal of the Promise: How South Africa Is Being Stolen" (Swilling, Bhorat, Buthelezi, Chipkin, Duma, et al., 2017; later extended and published under the title "Shadow State" see Chipkin and Swilling, 2018). One day, while I was in the midst of writing this depressing analysis of how our dreams of democracy were dashed on the rocks of corruption and inappropriate economic policies over a 20-year period that left 95% of all asset wealth in the hands of 10% of the population, I took a group on a tour of the Lynedoch EcoVillage (which I had not done for some time). I was over-whelmed by the huge contrast between what I was writing about and the exquisite beauty and 'alive energy' of the place. As I led the group around I described – as I always do on these so-called 'tours' – how over nearly two decades and more we experimented with low-cement construction, resulting in many different build-ing systems (from four different kinds of adobe brick, to recycled bricks reclaimed from landfills, to wood, to sandbags, to light steel frame clad with fibre cement and bricks made from waste construction materials); many different building designs (with shortcomings when it came to orientation); renewable energy starting with solar hot water heaters/no electric stoves (which halved the cost of the internal electrical cabling system and monthly electricity expenses of each household) and ending with a fully fledged so-called smart mini-grid system funded by ESKOM and installed in 2017; worm filters and vertical/horizontal wetlands to treat the sewage; wind chimneys and a rock store to cool the main building; septic tanks for every two to three houses (not for treatment purposes but to prevent inap-propriate materials being flushed down toilets that then become the community's problem); small plots to maximize shared space; two prices for every plot – one if you qualified for a government housing subsidy and the other if you did not, thus ensuring a social mix (instead of allocating the poor to a specific area and the rest to another); a democratically elected home owners association that approves who can buy a property and the designs of new houses; an innovative financing mechanism that enabled poorer households to borrow funds internally and get three forms of cross-subsidization (costs of labour, land and infrastructure); and the placement of an innovative primary school and Montessori preschool at the very centre of the community – what we have often called child-centred urban planning (i.e. an urban design governed by the principle that from any point in the neighbourhood 'you should be able to see a child').

All these innovations, and many more, emerged from a process of experimenta-tion that was uninformed by a particular theory of change or, indeed, of experi-mentation, nor were we aware of the literature on the governance of the commons. In *Just Transitions* it was referred to as "adaptive design" – how to create a material environment that orients people towards resolving their own problems 'adaptively' in collective ways rather than depending on technical expertise to resolve seem-ingly technical problems (Swilling and Annecke, 2012: Chapter 10). With hind-sight, what we did was profoundly incrementalist, and the lesson is clear – systemic change takes time, and what is possible is context specific. Innovation, experimenta-tion and incrementalism are themes explored further in this book.

Undoubtedly enraged by the evils of the past, the founders of the Lynedoch EcoVillage (primarily the members of the first board of a non-profit company called Lynedoch Development set up in 1999 to drive the development) were inspired by a sense that the grand dream of a post-apartheid non-racial South Africa could be miniaturized and lived out in reality. In practice, we took ourselves into the future and never escaped what it meant to be human, including attributing the cause of nearly all legitimate and illegitimate conflicts to the persistence of racism of various kinds. For some, this will prove why we cannot aspire to live relationally – "it is just not how we are" is an oft-repeated refrain I've heard over the years. On the other hand, if we are serious about conscious evolution, we have to find another way of being (post-)human by being a part of experiments that call forth a very different set of desires and passions to those encouraged by individualistic consumer-driven security-oriented urbanism that so many aspire to achieve (in particular the bizarre phenomenon called 'suburbs' built as they are around spending in malls, surrounded by security fences, with schools built like prisons on the margins – and then we wonder why things don't change).

In 2007, the South African government's Department of Science and Technology nominated me onto a newly established expert panel initiated by what was then called the United Nations Environment Programme (UNEP). My term ended in late 2019. During this time, I participated in the co-authoring of several global reports on material resource consumption and urban transitions. In many different ways, my contributions were shaped and influenced by what I had learnt in microcosm during the preceding decade and a half of experimentation and failure.

Up until this point, my description of the Lynedoch EcoVillage and the SI is completely depersonalized and thus somewhat ahistorical. It leaves out the obvious fact that I am a white male, brought up in a middle-class home with a private school education (albeit in anthroposophically based Waldorf Schools, with parents who were non-practicing Jews and committed vegetarians – unusual for the 1960s!). They got divorced when I was 17, and my gay mother has lived with her current partner – whom she married when this was legalized after 1994 – for 32 years. And it also leaves out the fact that what we were trying to do at Lynedoch is to go up against the harsh realities and logics of one of the most violent, racist, misogynistic and unequal societies in the world. With 95% of all wealth in the hands of 10% of the population (Orthofer, 2016) (most of whom, of course, are white, and then most of those whites who own property are males), the psycho-emotional implications are so horrifying that it is as if it is collectively unthinkable by all South Africans in any terms – race, gender or class. As the book by a new generation of young black writers – *Writing What We Like* – makes so clear, rage permeates this conundrum (Qunta, 2016). In a contribution to this volume, Mathe reveals what most fans the flames of rage when he writes how "[I]t irks me when white people see the flames of black people's anger, then use Mandela's ideas of peace and reconciliation to extinguish them. Firstly, this anger has value" (Mathe, 2016: l. 1088). Herein lies the conundrum: if South Africans agreed to become a 'rainbow nation' without addressing the real cause of black rage (and, indeed, women's rage), where does that rage go? And from this, what must a white male do when so little changes and so much rage remains suppressed?

I started asking these questions in the 1980s, preferring the writings of Robert Sobukwe and Steve Biko (and internationally the life of Malcolm X rather than Martin Luther King Jr. or Ghandi) to the then more fashionable Marxist texts about class that the 'white left' loved so much – texts that seemed to me to ignore consciousness and, therefore, race. But the answers to my questions were few and far between mainly because, I realized many years later, of the consequences of reductionism (i.e. don't worry about the effects, what is really relevant is the cause, that is, the economics of capitalist dynamics and class). As the Marxist historian Jeff Guy once said to me when I told him my PhD thesis includes a chapter that reconstructs the actual day of the Langa Massacre, which occurred on 21 March 1985 in the Eastern Cape town of Uitenhage, "That sounds like an excellent exercise in irrelevant detail". What a Marxist like Guy cannot see is that the killing of 43 people (most of whom were shot in the back) changed the consciousness of that region and the politics of resistance aimed at changing the material conditions. Eventually, I moved beyond Marxism and found in complexity theory (via systems theory) an escape from the kind of reductionism that enabled Jeff Guy to say what he did – an approach that relegates the realities of experience to the margins of analysis (or to the status of an 'epiphenomenon' in academic language).

We, the group that initiated the Lynedoch EcoVillage and the SI, shared a dream of creating a community within which we could bring up our children in ways that reflected our values and visions of the future. After leaving the EcoVillage in 2015, I've returned to the search that began so long back, but this time it is not just about the implications of being white in a racist society but what it means to be a man in a deeply misogynistic world. After several women shared with me how they were sexually assaulted, this became especially poignant and painful – the rapist's rage and their counter-rage became the coordinates of my explorations. And here I'm not just referring to society in general but South African society in particular, which is premised on a profoundly unsustainable highly extractive economy that defined natural resources, black bodies and women as inherently exploitable objects. This is what created the obscene inequalities in wealth that nearly 25 years of democracy has done little to change, not to mention persistent violence against women and pervasive racism (see Chapter 9).

Put simply, it means facing rage: the 'thousand years of rage' that women and black people carry because of the accumulated damage wrought by white (mainly heterosexual) men. But that is not enough: if it stopped there, the job would simply be 'anger management' – how to make sure the rage does not become destructive. This is the kind of thinking that "irks" Mathe so much. It is also what further inflames women's rage – as Manto Khumalo tweeted after Winnie Mandela died in April 2018:

> I am angry, very angry. I am angry for Mme Winnie, my great grandmother, my grandmother and my mother. I am angry for all the women across Africa

who are the butt of your jokes and the focus of your exploitation. I am fucking angry. I am left defenceless and vulnerable.

And when confronted by these men about her anger, she cries, men will "dumb down by tweets because 'umubi vele' or 'I am not like that'".

Instead, the challenge for me – and indeed all men both black and white – is to discover within myself what it is that makes me misogynistic and racist. At least with respect to misogyny, this will certainly mean going beyond the biology versus culture debate (or what some refer to as the 'essentialism versus constructionism' debate) to accepting that it is both – the 'othering' of women is not simply a cultural outcome of the way we humans – and particularly men – get socialized by the societies we are born into. When we boys are hit by a testosterone tsunami at puberty, nothing and no one prepares us for the way our world is sexualized and transformed almost overnight. This is biology, not just culture. Noting the substantial and convincing literature on the historical existence of matriarchal societies (Eisler, 1996), surely the stubborn persistence of misogyny (and maybe also racism) must have deeper evolutionary roots in the way the post-matriarchal individual psyche gets constructed when separating from the mother, from nature, from the 'other sex' (whoever that may be) and (maybe even) from 'the other' race without an appropriate matriarchal nature-centred relational culture to nurture a different outcome. But as the author of the book *Why Men Hate Women* put it, "I do not believe that biology causes gender, but that gender provides significance to biology" (Jukes, 1993:20). In other words, biological drivers evolve within certain gendered and – by extension – racialized cultural formations that become mutually reinforcing.

Whatever the balance is between nature and culture as determinants of sexuality, the 'other' seems to persist against all rationality as an inherent threat to this male individual-ego at both the 'cellular' and 'cultural' levels. This must play a key role in ensuring that misogyny (as femicide), racism (as genocide) and the destruction of nature (as 'eco-cide') persist. Recognizing this means accepting that these destructive evolutionary outcomes are now a fundamental threat to everyone's survival, not just the multi-species that have been destroyed by these inherently destructive forces.

This has, in my view, got a lot to do with our definition of what it means to be human. Unless this changes, not much progress will be made (see subsequent pages). But for this to work, I must also tap into my own rage. Rage needs to be rescued from its place in global culture where it gets branded as a dangerous force that needs to be contained, either forcefully or by the new modes of collaborative or partnership-based governance that seeks to 'manage rage' (Sloterdyk, 2006). In contrast, rage, in the words of Nigerian feminist novelist Chimamanda Adichie, needs to become a "positive force" – or, in Mathe's words, "anger has value".

All the themes reflected in the experiences of my personal journey (from democratization to experimentation, South Africa's transition to global transition, how state institutions work at different levels, resource extraction and sustainability, transdisciplinary knowing) will be elaborated and further explored in various ways in the chapters that follow. There has been a constant interaction between my experience of incrementalism in practice at the microlevel in the Lynedoch EcoVillage (including race and gender issues within my own community and family); at city-wide levels working closely on the transformation of Johannesburg, Cape Town and Stellenbosch; at national levels (with respect to various sectoral policies and state capture); in the development finance field via my board membership of the Development Bank of Southern Africa; and also at the international level working with the International Resource Panel (IRP), where I have gained an understanding of our changing world. This is my vantage point.

Thymotics of the relational self

Before proceeding to the next section, which summarizes the core argument developed in this book over the course of the chapters that follow, a fundamental axiological point of departure needs to be made explicit. One of the great ethical questions that generations of thinkers have posed over the ages is this: what does it mean to be human? Today, the question becomes: now that we are a geophysical force of nature with extraordinary informational powers, what does it mean to be human in the Anthropocene? And from this, my next question is what 'passion for change' best equips us for the making of a just transition?

Following a well-established tradition within Sub-Saharan African philosophy and the post-humanist turn in the social sciences, redefining what it means to be human belongs up front in this introductory chapter. This is key to an understanding of the transitional and relational dynamics discussed in the rest of this book.

Leonardo da Vinci's *Vitruvian Man* was the iconic image that most powerfully expressed the naturalistic universalized assumption about what it means to be human in 'modern times' (even though his enormous lifetime oeuvre [1452–1519] made him the first holistic systems thinker (see Capra, 2008)). *Vitruvian Man* defined what it meant to be human – as someone who is male, white, rational, alone, perfectly proportioned, disconnected from nature and free of any hierarchy (either social or cosmological). Following Braidotti (2013), everything and everyone else was 'othered': woman via sexualization, people of colour via racialization and nature via naturalization. Complementing this image of the disconnected male self, we have the relentless quest of Rene Descartes to find out what defines him as human. He concluded: "I think, therefore I am". This ultimate foundational statement from the dawn of contemporary Western industrial culture established binary thinking as the ultimate way of knowing – because A is not-A, and A cannot be A and 'not-A' at the same time (Nicolescu, 2002). Hence, we have the mind-body binary but also the other disconnects that follow: self and other, self and nature, self and things. By contrast, Sub-Saharan African philosophy has always emphasized 'relatedness' between all things (animate and inanimate) and beings (humans and non-humans) (see the following discussion).

The underlying science that substantiated the nature-culture binary is questioned across many disciplines. An alternative is offered by the notion of 'complex adaptive systems' (CAS) (see Preiser, Biggs, De Vos and Folke, 2017 and see also the more extensive discussion of Metatheory 2.0 in Chapter 2). Seeing reality as a CAS is remarkably similar to the relational worldview found in Sub-Saharan African philosophy (Murove, 2009a; Coetzee, 2017). The interconnected relational nature of all reality sits at the very centre of the worldview expressed in both these systems of thought. This, in turn, has major implications for our understanding of what it means to be human in the Anthropocene. In short, it results in the replacement of Vitruvian Man with a relational self that does not depend on the 'othering' of anyone or anything. In the words of Rosi Braidotti, one of the chief exponents of the Western post-human turn that is infusing many natural and social sciences,

> The human of [classical] Humanism is neither an ideal nor an objective statistical average or middle ground. It rather spells out a systematized standard of recognizability – of Sameness – by which all others can be assessed, regulated and allotted to a designated social location. The human is a normative convention, which does not make it inherently negative, just highly regulatory and hence instrumental to practices of exclusion and discrimination. The human norm stands for normality, normalcy and normativity. It functions by transposing a specific mode of being human into a generalized standard, which acquires transcendent values as the human: from male to masculine and onto human as the universalized format of humanity. This standard is posited as categorically and qualitatively distinct from the sexualized, racialized, naturalized others and also in opposition to the technological artefact. The human is a historical construct that became a social convention about "human nature".
>
> My anti-humanism leads me to object to the unitary subject of Humanism, including its socialist variables, and to replace it *with a more complex and relational subject framed by embodiment, sexuality, affectivity, empathy and desire* as core qualities.
>
> *(Braidotti, 2013:26 – emphasis added)*

Together with the science that subverted the foundational nature-culture dualism (Capra, 1996; Peat, 2002) and the rise of the 'complex adaptive systems' theory (Preiser, Biggs, de Vos and Folke, 2018), the combined impact over the past three decades of feminism (that challenged sexualization), post-colonial studies (that challenged racialization), political ecology (that challenged naturalization) and actor-network theory (that challenged technicism) has resulted in the collapse of the Vitruvian Man. In its place is the relational self, where connectedness rather than separateness is valued in a way that reconstitutes what it means to be, know, learn and act. This includes the connectedness between all sexual and racial identities, as well as between humans and nature and humans and technological artefacts. In this way, radical 'multi-species' relationality subverts classical Humanism (Haraway, 2008).

As already mentioned, the conception of the relational self we see emerging out of Western post-humanism is remarkably similar to notions at the very centre of a long-established and substantial body of Sub-Saharan African philosophy with deep roots in pre-colonial cosmologies and cultural practices (Murove, 2009a; Coetzee, 2017). For Murove, the two most important concepts in this body of knowledge are *Ubuntu* and *Ukama: Ubuntu* is about a relational humanness, but *Ukama* means relatedness of everything (Murove, 2009b).

Significantly, *Ukama* is not merely about relatedness between people but also relatedness between people and nature (including inanimate objects) and between people and ancestors (who are, in turn, experienced in very real ways in the everyday lives of many rural and urban Africans) (Murove, 2009b). Indeed, extending relatedness to include ancestors and inanimate objects is African philosophy's distinctive contribution to environmental ethics (Behrens, 2014). Based on an extensive review of African philosophical traditions and Western scientific thought, Murove concludes that Western conceptions of science have "been applied in a way that has severed the *Ukama* between environmental well-being and humanity" (Murove, 2009b:327). By contrast, trends like the 'new systems science' and Western post-humanism are regarded by Murove as a "vindication of African values of interconnectedness" (Murove, 2009b:326). He concludes by arguing that

> African ethics, as espoused in the concepts of *Ukama* and *Ubuntu*, offers a plausible paradigm that can help the present generation, and humanity at large, to harmonise its behaviour with the natural environment. . . . [T]hrough *Ukama*, . . . an authentic understanding of human existence should embrace human togetherness in all spheres of existence – social, spiritual, economic and ecological. In *Ukama* with all these dimensions of human existence, the individual derives personality and character.
>
> *(Murove, 2009b:329)*

Braidotti invokes what is specific about the present context: "We need a vision of the subject that is 'worthy of the present'" (Braidotti, 2013:51). It is this clarion call that I want to invoke here in the introductory chapter, as the explicit entry point into the explorations that follow in subsequent chapters. In the spirit of *Ukama*, are we prepared for the future with an appropriate sense of what it means to be human in this complex age of sustainability?

The status quo – the persistence of the 'unitary subject of Humanism' – will ensure that the knowledge, ingenuity, creativity and inspiration required to make a deep and just transition happen will remain suppressed and constrained. The sexualization, racialization and naturalization associated with this version of being Human excludes women, people of colour, and the natural world we are embedded within from being active partners in the production of the transformative knowledge we need to address the polycrisis we face. This systemic exclusion

> result[s] in the active production of half-truths, or forms of partial knowledge about these others. . . . *[O]therness induces structural ignorance* about those who, by being others, are posited as the outside. . . . The reduction to sub-human status of non-Western others is a constitutive source of ignorance, falsity and

bad consciousness for the dominant subject who is responsible for their epistemic as well as social de-humanization.

(Braidotti, 2013:28 – emphasis added)

And this 'dominant subject' is to a large extent the white heterosexual man.

This fusion of Sub-Saharan African philosophy of relatedness and Western post-
humanism makes it possible to identify and challenge the dominant values of mas-
culinism, racism, the dogma of the superiority of scientific reason, the devaluation
of indigenous knowledge and the worship of the ever-rational individual *homo
economicus*. If these values remain dominant, they will prohibit the emergence of
the 'transformed world' envisaged in the Preamble to the SDGs. Where they are
expressed most beautifully is via the notion of *Ukama* – a multi-species relationality
that is allowed to flower within an ecocultural commons.

For Braidotti, the relational self that emerged from the ruins of Western
Humanism to resist the "human of Humanism . . . a systematized standard of
recognisability – of Sameness – by which all others can be assessed, regulated
and allocated to a designated social location". The alternative is the "*complex and
relational subject framed by embodiment, sexuality, affectivity, empathy and desire* as core
qualities" (Braidotti, 2013:26 – emphasis added). Similarly, for the South African
philosopher Coetzee, the "the main ordering principles of sub-Saharan African
thought and world senses are *relationality, multiplicity, fluidity and difference, rather than
sameness, exclusion and stability*" (Coetzee, 2017 – emphasis added). The overlap from
very different vantage points is striking!

All well and good. But are these qualities and principles sufficient to fire up mass
action against environmental and social injustice? To sustain collective imaginaries
and actions by communities to build the commons, against all odds? To inspire
entire cities to reimagine their futures and act accordingly? To build an entirely
new generation of institutions that value well-being rather than GDP growth when
economic power is held by those who think the opposite? To inspire national leaders to see futures that their electorates may desire but not want? I don't think so.
What is missing is the gritty gut-wrenching subversive (and often divisive) power of
rage. This is the lesson I take from my Lynedoch experience and my experience of
the unresolved South African democratic transition that has left race and misogyny
largely unreconstructed (see Chapter 9).

From an African perspective, surely it's time to build on the rich philosophical
tradition of relational subjectivities cited previously and then invoke the fighting
spirit of Fanon when contemplating what it will take to translate decolonized
conceptions of self into new real modes of collective action and social organization
within society (Fanon, 1963). As a founding contributor to post-colonial studies, he
was obsessed with what it meant to contest, resist and transcend the 'othering' of
the colonized subject and, therefore by extension for this discussion, all who have
been 'othered'. Fanon and others in his tradition, such as South Africa's Steve Biko
(Mamdani, 2012), were interested in the dismantling of the psycho-social apparatuses that the colonizer uses to control the minds of the colonized. A key goal of the

colonizer was to diminish via culture/religion and – where necessary – forcibly kill off (via torture, imprisonment, concentration camps, genocide, mass rape, forcible removal of children, slavery) the rage felt by the colonized. The goal was always docility. As far as the colonizer was concerned, the colonized become human when they become docile. It is as simple as that. Docility is the condition that allows the colonized to accept the dominant colonial discourse as normal or even 'natural' or, as the missionaries insisted, as 'God-given'. Writ large, the claim that docility defines what it means to be human is true for all who are 'othered' to this day. The granularity of everyday micro-power dynamics that reproduce the dominant patriarchal definition of what it means to be human by insisting on docility is today all pervasive.

For Fanon and Biko, rekindling rage was a precondition for action to change. Castells has captured this with respect to contemporary protest, including the Arab Spring, Occupy movement and so forth. Drawing on neuroscience, he demonstrated that "outrage" instigates solidarities that are necessary to overcome the fear instilled by dominant elites to prevent collective action (Castells, 2012). Fear is what ensures docility; rage is what overcomes fear, thus preparing the way for collective action. Unsurprisingly, as Malcolm X understood so well, dominant cultures needed ways to contain and tame the power of rage.

Rage, however, is now blamed for many social evils. In his remarkable book about the "history of the present" moment entitled *Age of Anger*, Panjak Mishra argues that "political dysfunction", "economic stagnation" and "climate change" are resulting,

> as [Hannah] Arendt feared, [in] a "tremendous increase in mutual hatred and a somewhat universal irritability of everybody against everybody else", or *ressentiment*. An existential resentment of other people's being, caused by an intense mix of envy and a sense of humiliation and powerlessness, resentment, as it lingers and deepens, poisons civil society and undermines political liberty, and is presently making for a global turn to authoritarianism and toxic forms of chauvinism.
>
> *(Mishra, 2018)*

In this increasingly popular view, rage is associated with the resentments of those who want to resist liberal tolerance, in particular a toxic mix of right-wing racists longing for a return to a golden age of white hegemony, neo-fascists who long for certainty, authoritarian populists justifying neo-patrimonial governance, misogynists of all kinds yearning for the resubordination of women and religious fundamentalists who yearn for the restoration of imperial theocracies (for further discussion, see Chapter 9).

Although Mishra is critical of neoliberal globalization and the associated enrichment of the few at the expense of the majority, he is the most recent exponent of a very long tradition in Western thought that regards rage as a threat and therefore as something that needs to be contained. For him, like many others in this tradition across the ideological spectrum, the real danger lies in the way populist demagogues who promise certainty can easily manipulate young men and women who are "eager to transform their powerlessness into an irrepressible rage to hurt and destroy" (Mishra, 2018:Kindle location 4599). Rage, in short, can catalyse the quest for certainty.

A very different tradition – including, of course, that of Fanon, Biko, Castells, radical feminists and many others – extols the virtues of rage as the passion that inspires the righteous fight against domination and injustice. In his book *Rage and Time*, German philosopher Peter Sloterdijk excavates this tradition, going back to its roots in the very first line of the *Iliad* – the famous Greek celebration of the hero Achilles. In the first line of the *Iliad* the storyteller calls on the Goddess to sing in celebration of the rage of Achilles – "*Of the rage of Achilles, son of Peleus, sing Goddess*". For Sloterdijk, these words reveal a completely different conception of rage to that which is generally accepted in the world today: in this tradition, rage is the passion that gets mobilized to fight injustice against all odds. Although ignored by Sloterdijk, surely it is significant that it is the *Goddess* to whom this is addressed – the symbol of the sacred feminine principle, now called upon to praise the passion that allows Achilles to annihilate the enemy in battle after he loses his lover. This is, in short, rage that is acceptable to the sacred feminine. Maybe that is how we can discern the difference between destructive and productive rage: can the Goddess – the feminine in us all – sing to our Achillean rage? If not, it is more than likely destructive.

Confirming a long tradition represented by Castells, Fanon, Biko and the contributors to *Writing What We Like* (Qunta, 2016), for Sloterdyk,

> The early heroes are celebrated solely as *doers of deeds and achievers of acts*. Their deeds testify to what is most valuable. . . . Because true actions have been done the accounts of them answer the question, "Why do human beings do something at all rather than nothing?" Human beings do something so that the world will be expanded through something new and worthy of being praised. Because those that accomplished the new were representatives of humankind, even if extraordinary ones, for the rest as well an access to pride and amazement opens up when they hear about the deeds and sufferings of the heroes. . . . Only because the terrifying rage of heroes is indispensable may the singer turn to the goddess in order to engage her for twenty-four songs. If this rage, which the goddess is supposed to help to sing to, were not itself of a higher nature, the thought to appeal to it would already be an act of blasphemy. Only because there is a form of rage that is granted from above is it legitimate to involve the gods in the fierce affairs of human beings. Who sings under such premises about rage *celebrates a force that frees human beings from vegetative numbness*. This force elevates human beings, who are covered by a high, watchful sky. The inhabitants of the earth draw breath since they can imagine that the gods are viewers, taking delight in the mundane comedy.
>
> *(Sloterdyk, 2006:4–5 – emphasis added)*

In short, rage enables heroes to be "doers of deeds", thus ensuring "the world will be expanded", which, in turn, "frees human beings from vegetative numbness". Without rage, nothing will happen. Rage, however, is not something that the hero feels and then acts. Instead, like the Prophet who is merely a medium for the transmission of the divine message, the hero becomes an instrument of rage – rage possesses the hero so s/he can act, not the other way round (Sloterdyk, 2006:8).

> It is not the human beings who have their passions, but rather it is the passions that have their human beings. The accusative is still untamable.
>
> *(Sloterdyk, 2006:9)*

If it is passion that possesses the hero, where does it arise? For the Greeks, rage resides in the *thymos* – the organ just below the upside-down V in the middle of the chest. From here it emanates outwards, beyond the control of the rational mind. For us moderns, the notion that we can be possessed by passions in service of a just cause determined beyond the bounds of rationality is almost impossible to accept. It is too much of a threat to our belief in the infinite power of the mind, especially if it is assumed to be the source of all rationality. It threatens the deliberative conceptions of dialogically based change. And yet, as Castells has argued using different terms, thymotic rage holds the key to collective action for change to a better society because it is the antidote for fear.

Thymotic rage, however, has been systematically tamed and repressed over the millennia. The construction of the classical notion of what it means to be Human – as symbolized by the Vitruvian Man – required by necessity the systematic spiritualization, pathologization, psychologization and intellectualization of rage. Rage had to be tamed in order to suppress its power to 'hurt and destroy'. The Vitruvian Man, now embedded in the routines required to reproduce civil urbane societies, could not be allowed to be possessed with rage. The routines of mass societies and colonies require docility – just think of Japanese urban culture and associated routines where docility is taken to extremes. And yet, rage can never disappear – it will always surface in some way and if suppressed it will surface in negative (often violent) ways.

I am not advocating the glorification of rage per se. Rage can, of course, cause mass destruction, suffering and humiliation. The escalation of male youth violence is a case in point. So is there a difference between negative and positive rage, and how can we tell the difference? Well, the clue may lie in those first lines of the *Iliad*: if the Goddess agrees to sing the praises of the enraged hero, then that is thymotic rage. In other words, a thymotic rage that aligns with the feminine principle of care rather than the male principle of control may well be the key passion for change. From this perspective, it may well be that thymotic rage is a vital quality of the relational self that emerges from post-humanism and Sub-Saharan African philosophical traditions. The thymotics of the relational self entangles the passions of rage and love in ways that empowers activists to overcome fear, mistrust and hopelessness.

How the argument unfolds

The argument, in summary, in the rest of the book unfolds in four parts. Part I – which includes this chapter – essentially elaborates the points of departure for this book, namely a summary of the personal and conceptual framing of the arguments that follow in subsequent chapters. This introductory chapter opens with key questions about transition and presents my personal vantage point resulting from the accumulation of my life experiences. It also articulates my axiological point of departure – the thymotics of the relational self. This is followed by Chapter 2, which is an elaboration of the metatheoretical framework which informs the way I have written the subsequent chapters, including the conceptual language that I have used. Part II is about the dynamics of the global transition. It contains four chapters that together provide a broad global overview of the dynamics of our changing world from vastly different vantage points, including transition from a long-wave perspective (Chapter 4), resource constraints of current development trajectories (Chapter 4), a theory of radical incrementalism (Chapter 5) and case studies of ecocultural communities that have emerged in the global South (Chapter 6). Part III discusses the way sustainability transitions are 'made and unmade'. Drawing on my South African experience, I discuss in Chapter 7 the relationship between developmental states and just transitions and an analysis of the global energy revolution and the prospects for building energy democracies in Chapter 8. Chapter 9 discusses the backlash against sustainability, with respect, in particular, to the political economic implications of the defence of fossil fuels and nuclear energy in a world where renewable energy is the cheaper option. The concluding chapters, in Part IV, draw out the implications of the arguments presented in the preceding chapters for learning (Chapter 10) and activism (Chapter 11) in the age of sustainability.

Chapter summaries

Chapter 2 provides a theoretical synthesis in three parts: firstly, with respect to new metatheoretical thinking (what will be referred to as Metatheory 2.0 for convenience) about the nature of knowledge/knowing (epistemology) and reality (ontology); secondly, new thinking about the governance of non-equilibrium economies that aligns with Metatheory 2.0; and thirdly, the new literature on the commons as a framework for thinking about an authentic 'third way' beyond statism and market fundamentalism. In my view, the most significant shifts in academic thinking about our increasingly complex and fragile world during the emerging sustainability age have been away from both the modernist and the postmodernist frameworks that evolved during the course of the nineteenth and twentieth centuries. Three traditions of intellectual thought have shaped this emerging, widely held understanding of the world that is consolidating at the foundational base of the sustainability age (with applications within an estimated 35–50 disciplines): critical realism associated with the writings of Roy Bhaskar, the complexity thinking associated with the works of Edgar Morin and integral theory as developed by Ken Wilber. The metatheoretical synthesis of these three traditions achieves four things:

- a break from reductionism: things are explained in terms of the relations between the parts and not by reducing all the parts to what is assumed to be the primary determinant irrespective of the context – this alternative to reductionism is, therefore, the 'relationality' of all beings/things in a complex world;
- a return to a way of thinking that aspires to understand the 'big picture' with respect to the forces that shape what is going on and therefore the bold solutions that may be required to achieve a just transition – this does not mean micro-dynamics are irrelevant: on the contrary, they are given greater meaning as part of larger unfolding processes;
- respect for a plurality of perspectives: this marks a break from the rationalist culturally oppressive 'one-size-fits-all' approach that dominated modernism (and in particular economics), and it is also a break from the judgemental superiority of the postmodernists who pride themselves in seeing the hidden discursive formations that others cannot see (so-called 'deconstructionism');
- acceptance that it is not only material conditions that matter but also the non-material dynamics associated with, in particular, the interiority of the individual and collective consciousness about inter-human and human-nature relations or – using more elegant language – 'multi-species co-flourishing' (Haraway, 2008).

At the centre of this emerging metatheory is the notion of complex adaptive systems (CAS) (drawing in particular from Preiser, Biggs, de Vos, and Folke, 2018). Without this notion, it would be impossible to imagine the transformed world referred to in the Preamble to the SDGs, nor would it be possible to conceive of an appropriate theory of change and an adequate understanding of what it means to be human in a world in transition.

The second part of Chapter 2 builds on Metatheory 2.0 in order to address in a fresh way the two inter-linked fundamental conceptual challenges that face anyone interested in a just transition to a more sustainable and equal world. The first is the continued dominance of neoclassical economics despite its clear failure to address the challenges of our increasingly complex crisis-ridden and ecologically unsustainable world. Drawing from the literature on 'complexity economics' or, as I prefer, 'non-equilibrium economics',[9] it is argued that the root cause of the problem is that neoclassical economic theory fails to recognize that economies are subject to the laws of thermodynamics. Once the notion of non-equilibrium economics has been established as an alternative, the next challenge is the elaboration of a conception of governance appropriate for the increasingly complex non-equilibrium economies facing serious resource and energy constraints.

Those interested in sustainability transitions tend to lack an effective theory of politics and governance (Johnstone and Newell, 2017). The penultimate part of Chapter 2 addresses this challenge. Drawing on Jessop (2016), I discuss the need to recognize the fact that there has been a shift from government (as structure) to governance (as relation). This reflects a weakening of the state and the de-centering of politics since the late 1970s. However, this conception of governance is problematic because it is inappropriate for the long-term challenge of guiding the socio-political *directionality* of a transition to a more sustainable world. The solution lies in recognizing the fact that the real challenge is the "governance of governance", or what Jessop calls "collibration" (Jessop, 2016). This refers to the emergence of a new generation of governance institutions that 'bring the state back in' but without reverting back to a Weberian golden age of centralized bureaucratic control. Partnering is important but on terms set by states controlled by political leaders committed to the realization of public value over the longer term (Benington and Moore, 2010).

The third part introduces the literature on the commons that has been largely ignored by a surprisingly wide range of research fields. On the whole, there is very little reference to this literature in the sustainability sciences, sustainability transition studies, governance writing, eco-socialist alternatives and the metatheoretical literature. Even the literature on energy democracy (ED), ecocultures and cooperatives has not engaged with the commons literature. What I will argue is that the commons literature provides a language for making sense of a vast range of alternative formations at local and global levels that are crucial for imagining futures that are neither statist (varieties of social democratic/socialist) nor market driven (varieties of neoliberalism). The remarkably creative institutional configurations discussed in this literature give concrete expression to the relational worldview elaborated in this chapter. It also gives collibratory governance a specific mission – to foster the partnerships required to expand the commons. This argument becomes particularly pertinent in Chapter 8, where it is argued that the renewable energy revolution could go either way – the driver of a new energy commons or replication of a top-down financialized corporate-driven renewable energy sector.

In short, Chapter 2 makes explicit the underlying metatheoretical worldview that informs my practice, research and analytical writing. We need metatheories that equip us to understand the challenges of our times and act accordingly. The first step is an appropriate metatheoretical framework. We all have these in one form or another (often just a jumble of frequently contradictory ideas) – what matters is whether we make them explicit or not. I prefer to make my metatheoretical assumptions explicit, because that frees the reader to engage critically with the ideas that follow.

The remainder of the book then proceeds to argue the case for a just transition from a range of different entry points.

Chapter 3 provides a synthesis of the research output of the IRP since its inception in 2007. I was nominated by the South African government to be a founding member of this Panel. The body of work that has been produced since 2007 essentially documents the resource limits of the industrial era and as such is suggestive of the socio-metabolic transition that needs to emerge. This body of knowledge is significant because it goes way beyond the mainstream focus on climate change.

Even if there was significant decarbonization, the planet would still fall to pieces as ecosystems degrade and resources are over-exploited. Sustainability is about far more than the changing nature of the carbon cycle.

Chapter 4 argues that a deep transition needs to be understood as the emergent outcome of the asynchronous interaction between four long-wave transitions: socio-metabolic transitions, socio-technical transitions, techno-industrial transitions and long-term development cycles. However, whether or not the directionality of the coming deep transition will be oriented towards a just transition that addresses the twin challenges of inequality and unsustainable resource use will depend on the outcome of the struggles within the polity between a wide range of organized formations that represent divergent interests, specifically those who want to replicate the status quo (plus some greening on the side) and those searching for a more collective commons-oriented alternative.

Chapter 5 proposes a way of thinking about experimentation and innovation as key components of a theory of change that will be described as radical incrementalism. Following the work of Roberto Unger (Unger, 1998), it will be argued that we undervalue the significance and power of incrementalism by overvaluing the role of structures – what Unger refers to as "structure fetishism". Being liberated from structuralist thinking, we clear the way for an understanding of the transformative power of radical incrementalism.

Chapter 6 extends the argument developed in Chapter 5 by demonstrating that there are a wide range of ecocultural commons that have emerged within the global South. Because they share a commitment to both social justice and ecological sustainability, they prefigure in practice what a just transition could be on a larger scale. They are what radical incrementalism is all about.

Chapter 7 discusses the necessity for a synthesis of the long-established literature on developmental states and the burgeoning literature on sustainability transitions. Both advocate the need for 'structural transformation' but obviously differ fundamentally about what this means. Nevertheless, a synthesis is possible. At the centre of this synthesis lies a particular conception of politics, power and the polity that is best captured in Jessop's notion of 'collibration' (Jessop, 2016).

Chapter 8 discusses the global renewable energy revolution from an ED perspective. It will be argued that the decentralized and distributed nature of renewable energy systems provides a unique opportunity for building a new progressive politics of the energy commons. Energy democracies are publicly and/or socially owned renewable energy systems that enhance human well-being, the autonomy of inclusive local economies and the integrity of nature. Energy democracies, in turn, are the most tangible and immediately realizable manifestations of an emergent just transition inspired in part by a sense of the commons. The more extensive energy democracies become, the greater the chances that the emerging deep transition will have a just transition orientation.

Chapter 9 draws on South Africa's recent political history to revisit the implications of the renewable energy revolution (Chapter 8) for the future of governance (building on the concepts developed in Chapter 7). In this chapter the 'dark side' of governance is explored by drawing on the new literature on 'petro-masculinity' and the well-established literature on neo-patrimonialism that has emerged to make sense of the apparent dualism of the African state, with relevance for states elsewhere. It is all very well for the governance and policy innovation literature to be sanguine about the emergence of relational dynamics and, indeed, 'collibration', but powerful 'patrons' have been able to successfully manipulate these dynamics to their own advantage. The South African case seems to confirm a new global trend: neo-patrimonial subversion within a neo-masculinist narrative in order to defend elite accumulation strategies based on increasingly costly fossil fuel and nuclear energy systems relative to the financial, social and environment costs of renewable energy systems.

Finally, Chapter 10 reflects on nearly two decades of learning within the SI and since 2016 the CST. This experience has shed light on what is referred to as the 'evolutionary pedagogy of the present' – a particular approach to teaching and research that has emerged from practice within the South African context. It is this experience, coupled with my experiences since 2007 as a member of the IRP and since 2014 as a member of the Board of the Development Bank of Southern Africa, that has shaped the perspectives on the global polycrisis and sustainability transitions explored in the chapters that follow.

Conclusion

The aim of this chapter was to provide the reader with an introduction and entry point into the chapters that follow. This was done by framing my questions about the dynamics of change in the age of sustainability and by elaborating my life journey as an activist academic rooted in the complexities of the South African transition. While the former establishes my perspective, the latter is my vantage point. Together, they explain why I am interested in a just transition.

My conclusions thus far are straightforward. We are living through a deep transition from the industrial epoch to a potentially more sustainable and socially just epoch. This is the sustainability age – a time when the coordinates of our future imaginaries are drawn from interpretations of the SDGs in particular and sustainability in general. However, we need a better understanding of the dynamics of change. The traditional choices between revolution and reform will not do. We need a theory of radical incrementalism that exploits the evolutionary potential of the present. But for this, we need to answer Latour's question about where are the passions for change. To answer this question, I have proposed that the post-human and sub-Saharan African conception of the relational self is useful but needs to be infused with the ancient Greek conception of thymotic rage.

The next chapter elaborates the metatheoretical framework that informs the meso-level analyses that follow in subsequent chapters.

Notes

1 For more information about the "Great Transition", go to www.tellus.org/
2 Although the (largely Latin American literature) on post-development provides the context for the use of the word 'transformation' in this discussion, it needs to be acknowledged that 'transformation' is also used by the largely European literature generated by the 'resilience community' and refers to large-scale non-linear systems change. For a discussion of the ways 'transition' and 'transformation' are used, see Hölscher, Wittmayer and Loorbach (2018).
3 These are allusions to the French Revolution, when the Bastille was stormed, and the Russian Revolution, when the Winter Palace was seized by revolutionary forces.
4 For more details on SI, see www.sustainabilityinstitute.net
5 Edgar Pieterse is now Professor and Director of the renowned African Centre for Cities; Naledi Mabeba is a Montessori pre-school teacher and lives in the Lynedoch EcoVillage with her mother and children; Roshieda Shabodien is well-known in Cape Town's political and Islamic circles with a history of work in the women's movement; and Adrian Enthoven is a leading South African business leader active in Business Leadership South Africa.
6 The land area is 7 hectares, located within a 15-minute drive of Stellenbosch in the Western Cape. There are 46 residential sites, with only three left undeveloped. The SI owns the large Guest House, another large residential house, the so-called Main Building that houses a primary school that caters to early 500 poor and middle-class kids (overwhelmingly all black), and the premises of the SI. Sixty-five per cent of the 7 hectares is open land used for organic food gardens and an indigenous arboretum (for more details, see Chapter 10 of Swilling and Annecke, 2012).
7 This failure is due to the fact that the property developer, Ruby Ovenstone, did not deliver on her legal obligations, causing a series of rifts that lasted over two decades.
8 The insights discussed in this paragraph emerge from discussions with Amanda Gcanga, a PhD candidate working on collaborative water governance.
9 Following Brian Arthur, one of the founders of the Santa Fe school of economic thinking, I prefer the term 'non-equilibrium economics' – (see Arthur, 2010).

References

Agnew, J. (2017) 'The tragedy of the nation-state', *Territory, Politics, Governance*. Taylor & Francis, 5(4), pp. 347–350. doi: 10.1080/21622671.2017.1357257.

Ahmed, N. M. (2017) *Failing States, Collapsing Systems: Biophysical Triggers of Political Violence*. Cham, Switzerland: Springer.

Alexander, C. J. (1977) *A Pattern Language: Towns, Buildings, Construction*. Oxford: Oxford University Press.

Annecke, E. (2013) 'Radical openness and contextualisation: Reflections on a decade of learning for sustainability at the Sustainability Institute', in McIntosh, M. (ed.) *The Necessary Transition: The Journey Towards the Sustainable Enterprise Economy*. Sheffield, UK: Greenleaf Publishing, pp. 41–51.

Barnosky, A., Brown, J., Daily, G., Dirzo, R., Ehrlich, A., et al. (2014) 'Introducing the scientific consensus on maintaining humanity's life support systems in the 21st century: Information for policy makers', *Anthropocene Review*, 1(1), pp. 78–109. doi: 10.1177/2053019613516290.

Bauwens, M. and Ramos, J. (2018) 'Re-imagining the left through an ecology of the commons: Towards a post-capitalist commons transition', *Global Discourse*, pp. 1–19.

Behrens, K. (2014) 'An African relational environmentalism and moral considerability', *Environmental Ethics*, 36(1), pp. 63–82.

Benington, J. and Moore, M. (2010) *In Search of Public Value: Beyond Private Choice*. London: Palgrave.

Berry, T. (1999) *The Great Work: Our Way into the Future*. New York: Bell Tower.

Bollier, D. and Helfich, S. (eds.) (1978) *Patterns of Commoning*. Amherst, MA: Off the Common Books.

Braidotti, R. (2013) *The Posthuman*. Cambridge: Polity Press.

Callinicos, A. (1992) *Between Apartheid and Capitalism: Conversations with South African Socialists*. Ann Arbor, MI: University of Michigan Press.

Capra, F. (1996) *The Web of Life*. New York: Anchor Books.

Capra, F. (2008) *The Science of Leonardo: Inside the Mind of the Great Genius of the Renaissance*. New York: First Anchor.

Castells, M. (2012) *Networks of Outrage and Hope*. Cambridge, UK: Polity Press.

Chipkin, I. and Swilling, M. (2018) *Shadow State: The Politics of Betrayal*. Johannesburg: Wits University Press.

Coetzee, A. (2017) *African Feminism as a Decolonising Force: A Philosophical Exploration of the Work of Oyeronke Oyewumi*. Stellenbosch: Stellenbosch University.

Crutzen, P. J. (2002) 'The anthropocene: Geology and mankind', *Nature*, 415, p. 23.

D'Alisa, G., Demaria, F. and Kallis, G. (2015) *Degrowth: A Vocabulary for a New Era*. New York: Routledge.

Ehrlich, P. (2002) *Human Natures: Genes, Cultures and the Human Prospect*. London: Penguin.

Eisenstein, C. (2013) *The More Beautiful World Our Hearts Know Is Possible*. Berkeley, CA: North Atlantic Books.

Eisler, R. (1996) *Sacred Pleasure: Sex, Myth, and the Politics of the Body – New Paths to Power and Love*. San Francisco: HarperCollins.

Escobar, A. (1995) *Encountering Development: The Making and Unmaking of the Third World*. Princeton, NJ: Princeton University Press.

Escobar, A. (2015) 'Degrowth, postdevelopment, and transitions: A preliminary conversation', *Sustainability Science*, 10(3), pp. 451–462. doi: 10.1007/s11625-015-0297-5.

Fanon, F. (1963) *The Wretched of the Earth*. New York: Presence Africaine.

Fioramonti, L. (2015) 'A post-GDP world? Rethinking international politics in the 21st century', *Global Policy*. doi: 10.1111/1758-5899.12269.

Flyvbjerg, B. (1998) *Rationality and Power: Democracy in Practice*. Chicago: University of Chicago Press.

Fry, T. (2012) *Becoming Human by Design*. London: Berg.

Gore, C. (2010) 'Global recession of 2009 in a long-term development perspective', *Journal of International Development*, 22, pp. 714–738.

Grin, J., Rotmans, J., Schot, J., Geels, F. and Loorbach, D. (2010) *Transitions to Sustainable Development: New Directions in the Study of Long Term Transformative Change*. New York: Routledge.

Haberl, H., Fischer-Kowalski, M., Krausmann, F., Martinez-Alier, J. and Winiwarter, V. (2011) 'A socio-metabolic transition towards sustainability? Challenges for another great transformation', *Sustainable Development*, 19, pp. 1–14.

Hajer, M., Nilsson, M., Raworth, K., Bakker, P., Berkhout, F., et al. (2015) 'Beyond cockpitism: Four insights to enhance the transformative potential of the sustainable development goals', *Sustainability*, 7, pp. 1651–1660.

Haraway, D. (2008) *When Species Meet*. Minneapolis, MN: University of Minnesota Press.

Hodson, M., Marvin, S., Robinson, B. and Swilling, M. (2012) 'Reshaping urban infrastructure: Material flow analysis and transitions analysis in an urban context', *Journal of Industrial Ecology*, 16(6). doi: 10.1111/j.1530–9290.2012.00559.x.

Hölscher, K., Wittmayer, J. M. and Loorbach, D. (2018) 'Transition versus transformation: What's the difference?', *Environmental Innovation and Societal Transitions*, pp. 1–3. doi: 10.1016/j.eist.2017.10.007.

Hopkins, R. (2018) *The Transition Starts Here, Now and Together*. Arles, France: Actes Sud.

Jessop, B. (2016) *The State: Past Present Future*. Cambridge: Polity Press.

Johnstone, P. and Newell, P. (2017) 'Sustainability transitions and the state', *Environmental Innovation and Societal Transitions*. Elsevier, 27(October), pp. 72–82. doi: 10.1016/j.eist.2017.10.006.

Jukes, A. (1993) *Why Men Hate Women*. London: Free Association Books.

Lang, M. and Mokrani, D. (2013) *Beyond Development: Alternative Vision from Latin America*. Quito: Rosa Luxemburg Foundation.

Latour, B. (2010) 'Where are the passions commensurate with the stakes?', *IDDRI Science Po: Annual Report 2010*, p. 3. Paris. Available at: www.iddri.org.

Lovelock, J. (2006) *The Revenge of Gaia*. New York: Basic Books.

Macy, J. and Brown, H. (1998) *Coming Back to Life: Practices to Reconnect Our Lives, Our World*. Gabriola Island, Canada: New Society Publishers.

Malm, A. and Hornborg, A. (2014) 'The geology of mankind? A critique of the Anthropocene narrative', *The Anthropocene Review*, 1(1), pp. 62–69. doi: 10.1177/2053019613516291.

Mamdani, M. (2012) 'A tribute to Steve Biko', *Transformation*, 80, pp. 76–79.

Maslin, M. and Lewis, S. (2015) 'Anthropocene: Earth system, geological, philosophical and political paradigm shifts', *The Anthropocene Review*, 2(2), pp. 108–116. doi: 10.1177/2053019615588791.

Mathe, S. (2016) 'White supremacy vs transformation', in Qunta, Y. (ed.) *Writing What We Like*. Cape Town: Tafelberg, pp. 1037–1114 (Kindle Location).

Mazzucato, M. and Penna, M. (2015) *Mission-Oriented Finance for Innovation: New Ideas for Investment-Led Growth*. London and New York: Rowman & Littlefield International.

Mishra, P. (2018) *Age of Anger: A History of the Present*. London: Penguin Random House.

Moore, J. (ed.) (2016) *Anthropocene or Capitalocene? Nature, History, and the Crisis of Capitalism*. Oakland, CA: PM Press.

Muhar, A., Visser, J. and van Breda, J. (2013) 'Experiences from establishing structured inter- and transdisciplinary doctoral programs in sustainability: A comparison of two cases in South Africa and Austria', *Journal for Cleaner Production*, 61, pp. 122–129.

Murove, M. (ed.) (2009a) *African Ethics: An Anthology of Comparative and Applied Ethics*. Durban: University of KwaZulu-Natal Press.

Murove, M. (2009b) 'An African environmental ethic based on the concepts of ukama and ubuntu', in Murove, M. (ed.) *African Ethics: An Anthology of Comparative and Applied Ethics*. Durban: University of KwaZulu-Natal Press.

Nicolescu, B. (2002) *Manifesto of Transdisciplinarity*. New York: State University of New York Press.

Orthofer, A. (2016) *Wealth Inequality in South Africa: Insights from Survey and Tax Data*. REDI3X3 Working Paper 15. Cape Town: University of Cape Town. Available at: http://web.archive.org/web/20180204212928/www.redi3x3.org/sites/default/files/.

Parenti, C. (2016) 'The catastrophic convergence: Militarism, neoliberalism and climate change', in Buxton, N. and Hayes, B. (eds.) *The Secure and the Dispossessed: How the Military and Corporations Are Shaping a Climate-Changed World*. London: Pluto Press, pp. 23–38.

Peat, D. (2002) *From Certainty to Uncertainty: The Story of Science and Ideas in the Twentieth Century*. Washington, DC: Joseph Henry Press.

Perez, C. (2016) *Capitalism, Technology and a Green Golden Age: The Role of History in Helping to Shape the Future*. WP 2016–1.

Preiser, R., Biggs, O., de Vos, A. and Folke, C. (2017) 'A complexity-based paradigm for studying social-ecological systems', *Tbc*, (March), pp. 1–18.

Preiser, R., Biggs, O., de Vos, A. and Folke, C. (2018) 'A complexity-based paradigm for studying social-ecological systems', *Ecology and Society*, 23(4), p. 46. doi: 10.5751/ES-10558-230446.

Qunta, Y. (2016) *Writing What We Like: A New Generation Speaks*. Cape Town: Tafelberg.

Regeer, B. J. and Bunders, J. F. G. (2009) *Knowledge Co-creation: Interaction Between Science and Society: A Transdisciplinary Approach to Complex Societal Issues*. Amsterdam: RMNO.

Robinson, B., Musango, J., Swilling, M., Joss, S. and Mentz-Lagrange, S. (2013) *Urban Metabolism Assessment Tools for Resource Efficient Cities*. Unpublished report commissioned by United Nations Environment Programme. Stellenbosch: Sustainability Institute.

Schot, J. and Kanger, L. (2018) 'Deep transitions: Emergence, acceleration, stabilization and directionality', *Research Policy*. doi: 10.1016/j.respol.2018.03.009.

Sloterdyk, P. (2006) *Rage and Time: A Psychopolitical Investigation*. New York: Columbia University Press.

Snowden, D. (2015) *The Evolutionary Potential of the Present*. Blog Post. Available at: https://cognitive-edge.com/blog/the-evolutionary-potential-of-the-present/.

Steffen, W., Broadgate, W., Deutsch, L., Gaffney, O. and Ludwig, C. (2015) 'The trajectory of the anthropocene : The great acceleration', *The Anthropocene Review*, 2(1), pp. 81–98. doi: 10.1177/2053019614564785.

Swilling, M. (1999) 'Rival futures', in Judin, H. and Vladislavic, I. (eds.) *Blank: Architecture, Apartheid and After*. Rotterdam: NAI Publishers.

Swilling, M. (2008) 'Tracking South Africa's elusive developmental state', *Administratio Publico*. Cape Town: HSRC Press, 16(1), pp. 1–29.

Swilling, M. (2015) 'How black is the future of green in South Africa's urban future?', in Mangcu, X. (ed.) *The Colour of Our Future*. Johannesburg: Wits University Press, p. 215.

Swilling, M. and Annecke, E. (2012) *Just Transitions: Explorations of Sustainability in an Unfair World*. Tokyo: United Nations University Press.

Swilling, M., Bhorat, H., Buthelezi, M., Chipkin, I., Duma, S., et al. (2017) *Betrayal of the Promise: How South Africa Is Being Stolen*. Stellenbosch and Johannesburg: Centre for Complex Systems in Transition and Public Affairs Research Institute.

Swilling, M., Musango, J. and Wakeford, J. (2015) 'Developmental states and sustainability transitions: Prospects of a just transition in South Africa', *Journal of Environmental Policy and Planning*. doi: 10.1080/1523908X.2015.1107716.

Swilling, M., Musango, J. and Wakeford, J. (2016) *Greening the South African Economy*. Cape Town: Juta.

Swilling, M., Robinson, B., Marvin, S. and Hodson, M. (2013) *City-Level Decoupling: Urban Resource Flows and the Governance of Infrastructure Transitions*. Available at: www.unep.org/resourcepanel/Publications/AreasofAssessment/Cities/tabid/106447/Default.aspx.

Unger, R. M. (1998) *Democracy Realized*. London: Verso.

van Breda, J. and Swilling, M. (2018) 'Guiding logics and principles for designing emergent transdisciplinary research processes: Learning lessons and reflections from a South African case study', *Sustainability Science*. doi: 10.1007/s11625-018-0606-x.

Weber, A. (2013) *Enlivenment: Towards a Fundamental Shift in the Concepts of Nature, Culture, and Politics*. Berlin: Heinrich Boll Foundation.

2

UKAMA

Emerging metatheories for the twenty-first century

Introduction

How we act in the world is conditioned by how we conceptualize it. Usually, our conception of the world is a jumble of different ideas drawn from what we have heard, read and seen. However, while we stumble about trying to grasp whatever comes our way to make sense of an increasingly incomprehensible reality, some ideas start to gel together in more coherent ways until eventually a small group of people assemble to formalize and systematize them. What follows are descriptions of this process of conceptual breakdown and synthesis, giving rise to new ways of seeing our world and the crises we face. If there is a word that can best sum up the emergent way of seeing the world, it would be the African notion of *Ukama*, meaning *relatedness* or *relationality*. And if there is a phrase that sums up how we go about changing the world that flows from this emergent *relational* perspective, it would be *radical incrementalism*. This chapter is about our emerging relational conception of the world, and how this translates into a radical politics of the commons by way of incremental actions. This provides the basis for constructing relational theories of economy and governance.

The first part of this chapter describes the emergence of a new metatheoretical framework for making sense of a rapidly changing reality and how a relational understanding of this reality has become a precondition for our survival as a species. Based on this metatheoretical framework, the second part proposes a framework for replacing conventional neoliberal economics with a theory of non-equilibrium economics that is aligned with the laws of thermodynamics. The third part of the chapter suggests a theory of 'collibratory governance' that is appropriate for managing non-equilibrium economies. The fourth part argues that the new literature on the 'commons' and 'commoning' provides a way of conceptualizing the types of post-statist post-market collective actions that could drive a just transition. The commons is relationality in practice. It really makes sense only from the perspective of non-equilibrium economics, and without collibratory governance it will remain marginal to those few believers in 'commoning'. Together, this chapter proposes a metatheoretical framework for making sense of a world in transition, with special reference to what will be referred to as the collibratory governance of non-equilibrium economies in general and the commons in particular. This conceptual framework sets the stage for the discussion in the chapter that follows which is about the next deep and potentially just transition.

Ways of seeing in the sustainability age

The transition to the 'modern era' starting with the industrial revolution 250 years ago was made possible by the rise of what came to be called 'modernism', which was itself the product of the Enlightenment (Peat, 2002). However, with the collapse of certainty originating in the decolonization and counter-culture movements of the 1960s, modernism as the intellectual 'high culture' of modernity has been replaced with what became known as 'postmodernism'. As will be argued, neither of these are adequate for comprehending the polycrisis we face (as portrayed in Chapter 3) and an appropriate theory of change (Chapter 5). What is required is an appropriate metatheory that is sufficiently comprehensive, integrative and transdisciplinary for comprehending and informing appropriate responses to the polycrisis.

With roots in quantum theory (Capra, 1996) and Whitehead's foundational work (Whitehead, 1929), sophisticated metatheories have emerged out of late twentieth-century Western science and philosophy that have begun to merge into a 'new' metatheoretical framework. These are integral theory (associated with the US-based social philosopher Ken Wilber), critical realism (associated with the English philosopher Roy Bhaskar) and complexity theory (associated with French social theorist Edgar Morin) (Bhaskar, Esbjorn-Hargens, Hedlund and Hartwig, 2016). Building on the synthesis achieved by Bhaskar et al. and the initial thinking in Chapter 1 of *Just Transitions* (Swilling and Annecke, 2012), what follows here is an interpretation of this emergent metatheoretical worldview from the vantage point of the African notion of *Ukama* referred to in Chapter 1.

Although far more influential across dozens of disciplines across all world regions, these three Western metatheories of relational ways of seeing have unwittingly caught up with a much longer and deeper African tradition of relational thinking (Murove, 2009a; Behrens, 2014). Whereas the Western metatheories break from Western reductionism and binaries to arrive at a more universally applicable relational epistemology and ontology, African cosmology and philosophy has always been a relational perspective, albeit suppressed below colonial narratives that devalued African contributions to knowledge (Murove, 2009a; on colonial 'denialism' and 'debasement' of African ethics, see especially Murove, 2009b; Coetzee, 2017).

Although the Sub-Saharan African literature is rich and diverse (Kagame, 1976; Masolo, 1994; various contributions to Wiredu, 2004, 1995; Bujo, 1997;

Oyewumi, 1997; Coetzee and Roux, 1998; Mama, 2001; Eze, 2009; Murove, 2009a; Coetzee, 2017), it has yet to be synthesized into a fully integrated metatheory that expresses the African perspective commensurate with the systematic breadth and methodological sophistication of complexity theory, critical realism and integral theory. Nevertheless, to rebut colonial "denialism" and "debasement" of the African contribution, Murove celebrates the vindicationist approach in post-colonial studies, which is clearly the first step towards developing this African perspective:

> Finally ... came the vindicationist approach which said that, while African values might have been debased by western-biased colonial scholarship, these values have been the source of African resilience against inhumane treatment. These post-colonial scholars maintain that the debasement of African values, in the guise of civilisation and modernity, was a western ploy to destroy the African personality and ensure that western values dominated all spheres of African existence. These scholars also maintain that the African ontology of a person as relationally constituted (made up of relationships with others), and the general understanding of reality as an interconnected whole, is currently *vindicated by the new sciences as commensurate with the nature of reality.*
>
> *(Murove, 2009b: 16–17 – emphasis added)*

As argued in Chapter 1, at the centre of this Sub-Saharan African perspective is the notion of *Ukama* – the profoundly African ethic of relatedness. As Murove puts it,

> An ethic that arises from a civilisation sensitised to relatedness among all that exist can only be an ethic about relatedness. Thus, in African ethics, relatedness is not restricted to human relations but extends to the natural environment, the past, the present and the future. This relatedness blurs the distinction between humanity and nature, the living and the dead, the divine and the human.
>
> *(Murove, 2009b:28)*

The three Western metatheories and the African *Ukamian* approach accept the need to break from what critical realists refer to as the "epistemic fallacy", namely the Western tradition (rooted in Hume, Kant and Hegel) that reduces ontology (our theory of reality) to epistemology (our theory of knowledge) (this is further explained in subsequent pages). They are explicitly critical of modernism, and the three Western metatheories seek also to go well beyond postmodernism. The three metatheories have sufficient in common to justify drawing these metatheories together into a metatheoretical framework that underpins the meso-level frameworks used in the rest of the book for 'seeing' the polycrisis and the dynamics of change underway. In so doing, a body of knowledge gets integrated in a way that can contribute to the African synthesis that must still happen.

Modernism and postmodernism

Modernism as a way of conceptualizing the world emerged from epistemological revolutions in so-called Western societies (mainly Europe and later North America): the Scientific Revolution (fourteenth to sixteenth centuries), the Renaissance (fourteenth to seventeenth centuries) and the Enlightenment that started to emerge in the eighteenth century. Inspired by Descartes' famous dictum "I think, therefore I am", modernism replaced the authority of the Church that had hitherto monopolized the power to conceptualize the world. Modernism offered a totally secular conception of the world that depicted individuals as inherently individualistic, materialistic and competitive. Society became merely a sum of the newly defined (ego-centred) individuals, without any reference to society's ecological or cosmological context. Da Vinci's 'Vitruvian Man' has survived as the epitome of this worldview – the perfect white male, rational and alone in his circle, with all else – woman, nature and people of colour – 'othered' (see Chapter 1 for a more detailed discussion).

The power of science to quantify and reduce to component parts became the primary source of knowledge (Capra, 1996), abrogating the right to deny all other forms of knowledge – a sine qua non for colonizing others and subverting their beliefs and knowledge frameworks. Rationality, individual liberty (initially only for white males), progress, universal laws and the conquest of nature became the organizing principles for modernism. Deeper gender and race assumptions rooted in pre-modern societies survived and morphed into their modern form but now denied by ideological notions like 'equality before the law'. *Liberty, egalite* and *fraternite* were the slogans of the French Revolution that modernism inspired and which became the foundations for modern democracies. In practice, modernity consisted of secularized nation-states, a market economy (more or less socially regulated), new class-based social divisions (including the reformation of pre-industrial racial and gendered divisions into permutations of these class structures), a secular and materialistic culture, a dominant role for scientific knowledge managed by a professional elite, the consolidation of the notion of 'the social' or 'society' as a cultural space distinct from economy and nature and the separation of society from nature (Hall, Held, Hubert and Thompson, 1996).

Although modernity came to be associated with democracy after the adoption of the Universal Declaration of Human Rights after World War II in 1948 and the subsequent decolonization period starting with Indian Independence in 1947, in reality modernity has been expressed in many non-democratic, semi-democratic and democratic forms. By the end of World War II, only 11 democracies existed (Keane, 2009). Since then, democracy evolved in waves of democratization, starting with Western Europe and Japan in the 1950s; some former colonies in the 1960s; Latin America, Southern Europe and chunks of Africa in the 1970s and 1980s and after the collapse of the Soviet Union and the ending of the Cold War, in Eastern Europe, chunks of Asia, South Africa and many former African dictatorships in the 1990s (most notably the Ethiopian Revolution that toppled the Marxist-Leninist dictatorship led by Mengistu, and the 'Arab Spring' in the 2000s prior to its tragic reversal). There is evidence, however, that since 2009 there are signs of a rollback of democracy in the face of rising reactionary populism (Mishra, 2018) and the expanding influence of China. As argued in Chapter 10, the endgame for fossil fuels is catalysing the rise of a new authoritarianism in both the developed and developing worlds.

Postmodernism emerged from across a wide range of geographical locales from the 1960s onwards (including significant contributions from the African context (Mbembe, 2002; Simone, 2004)) and became highly influential during the last quarter of the twentieth century. Inspired by the French schools of thinkers associated with Jacques Derrida and Jean-Francois Lyotard, it became a productive academic field that spanned many disciplines (especially cultural studies, literature, art, philosophy of science, post-colonial studies and feminist theory). Often referred to as the 'semantic turn' in the social sciences, the emphasis was on our 'discursive constructions' of reality – or in simple terms, the stories we share about reality. Instead of explaining oppression and exploitation in terms of the accumulated power of a political or capitalist elite, postmodernism focused attention on the way a shared narrative ensures the reproduction of a given social structure. Postmodernism questioned the Enlightenment intellectual project, especially modernist universalist notions of objective truth, rationality, 'laws of history', progress, morality, structure and the nature of language. Modernist universalism was a key discursive rationale for colonial domination of the world until the 1960s. Unsurprisingly, postmodernism was attractive to those interested in post-colonial cultural constructions in art, literature, cultural production and politics. Using irony and scepticism, postmodernists have always been deeply suspicious of anything that claims to be – or inadvertently sees itself as – a 'grand narrative'. A grand narrative is a set of constructs that depict reality in a specific structured way that usually suits the way elites would like the world to be understood. However, these purposes are hidden and denied by the ideological claims of modernism, in particular the claims about rationality, progress, universal truth, justice and individual choice.

Whereas the method of modernism was 'positivism' (the facts of quantitative knowledge), the predominant method of postmodernism was discursive 'deconstruction', or the 'semiotic tropes' revealed by 'discourse analysis'. What deconstructionism succeeded in achieving is the delegitimization of any attempt to create a metatheory with universalist claims, because for postmodernists this contains the seeds of another 'grand narrative'. An unintended consequence is an implicit reluctance to confront the systemic nature of the polycrisis in ways that suggest actions for radical change.

As argued by Bhaskar, Esbjorn-Hargens, Hedlund and Hartwig (2016), neither modernism nor postmodernism is adequate for understanding and responding to the polycrisis:

> Yet, while there are some countervailing trends…, much of the contemporary
> academy remains hypnotized by either the hyper-analytic, hyper-specialized,

and fragmented gaze of late modernity, or the sliding scale of postmodern rel-
ativism and its antipathy to integrated knowledge and meta-level understanding. Together these two orientations offer inadequate understanding(s) of our
many complex problems and their root causes, let alone the socio-ecological
crisis at large. Without being able to adequately illumine such root causes, the
academy remains largely impotent to address and help transform them.

(Hedland, Esbjorn-Hargens, Hartwith and Bhaskar, 2016: 2)

We need a theory of change commensurate with our times. Neither modernism
nor postmodernism can deliver that.

Re-emergence of metatheory

Writing from within a broad sweep of Western scholarship, Bhaskar, Esbjorn-Hargens, Hedlund and Hartwig (2016) convincingly demonstrate that integral theory,
critical realism and complexity theory are currently the most sophisticated and
influential *meta*theories, with active applications across 35–50 different disciplines
in the natural and social sciences. The strengths of all three are that they transcend modernism and postmodernism, refute reductionism, contextualize the social
within the ecological, promote individual and social emancipation, create space for
interiority/spirituality and give primacy to relationality. While Chapter 1 of *Just
Transitions* depended on the writings of Morin (1999) and Cilliers (1998)[1] to posit a
complexity-oriented metatheory as the basis for understanding sustainability challenges, driven by post-colonial narratives I had already begun to integrate critical
realism into my desire to go beyond constructionism.

This, however, is not the place for taking the next logical step by following Esbjorn-Hargens (2016) and Marshall (2016), who distil what emerges from a synthesis
of these metatheories to propose what they refer to as a new metatheory of *complex
integral realism.*[2] What matters for our more limited purposes in this book is how
leading-edge paradigmatic work that emerges from mainly Western traditions that
have had little connection with one another (until recently) are all calling for ways
of conceptualizing reality that have major implications for how we think about the
dynamics of change in the twenty-first century. In particular, by transcending the
traditional obsessions with structure, reductionism and one-dimensional notions
of what it means to be human, these three metatheoretical traditions and the Sub-
Saharan African tradition of *Ukama* create a wide-open space for critically rethinking our assumptions about agency, relationality and the changing nature of complex
adaptive systems. The meso-level conceptual themes addressed in the chapters that
follow which address the dynamics of transition at different temporal and spatial
scales would make little sense without grasping the intellectual trends that come
together in this emergent meta-theoretical perspective that so aptly vindicates the
African perspective. That said, this has nothing to do with building another grand
'metatheory of everything' that somehow trumps all others, which is what modernism strove to achieve. Rather, as Edwards puts it, it is about "a balance between
an integrative synthesis and a respect for the pluralism of perspectives" (Edwards

quoted in Esbjorn-Hargens, 2016:111). This creates space for an African metatheoretical framework to take its rightful place in the global pantheon of future-oriented sensemaking frameworks.

The three Western meta-theoretical frameworks are described herein, followed
by a distillation of a set of themes that are common to all three paradigms.

Integral theory: Wilber's four quadrants

For Wilber, every aspect of reality is a holon – each aspect is simultaneously a
whole in itself and a part of something else. As represented in Figure 2.1, every
holon has the following four dimensions: intentional ("I", subjective), behavioural
("it", objective), cultural ("we", intersubjective) and social ("its", interobjective).
Scientific disciplines, however, tend to focus on one dimension or the other, and
in so doing the reality we face gets cognitively fractured. To grasp the full extent of
the polycrisis we face, it will be necessary to accept the validity of knowledge fields
within all four quadrants and how they are interrelated.

Bhaskar's four-planar social being was central to his conception of 'dialectical
critical realism'. Each of the four planes represented in Figure 2.2 corresponds
to the four dimensions of social life. They are "dialectically interdependent" and
yet conceptually distinct. Corresponding to Figure 2.2, the four planes are (a) the
plane of material transactions of nature, (b) the plane of social interactions between
people, (c) the plane of social relations or social structure and (d) the plane of the
stratification of the embodied personality or agency.

As Esbjorn-Hargens notes, Wilber's four quadrants and Bhaskar's four planes are
"talking more or less about the same dimensions" (Esbjorn-Hargens, 2016:106).
Without implying an exact 1:1 correlation in meanings between the planes and
quadrants, Esbjorn-Hargens does represent the similarities listed in Table 2.1.

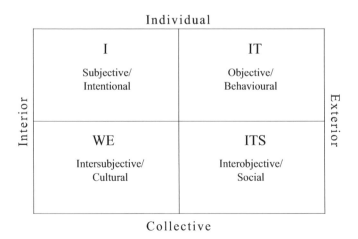

FIGURE 2.1   Integral Theory: Wilber's Four Quadrants.

Source: Adapted from Esbjorn-Hargens (2016).

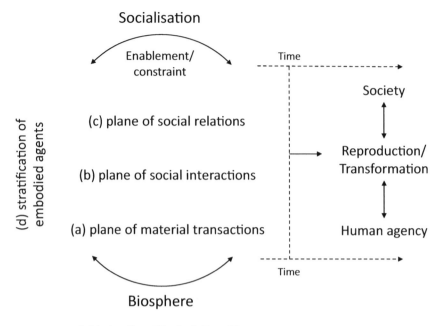

FIGURE 2.2 Critical realism: Bhaskar's Four Planes.

Source: Adapted from Esbjorn-Hargens (2016).

TABLE 2.1 Wilber's four quadrants and Bhaskar's four planes

Bhaskar's four planes	*Wilber's four quadrants*
"The plane of material transactions with nature"	Upper right quadrant of objectivity
"The plane of social interactions between people"	Lower left quadrant of intersubjectivity
"The plane of social structure *sui generis*"	Lower right quadrant of intersubjectivity
"The plane of the stratification of the embodied personality"	Upper left quadrant of subjectivity

Source: Esbjorn-Hargens, 2016

Complexity: Morin's recursivity

In remarkably similar ways to Wilber and, indeed, similar to the Sub-Saharan African notion of *Ukama*, Morin attempts to grasp the interconnectiveness of "I", "we" and "they":

> In every human "I" there is a "we" and a "they". The I, therefore, is not something pure, nor is it alone. The I could not speak were it not for "they".
> *(Esbjorn-Hargens, 2016:108)*

This is why the Cartesian notion that "I think therefore I am" is anathema for both complexity theory and *Ukamian* thinking: both would then ask "yes, but where and with whom?" Until those recursive questions of context and community are answered, "I think therefore I am" makes no sense. They are recursive not simply because they must be repeatedly asked because the context and community always changes but because as each constituent element changes, so does the whole, and the whole, in turn, interacts with the parts.

Although the co-constitutive recursive nature of the (inter-)subjective and (inter-)objective are present in this formulation, it becomes clearer in Morin's notion of the "computo-cogito loop". Morin is interested in the relationship between brain (*computo*) and mind (*cogito*) and how the socially-culturally embedded computo-cogita recursively interacts with each other and with this wider social-cultural context. Morin explains how the recursive interaction between brain and mind within an individual is a loop that is also recursively related to culture (intersubjectivity for Wilber) and society (interobjectivity for Wilber). As Esbjorn-Hargens puts it, "Morin seems to be highlighting that the subjective mind, objective brain, intersubjective culture, and interobjective society recursively create each other in an ongoing loop" (Esbjorn-Hargens, 2016:109).

Esbjorn-Hargens concludes his discussion of the commonalities:

> What is striking about all the tetradynamic examples above is they each in their own way find a type of integral wholeness through the inclusion of these four dimensions. Each theorist includes these four dimensions in a signature way.
>
> *(Esbjorn-Hargens, 2016:109)*

There are so many remarkable similarities between this observation and the worldview emerging from the gradual consolidation of an "African Ethics" (for the best overview, see Murove, 2009a).

Hedland, Esbjorn-Hargens, Hartwith and Bhaskar (2016) describe these three metatheories (and various versions of how they can be synthesized) as the foundations of *Metatheory 2.0*. They distinguish *Metatheory 2.0* from *Metatheory 1.0* by describing the aim of the latter as "theoretical monism". By this they mean the desire to identify general fundamental laws of motion (e.g. class struggle for Marx, the market for Adam Smith, *geist* for Hegel, bureaucratization for Weber, gravity for Newton, tendency towards equilibrium in classical pre-quantum physics and neoclassical economics) that explain all other phenomena via a grand narrative of one kind or another. This is the essence of reductionism – explaining a complex reality in terms of a primary determining factor or a set of 'building blocks. The alternative is a relational perspective that sees interactive context-specific relations between all parts of the social-ecological system.

Towards a synthesis

Metatheory 2.0 is pluralist and devoid of explanatory claims via any form of reductionism. Instead, the emphasis is on transparency about methodologies/methods used by researchers, ethical responsibility for the consequences/impacts of research, admitting the 'positionality' of the researcher, acceptance that reality is more than a mere construct of the mind (ontological realism) and finally absorbing a plurality of perspectives (epistemological plurality) and a plurality of contexts (ontological plurality) (Hedland, Esbjorn-Hargens, Hartwith and Bhaskar, 2016:7–8).

Following Marshall, there are nine themes (the first eight drawn specifically from Marshall) that are common to all three metatheories (Marshall, 2016:141–143). It can be argued that the following nine themes can be found in different forms in all three metatheories:

- integrative, maximally inclusive and non-reductionist: by refusing to be reductionist in all respects, all three metatheories clear the way for integrating a vast array of disciplines into the way we go about making sense of the world, underpinned by a strong desire to be inclusive of a wide range of interpretations;
- 'post-formal' cognition: complex normative dialectical thinking makes it possible for all three metatheories to go "beyond both the atomism of analytical thinking (reduction of wholes to parts) and the holism of systems thinking (reduction of parts to wholes)" – the result is a complex adaptive systems perspective that grasps the rich and ever-changing relational complexities of social and natural phenomena, including an openness to ways of knowing that includes spiritual and indigenous knowledge;
- a 'realist' ontology: rejecting what critical realism refers to as the 'epistemic fallacy', all three metatheories accept the "existence of a mind-independent reality" and by doing so they break from a long tradition in Western thought that has consistently prioritized epistemology (as theory of knowledge – or knowing/ cognitive constructions) over ontology (as theory of reality – or what reality is and how it works);
- stratified vision of reality: given the re-balancing of epistemology and ontology, all three conceptualize reality as comprising irreducible component parts, that is, no one part can be deemed a priori to be more causal than another: the "physiosphere" (atoms, molecules), "biosphere" (cells, organisms) and "noosphere" (human beings and their capacity for consciousness) are in instant interaction without one another;
- inter- or trans-disciplinarity: resulting from these commitments, all three advocate the deployment of interdisciplinary and transdisciplinary research methods for making sense of a complex, stratified and multidimensional reality – a complex reality that cannot be understood via mono-disciplinary research;
- unitas multiplex: all three want to avoid reducing "interiors to exteriors" (modern science), the erasure of particularity/singularity by an assumed universality

(modernity) and the reduction of "the individual subject (and thus agency) to intersubjective networks" (postmodernism) – instead, they emphasize the existence of a "unitas multiplex" where the "'relative' autonomy of the subject with its own emergent powers and transformative agency" is firmly recognized and promoted (and appears in Chapter 5 as the underlying assumption for the notion of 'radical incrementalism');

- "spirituality" and an ethics of "emancipation": all three in different ways explicitly acknowledge the salience of the spiritual dimension in the formation of an ethics of individual and collective emancipation (axiology);

- homo complexus: there is a shared conception of human nature that breaks from essentialism (i.e. 'there is only one human nature') and clears the way for an emancipatory and transformative conception of human agency (and its multiple context-specific expressions) in a complex world;

- contextual specificity: all three metatheories deeply respect the idea that context matters, and therefore context is neither merely constructed (postmodernism) nor is it a derivative of a universal reality (modernism) – in short, what is possible will always be co-determined by what is specific to each particular context.

Synthesizing these nine themes drawn from the three metatheories (using the keywords from the above-listed nine themes), we arrive at a particular understanding of reality (ontology) and ways of understanding that reality (epistemology) that restores an appropriate balance between the two. To re-establish the significance of ontology relative to epistemology from a complexity perspective, Preiser et al. argue that the "emergent properties and patterns of behaviour [of complex adaptive systems] *are real*" (Preiser, Biggs, de Vos and Folke, 2018 – emphasis added). The conceptual foundation provided by complexity thinking makes this balance possible:

> The deeper paradigm of complexity spans the ontological, epistemological and ethical domains of scientific inquiry to form a new conceptual framework through which humans can understand and reflect on the nature of the world and what it means to be human.
>
> *(Preiser, Biggs, de Vos and Folke, 2018)*

From this 'new conceptual framework' flows the key dimensions of Metatheory 2.0 drawn from the work of Bhaskar et al. and Marshall's particular contribution already summarized. Using the specific terms from the nine themes, it is possible to provide the following summation: complex systems are not mere cognitive constructions to comprehend reality, they exist in reality as well – this is what was referred to earlier as *'ontological realism'*. This reality, however, is stratified into various irreducible component parts that all interact with one another in dynamic non-deterministic ways that are usually context-specific (*'stratified vision of reality'*). The bio-physical context, the socio-cultural context and the knowledge context (i.e. the body of knowledge that is drawn on to make sense of the world) co-exist and co-constitute one another. This means these contexts have a mind-independent existence that sets the arenas for action by subjects in the real world who enjoy actual agency and relative autonomy to change things (*'unitas complex'*). That said, it is not possible to assume that individuals/collectives will act in accordance with a single fixed predictable rationality such as the 'self-interest' that is so central to neoclassical economic theory. Instead, echoing Ehrlich (2000), individuals are – their 'human nature' – both culturally specific and contextually changeable (*'homo complexus'*). Each context, in turn, has its own specific character, which means no a priori assumptions derived from universal imperatives can be made about a given reality – and how, in particular, it can change – prior to actually gaining an understanding of that particular context (*'contextual specificity'*). To grasp this reality, we need ways of knowing that are, first and foremost, profoundly anti-reductionist (*'anti-reductionism'*). This means wholes cannot be explained in terms of particular parts, and parts cannot be explained purely in terms of wholes. Explanations, therefore, are more or less valid depending on the dynamics and nature of the context.

Simultaneously, we must accept that to take action a choice always has to be made which at that moment in space and time will require a conclusion that will inevitably reduce complexity to justify what needs to be done. As long as this claim is admitted reflexively (i.e. admitting positionality and partial understanding of the system), it is consistent with this overall framework of understanding. By breaking with reductionism, we can build explanations that focus on the relationships between the component parts of a system rather than on the discreet parts themselves. By looking for dynamic relational interactions at all levels and dimensions (rather than making atomistic or holistic assumptions about reality), new understandings open up about the dynamics of personal and social change that were not apparent when only quantitative knowledge was regarded as valid knowledge (*'post-formal cognition'*). Consequently, indigenous knowledge, intuitive knowledge, process knowledge, cultural practices and the insights from depth psychology become significant. This profoundly relational epistemology clears the way for recognizing the spiritual dimensions of interior and collective life and, therefore, the potential for a relational axiology that is responsive to more than mere material realities (*'spirituality'*). To operationalize these emergent ontological realities and relational epistemologies, transdisciplinary research methods will be required. Transdisciplinary research can be defined as interdisciplinary research to co-produce knowledge *with* societal actors that helps generate solutions that address real-world problems (*'inter- and trans-disciplinarity'*).

The preceding two paragraphs capture the essence of what is common to the three metatheories that have been integrated by Bhaskar, Esbjorn-Hargens, Hedlund and Hartwig (2016). I find their contention convincing that this synthesis is, indeed, emerging as a way of seeing that has become increasingly influential across many disciplines and socio-cultural movements, especially since the dawning of the sustainability age and since the adoption of the SDGs. However, this emerging 'way of seeing' is hardly ever articulated in this integrated and systematic way. Most of the time only some of the various elements are implicit in various articulations of (largely partial) alternatives. For example, anticipatory thinking, transdisciplinary research, systems thinking, complexity economics, political ecology, deep ecology, transitions approaches, autonomous communities, new age spiritual perspectives and some radical social movements may all go beyond modernism and postmodernism in practice, but they rarely make systematically explicit their underlying meta-assumptions.

To avoid the risks of leaving implicit what needs to be made explicit, the metatheoretical synthesis articulated in this section provides the foundation for this book. Subsequent chapters draw on this core conceptual logic, but the emphasis will shift depending on the substance and focus of a particular chapter. The aim, however, is not to prove the correctness of this metatheoretical synthesis (or even the correctness of a complex adaptive systems perspective which is my bias) but rather to use this as a 'way of seeing' through the issues addressed at a meso-level and at an empirical level in subsequent chapters. Explicit referencing back to these nine themes will not always be necessary – unless specifically required, they will remain implicit in the unfolding logic of the argument.

Towards a non-equilibrium economics

What the emerging metatheoretical synthesis discussed earlier invites is a new way of thinking about the political economy and political ecology of social-ecological systems in the sustainability age. Unfortunately, an integrated body of knowledge that achieves this synthesis has yet to be developed. However, over the past 20 years, there has been a remarkable convergence of meso-theoretical thinking about the nature of the state, economy and energy that takes complexity as a point of departure. Three recent major texts that attempt to recast the conceptual foundations of each of these three intersecting dimensions of (social-ecological) reality are worth discussing, namely *Energy, Complexity and Wealth Maximization* by Robert Ayres, *The State: Past, Present and Future* by Bob Jessop and *Origin of Wealth: Evolution, Complexity, and the Radical Remaking of Economics* by Eric Beinhocker. All three synthesize broad swathes of related literatures in their respective spheres of thought, but only Ayres directly addresses the social-ecological reality we are mainly concerned with in this book. While Beinhocker focuses on the more superficial complex dynamics of markets and Jessop's focus is on the profound dynamics of the political economy of governance, both of them do recognize the significance of the ecological crisis. The discussion that follows in this section deals mainly with the emergence of non-equilibrium economics (drawing mainly from Ayres but also from Brian Arthur), while the next deals with governance (drawing on Jessop). Read together, these two sections are merely suggestive of how a convergence could possibly be taken further into a full-blown synthesis, with Metatheory 2.0 as the point of departure. This grand synthesis, however, is not what will be attempted here. What emerges, of course, from what follows is how this meso-level convergence aligns with the relational metatheory discussed earlier and contextualizes the discussions that follow in subsequent chapters.

There is a fundamental contradiction between the economic world that neoclassical economics portrays and the realities that complexity economics addresses (Arthur, 2015). While the former underpins the models used to generate the economic policies of most governments today and many global governance institutions, the latter underpins the emerging alternatives that will be required to guide a just transition to a more sustainable epoch. Contrary to what Arthur has argued (Arthur, 2010), these two paradigms are fundamentally incompatible. For neoclassical economics, the fundamental principles are essentially greed, rationality and equilibrium (Ayres, 2016). For complexity economics, the fundamental principles are differentiate, select and amplify (Beinhocker, 2006). These value sets cannot be reconciled to create a complexity-oriented neoclassical economics. As long as neoclassical economics remains hegemonic, a just transition is inconceivable. Complexity economics provides a way out because it enables a theory of a just transition. It is, in my view, that simple.

Echoed across many different literatures (from systems theory to political ecology, ecological economics, resilience theory and environmental science), there is consensus that the fundamental premise of neoclassical economics contradicts the laws of non-equilibrium thermodynamics. As will be argued later, non-equilibrium thermodynamics is a conceptual framework rooted in late nineteenth-century physics but not fully consolidated until Prigogine's Nobel Prize winning work in the 1970s/1980s (Prigogine, 1997). By contrast, however, for neoclassical economists real economies tend towards equilibrium. As will be argued later this, in turn, is why neoclassical economists regard state intervention as economically irrational. And without state intervention, a just transition is inconceivable.

Non-equilibrium thermodynamics is the point of departure for 'complexity economics' (which means the same thing as 'non-equilibrium economics'). Indeed, non-equilibrium thermodynamics (following Prigogine) is the basis for ontological complexity in general. As a result, complexity economics (and, by the way, much heterodox economics) accepts that economies (understood now as embedded within wider social-ecological systems) tend towards disequilibrium. This, in turn, is why 'state intervention' is a necessity and therefore economically rational. What form it takes, however, is contextually specific. A 'once-size-fits-all' logic does not apply.

To further elucidate this argument, it is necessary to describe what non-equilibrium thermodynamics is about, followed by a discussion about why mainstream economics went off in a very different direction.

The first law of thermodynamics states that energy is neither created nor destroyed, otherwise known as the law of conservation of energy. Any process – chemical or mechanical – that involves the conversion of one form of energy into another (e.g. a burning coal into steam, the impact of a rock rolling down a hill into a dam) will result in an amount of energy gained (by the steam engine, the dam) equal to the amount of energy lost (by the coal, the falling rock). Given that these amounts are equal, energy is neither created nor destroyed. However, the second law of thermodynamics states that *in a closed system* over time, the amount of useful energy available to do useful work (otherwise known as exergy) diminishes with every process that occurs. As exergy diminishes, that component of energy that is not useful for work increases with each process that occurs – this increasing quantity of non-useful energy is called entropy. Eventually, the exergy component of energy (sometimes referred to as 'order') will run out and all that will be left is the entropy component (i.e. 'disorder') – the first law, therefore, remains valid (total energy defined as exergy plus entropy is not lost), but the second law explains the changing nature of the system (why exergy diminishes over time). As a result, a closed system will always tend towards thermodynamic equilibrium. A system in thermodynamic equilibrium is effectively a dead system because there is no potential for change.

Well, of course, economies – like ecosystems and biochemical systems – are not closed systems. They are open systems. In other words, they are part of systems of systems that interact and interrelate with one another. They do not exist in reality as isolated systems. Even the planetary system is dependent on radiation from an external system, namely the sun. The 'openness' of systems is why a given system can access exergy from other systems and in so doing counteract the decline in exergy and the rise in entropy that occurs within itself. A human being, for example, is a system that depends on food supplies from natural systems (which, in turn, have been organized in agricultural systems to produce food). Without food from outside itself, the biological systems of this human will tend towards equilibrium and eventually s/he will die. For Prigogine and his colleagues, this process of resisting entropy by accessing external sources of exergy is achieved through "dissipative structures" which are effectively complex adaptive clusters of systems that operate far from equilibrium (Prigogine, 1997). An open system, therefore, can access and use energy from related systems to counteract rising entropy, thus creating more order and structure for a contextually specific period of time. As Beinhocker puts it,

> Our planet, for example, is an open system; it sits in the middle of a river of energy streaming out from the sun. This flow of energy enables the creation of large, complex molecules, which in turn have enabled life, thus creating a biosphere that is teaming with order and complexity. Entropy has not gone away; things on the earth do break down and decay, and all organisms eventually die. But the energy from the sun is constantly powering the creation of new order. In open systems, there is a never-ending battle between energy-powered order creation and entropy-driven order destruction. Nature's accounting rules are very strict, and there is a price to be paid when order is created in an open system. For order to be created in one part of the universe, order must be destroyed somewhere else, because the net effect must always be increasing entropy (decreasing order).
>
> *(Beinhocker, 2006:68–69)*

The conception of complex adaptive systems that stems in part from the epistemological consequences of non-equilibrium thermodynamics makes it possible to reconceptualize economics. Economic systems are real: they are sustained by through-flows of exergy, resources and information at points far from equilibrium (see Chapter 4 for a breakdown of the real material flows that all economies depend on). The greater the complexity of the system (or, more precisely, the system of systems or 'dissipative structures'), the more sophisticated its capacity to resist entropy (i.e. disorder) by retaining and using energy, resources and information accessed from other systems. Over time, as demonstrated by Giampietro et al. (who build on the work of the economist Georgescu-Roegen[3] (1971)), an increasing quantity of energy and resources are invested in industrial techno-economic infrastructures (on terms aligned with the elite interests of those who own them). Highly specialized governance institutions are required to regulate these resource flows and capital stocks (the internal system). However, as this happens, increasing quantities of energy and resources need to be extracted from the biophysical environment (the external system) to sustain the system as a whole (Giampietro, Mayumi and Sorman, 2012). However, the dissipation of exergy (through over-exploitation) in the external environment reaches a point where it constrains the further evolution of the internal environment (most notably by not being able to provide cheap oil in abundance), no matter the level of development that has been achieved. The result is more conflicts over resources and the polycrisis that is a central focus of this book. The solution lies in a socio-metabolic transition that will create new thermodynamic flows of exergy and resources between the internal and external systems. That is what will be referred to in Chapter 2 as the 'deep transition'. For this to be a just transition, however, much more than the thermodynamics of exergy and resource flows will need to change. Whereas 'deep transition' implies a transition to a sustainable set of resources and flows, a just transition implies radical changes in who owns these resources and flows.

A just transition is unlikely to happen without resource conflicts (Ahmed, 2017). However, these could build up into wider social revolutionary action against injustice. Recent research shows that since 1500 energy revolutions have coincided with the emergence of social revolutions organized around a specific set of new revolutionary ideas about the nature of reality, political power and the future (Fischer-Kowalski, Rovenskaya, Krausmann, Pallua and Mc Neill, 2019). This has direct relevance for the current historical moment: as Chapter 8 shows, there is a global energy revolution underway, and this is having major political repercussions and impacts. As significant, the global energy revolution confirms that a 'deep transition' is underway. As argued in Chapter 8, however, the current trajectory is away from the 'energy democracies' that would ensure that the 'deep transition' is also a 'just transition'.

> Robert Ayres (2016), Eric Beinhocker (2006) and Arthur (1999) have similar goals. They want to redefine economics from a complexity perspective by critiquing and transcending neo-classical economics. They want to propose a conception of wealth that is consistent with a more sustainable and equitable world and, therefore, is not reducible to equations about supply and demand, or rational self-interest. They want to avoid the common and very misleading

epistemological mistake of transposing metaphors about the complexity of natural systems onto social systems: instead of un-useful statements such as "economies are *like* complex systems", they prefer statements like "economies *are* complex systems". They take non-equilibrium thermodynamics as a point of departure because this is the best way to conceptualize the embeddedness of economies within real-world flows of exergy and resources. They agree with the basic contention that neo-classical economics ignores the fact that economies are open *real-world* systems, and therefore they cannot escape the first *and* second laws of thermodynamics that condition the workings of the biophysical world that mainstream economics cannot grasp.

(Sekera, 2017)

The belief that economies tend towards equilibrium lies at the very centre of neo-classical economics and justifies the dogmatic belief in free markets and minimal state intervention (i.e. 'if markets *naturally* tend towards equilibrium, state intervention is irrational'). The question, of course, is where does this belief within economics originate from? The answer lies in the formulations developed by nineteenth-century economists. French economist Leon Walras and his fellow so-called 'Marginalists' succeeded in applying the mathematical techniques of differential calculus to the hitherto largely philosophical discipline of economics. Walras achieved this in his famous book published in 1872 entitled *Elements of a Pure Economics*. The claim that economics is a 'science' – and therefore by implication 'pure' – is a profoundly Walrasian assumption. It is also indelibly modernist and profoundly wrong from a complex adaptive systems perspective.

For the mathematics to work, Walras needed quantitative predictability and more specifically the equivalent of the kind of equilibrium found in the first law of thermodynamics. The result was the imagined balance between supply and demand in an *economic equilibrium* as equal to the balance of physical forces in a *physical equilibrium*. For Walras, the perfect indicator of this balance – this point of equilibrium – would be a stable price – the ultimate 'signal' that this ideal state (i.e. a balance between supply and demand) had arrived. As Beinhocker shows, Walras then borrowed the mathematics of physicist Louis Poinsot (*Elements of Statistics* published in 1803) in order to demonstrate in detailed mathematical terms for the first time how supply, demand and prices align in markets under conditions of *general equilibrium* (Beinhocker, 2006:31–32). This is when resources are most efficiently used and welfare is 'Pareto optimal'. The dye was cast. The general equilibrium models that are used to run economies today replicate this basic founding logic. Beinhocker sums up the resultant epistemological revolution that became one of the cornerstones of mainstream economics and, ultimately, of economic modernism:

Walras declared that his "pure theory of economics is a science which resembles the physico-mathematical sciences in every respect". Jevons believed that he had created a "calculus of moral effects". And Pareto proclaimed, "The theory of economic science thus acquires the rigor of rational mechanics".

In their view, the Marginalists had succeeded in their dream of turning economics into a true mathematical science.

(Beinhocker, 2006:36)

From Samuelson and Arrow in the 1940s/1950s to Milton Friedman and the Chicago School in the 1960s/1970s, this basic logic was consolidated into the late twentieth century as a fully fledged theoretical and methodological framework that became known as the neoclassical economic consensus. This theory has been formalized into the highly sophisticated general equilibrium models that are used – with few exceptions – to support economic policy formation in the world today. However, as Beinhocker shows, it was premised on a conception of equilibrium that was borrowed from physics at a time when the second law of thermodynamics was still in the very early stages of construction. The full theoretical elaboration of the second law of thermodynamics within closed/open systems was only finally realized with Prigogine's Nobel Prize winning work in the 1970s.

Contemporary non-equilibrium thermodynamics is the emergent outcome of a century of scientific endeavour that took physics to a place very far away from the equilibrium-oriented worldview (the first law) that predominated in Western scientific circles in the late 1800s. Unfortunately, economics remained ignorant of these trends, getting stuck in a nineteenth-century epistemology that was applied with great force and arrogance during the last quarter of the twentieth century to the detriment of billions of people and nature. For Beinhocker, neoclassical economics (what he refers to as the "Traditional Model") was built on a profound conceptual flaw that resulted in the "misclassification" of the economy:

> The Traditional model was created with the implicit assumption that the economy is a thermodynamically closed equilibrium system, even though, at the time, Walras, Jevons, and their fellow Marginalists did not know that they were building this assumption into their theories. For the next one hundred years, as economics and physics each went their separate ways, this assumption lay buried in the mathematical heart of Traditional Economics.
>
> *(Beinhocker, 2006:70)*

John Reed, Chairman and CEO of the giant global bank Citicorp in the 1980s, was dissatisfied with how disconnected economists were from the crisis-ridden real-world dynamics of the actual economy. He eventually funded a highly significant meeting in 1987 – at the Santa Fe Institute in New Mexico – between the world's leading economists (including former US Presidential Advisor Larry Summers) and the world's leading complexity thinkers, including leading physicists (who had recently founded the Santa Fe Institute). It was here that the economists were confronted by physicists who were extremely surprised to discover a group of renowned highly influential intellectuals who had absolutely no idea that their underlying assumptions about the nature of reality were no longer theoretically valid. This, of course, did not collapse the dogma because by then the neoclassical revolution was well underway, supported by the world's major business coalitions and unleashed by the earlier elections of Ronald Reagan as President and Margaret Thatcher as Prime Minster, respectively, of the United States and the United Kingdom. Keynesian economics was being discredited, welfarism dismantled and financialized globalization unleashed via deregulation. This was not the moment that neoclassical economists were about to fall on their swords simply because a bunch of physicists thought their theoretical assumptions were wrong.

Ascendant support for neoclassical dogma – coupled to newly acquired computing power to drive ever-more complex general equilibrium models – seemed to vindicate the economists who were arrogant enough to argue at the Santa Fe meeting that the test of the adequacy of their assumptions was not whether they accorded with reality or not (as per normal scientific practice) but whether their assumptions made it possible to make useful predictions about what is possible. As long as their self-fulfilling prophesies seemed to be delivering economic growth (which seemed evident from the late 1980s for about a decade), they felt vindicated. That they were ignoring deeper underlying dynamics that would later result in the crash of 2007/2009 (that none of them predicted) and a wider systemic polycrisis (that none of them cared about) did not seem to bother them, and since then there seems little self-recognition from within mainstream economics that the entire theory is problematic.

However, some economists trained in the neoclassical tradition – such as Georgescu-Roegen – defected, fused together thermodynamics and economics and then became bitter critics of their discipline and former colleagues (Georgescu-Roegen, 1971). Over subsequent years, Brian Arthur coordinated the collaboration at the Santa Fe Institute between economists and scientists that generated what can now be called 'non-equilibrium economics' (Arthur, 2010). Beinhocker has inherited this tradition. It is a tradition, however, that is overly obsessed with modelling market dynamics rather than the deeper structural dynamics of capital flows and property ownership. Ayres breaks from this obsession with markets by focusing on resource flows and ownership – a perspective that animates the rest of this book.

The magnum opus *Energy, Complexity and Wealth Maximization* by Ayres is firmly within the theoretical tradition initiated by Georgescu-Roegen (subsequently elaborated by Giampietro et al.). It is by far the most comprehensive theoretically integrated reinterpretation of geological history, human history, the history of technology (especially energy technologies) and ideas about energy and economics that we have today. As the integration of his life's work as a physicist and ecological economist, Ayres' book marks the coming of age of non-equilibrium economics. It provides the platform for building the economics of a just transition.

In many ways, the story Ayres wants to tell is quite simple: prior to the arrival of humans, the natural earth system had evolved over millennia in a way that accumulated natural resources ('natural wealth') in increasingly complex geological and biological systems and formations. Evolution was the thermodynamic process that gave rise to these biological and geological resources – what Ayres refers to as "the history of material differentiation, and increasing diversity and complexity" (2016:12) – or what Beinhocker referred to as the evolutionary process of "differentiate, select and amplify". Hence, Ayres wrote: "Complexity is a form of natural wealth" (2016:6). What this means is that natural wealth is embodied exergy that evolved thermodynamically within complex adaptive biological and geophysical systems over millennia.

Humans, according to Ayres, then arrived and built increasingly complex social systems based on the extraction of exergy and natural resources at ever-accelerating rates as a means of accumulating 'economic wealth' within a new set of socio-economic systems based on ownership. Following the laws of thermodynamics and the dynamics of evolution, the complex adaptive systems humans built to harness and hold exergy corresponded with rising entropy in the natural systems. While imperceptible at first, over the long run this transformation of natural wealth into economic wealth became increasingly unsustainable because there are limits to what humans can extract from the natural systems that humans depend on. This is the Anthropocene – an era where humans (in particular the 20% who consume 80% of the resources) have become a geophysical force of nature (Crutzen, 2002). Nature, however, will survive but mutate to resist entropy in ways that will cease to be useful for humans – a condition that could herald the 'post-Anthropocene'. As Ayres puts it rather gloomily,

> Resource exhaustion in human civilization bears a certain resemblance to the process that led to supernova explosions. The explosion creates a bright but brief light, and what follows is devastation.
>
> *(Ayres, 2016:9)*

The unsustainability of the accelerating transformation of natural resources and exergy into economic wealth is only one dimension of the problem. The other is the fact that social wealth is a very particular human construct that is reproduced by our collective capacity to live in imaginary worlds and organize ourselves accordingly in large numbers (Harari, 2011). Over time, this mode of organization that transformed wealth into a human construct evolved into the global capitalist system through the violence of property ownership, slavery, colonization and exploitation of wage labour. This system now dominates the global economy, with the bulk of assets owned by an increasingly smaller and richer elite (Picketty, 2014). As Ayres puts it,

> Wealth is a word that captures the notion of material possessions with value to other humans. Material possessions imply ownership, and ownership implies rights of use and rights to allow, or prohibit, rights of use by others. Owners may exchange these rights for money by selling the possession for money. But money is only valuable if there is a choice of goods or services available to purchase.
>
> *(Ayres, 2016:1–2)*

Under capitalism, these transactions occur via markets. Economists, however, have been content to focus on these market dynamics and not on the wider exogenous material dynamics and structures of ownership within which markets are embedded.

Ayres then proceeds to elaborate on a grand scale his basic thesis over several very detailed chapters about how the natural world evolved. He commences with the evolution of the cosmos, the sun and the coming into being of the earth (Chapter 4). He then goes on to give an account of the origin of life as we know it and how it evolved over time (Chapter 5). This is followed by a description of the emergence and reproduction of natural energy, water and climate cycles over time (Chapter 6). He then elaborates in as much detail the evolution of human civilization: Chapter 8 is about the evolution of energy and technology, which includes a recapitulation of the evolutionary history of the human species and how it gets organized to access and use resources. This is followed by an account of the New World and the evolution of contemporary Science (Chapter 9). And Chapter 10 is about energy, technology and the future, focusing in particular on declining exergy or exergy invested at a time when more and more exergy per unit is required to extract and use resources.

For Ayres, the purpose of this extraordinary synthesis of contemporary scientific knowledge of our natural and social systems is to build up a rich and detailed empirical case against the obfuscations of neoclassical economics. Ayres wants to show repeatedly and forcefully that economies are embedded in real flows of resources and exergy that originate from within natural systems, flow through economic systems and end up either as built stocks or as waste outputs in the natural systems (including the air, rivers, land and sea).[4] "It is high time", he argues,

> to confront the single major problem with neoclassical economics, which is the dominant paradigm today. . .: economic theory, especially as applied to economic growth, has grossly under-estimated the importance of "useful" energy (exergy) and "useful work". As already pointed out in several places, most of the useful energy consumed in the world economy today comes from fossil fuels. Not only that, most of the industrial technology that supported economic growth in the nineteenth century and most of the twentieth century was "invented" and developed to utilize natural exergy resources, notably coal and petroleum.
>
> Yet the standard theory of economic growth assumes that growth is essentially automatic ("the economy wants to grow") and that it happens smoothly thanks to the accumulation of capital per worker, although the precise mechanism is unclear. The role of increasing returns, complexity and path-dependence is not widely understood. . . . The standard economic theory since the 1950s says (in effect) that natural resources are not essential because there is always a way to substitute ([with] a little more capital or a little more labour) for any scarce physical resource.
>
> *(Ayres, 2016:413–414)*

The faith in substitutability (i.e. 'if there is a shortage of something, a substitute that achieves the same/similar outcome will emerge via market dynamics if the price is right') makes it impossible for neoclassical economists to see two obvious empirical realities (discussed in detail in Chapters 4 and 8). Firstly, as Ayres shows, they cannot grasp the significance of the fact that every major spike in the oil price has triggered a recession (2016:185). There can be no clearer empirical proof than this of the dependence of modern industrial economies on exergy. Indeed, although recessions are conventionally understood to be merely a rebalancing of supply and demand, the underlying reality is that they are a reset of the complex institutional and economic systems humans have constructed to extract exergy from natural systems for combusting to drive their economic systems. The resultant entropy is exported back into the environment resulting in climate change. Climate change–related taxes and expenditures are the latest additions to these economic systems to counteract the entropic consequences of greenhouse gas (GHG) emissions.

Secondly, the 'energy return on energy invested' (EROI) ratio (i.e. the amount of energy needed to produce a particular unit of energy)[5] has been steadily declining for nearly a century (see Chapter 8 for a detailed description of the EROI). As far as oil is concerned, whereas one barrel in the 1930s was needed to generate 100 barrels of oil, by 2010 one barrel generated between 20 (Ayres, 2016) and 10 barrels (Ahmed, 2017). In other words, the EROI ratio declined over this period from 1:100 to 1:10/20. This is the primary reason why the cost of oil has escalated. To cope with this, production levels have increased over time with oil production reaching a plateau in 2005 (Murray, 2012).

In other words, rising production levels (often celebrated as proof that 'oil peak' is a myth) is an indicator of declining EROI. This is why, Ahmed argues, we should be talking about "peak EROI" and not "peak oil" (Ahmed, 2017:22). The reason is obvious: as the EROI declines, the price of oil needs to rise to cover the rising cost of energy as the amount of exergy per barrel declines. But as the price of oil goes up, growth rates of oil-dependent economies tend to go down (see Chapter 8 for the details of this trend). Logically, a point is reached where either prices are too low to cover the costs of production (causing a production-driven decline) and/or the demand for oil declines as cheaper energy sources are found to meet rising energy demands.

In short, the thermodynamics of oil-dependent economies is pushing them further and further away from a stable equilibrium regulated purely by prices. There is no other source for the type of exergy oil can provide. A tipping point will be reached that will force policymakers to realize that the entire energy system has to change. Following similar patterns since 1500 (cited earlier), this is unlikely to happen without a political crisis (see Chapter 9) and the replacement of neoclassical economics with non-equilibrium economics.

This brings us to the very core of Ayres' concern about the current structure of the global economy, what he calls the "double whammy" of resource depletion and declining EROI. Resource depletion means that more needs to be extracted from the crust of the earth to produce the same useful output (from metals to nutrients, to nearly everything else) (International Resource Panel, 2019). This, however, requires more and more exergy. The declining EROI means that the average cost of exergy over time will rise (Ayres, 2016:20). This may generate price signals that result in the rise of alternative renewable energy resources, but renewables generate a completely different kind of exergy via a vast decentralized and distributed network of energy generation systems (with political implications further explored in Chapter 8). There is, in short, no substitute for the exergy oil can provide. This is what underlies Ayres' relentless critique of neoclassical economics: it is an ideological framework that masks the fundamental thermodynamically non-equilibrium causes of the polycrisis and, therefore, the solutions we need today.

Towards collibratory governance for contemporary non-equilibrium economies

If economies tend towards disequilibrium as argued by non-equilibrium economics (and some heterodox economists (Lavoie, 2006)), then we need an appropriate conception of the state and governance that aligns with this post-neoclassical conception of political economy. The nineteenth-century conception of the state as a function of a defined territory, a specific population and a distinct uni-centric institutional apparatus (referred to as 'government') no longer reflects the reality of actually existing political dynamics and statecraft. Instead, following Jessop (who draws from a fusion of Foucaldian and Gramscian traditions), "'the [contemporary] state in its inclusive sense' can be defined as 'government + governance in the shadow of hierarchy'" (Jessop, 2016:176). Jessop argues that because "the state is just one part of a complex social order" (Jessop, 2016:86), it is necessary to adopt what he calls a "strategic-relational approach" (SRA) that recognizes "states are polymorphic, displaying different forms depending on changing principles of societal organization or on specific challenges and conjunctures – if not on both" (Jessop, 2016:8). In short, Jessop advocates a relational – as opposed to a structural – conception of 'the state' that is appropriate for the complex realities of our times in both developed and developing countries. He argues,

> In strategic – relational terms, state power is an institutionally and discursively mediated condensation (a reflection and a refraction) of a changing balance of forces that seek to influence the forms, purposes, and content of polity, politics, and policy.
>
> *(Jessop, 2016:10)*

Hence, the state

> is a complex ensemble (or, as some scholars put it, assemblage) of institutions, organizations, and interactions involved in the exercise of political leadership and in the implementation of decisions that are, in principle, collectively binding on its political subjects. These institutions, organizations, and

interactions have varying spatiotemporal extensions and horizons of action and mobilize a range of state capacities and other resources in pursuit of state objectives.

(Jessop, 2016:16)

At the centre of this ensemble is the "polity", which Jessop defines as

the institutional matrix that establishes a distinctive terrain, realm, domain, field, or region of specifically political actions. . . . Further, while the polity offers a rather static, spatial referent, politics is inherently dynamic, open-ended, and heterogenous.

(Jessop, 2016:17)

Since the 1970s, there has been a shift in the balance of forces and therefore the composition of the polity across all world regions, making it nearly impossible (with some obvious exceptions) for governments to govern in traditional state-centric ways that can be regarded generally as enjoying popular legitimacy. In many ways, eighteenth- and nineteenth-century Weberian conceptions of government never anticipated the degree and extent of the complexities of late twentieth-century socio-economic and political realities. The result was the gradual emergence of the notion of 'governance' that seemed to capture this shift in the dynamics of state-society relations from state-centrism (both social democratic and socialist varieties) to a complex set of reciprocal – albeit unequal – *relational* (and quite often dialogical) configurations and processes that unfolded in the 'shadow of hierarchy'. It takes many different forms, but the shift from government to governance is in essence part of a general cultural-aesthetic shift from object to subject, thing to relation, passive to active, structure to process, top-down to bottom-up, from 'planning for' to 'planning with', instruction to dialogue, teaching to learning, dependence to interdependence and action to interaction. The upshot is what Hajer refers to as the emergence of the self-activated "energetic society" (Hajer, 2010). More specifically, this shift is above all else a recognition of the shift from abstraction to context. As Jessop argues,

Indeed, only when we abandon the reified notion of "the state" can we begin serious study of the state system in all its *messy complexity* and undertake a serious critique of different state ideas. . . . Only then can we hope to transcend the misrecognition of the state in the "state idea" and to examine the state as it *actually exists and operates*, on its own terms and in its wider political and social contexts.

(Jessop, 2016:18 – emphasis added)

By emphasizing context (which is especially important from an African perspective) and the embeddedness of the state conceived as a complex ensemble of unevenly developed institutions that express the social dynamics of particular contexts, the SRA creates the space for addressing a crucial problem in sustainability research. In a recent article, an eminent group of sustainability scientists argued that there is an implicit mistaken assumption within sustainability science circles that a kind of decision-making "cockpit" of key decision-makers exists somewhere that can somehow be reached and given the correct empirical message about what is wrong with the current socio-economic trajectories (Hajer, Nilsson, Raworth, Bakker, Berkhout, et al., 2015). This politically naïve perspective – critiqued here by Hajer et al. – is always deeply shocking when I come across it, which is often. Even if a cockpit did exist and such a message could be delivered into it, the decision-makers in this mythical cockpit would have received very little guidance as to *how* they should change the system and what exactly the alternatives are.

The problem with "cockpitism" is that it fails to recognize what has been achieved by at least four decades of research on the nature of politics in largely capitalist societies. For Jessop (who may well be the most significant contributor to this literature), his conclusion is as follows:

> First, there can be no general, let alone transhistorical, theory of the state. . . .
> Second, as a complex political association, apparatus, *dispositive*, ensemble, or assemblage (language varies) linked to a wider set of social relations, the state system can be studied from many theoretical entry points.
>
> *(Jessop, 2016:5 – emphasis in original)*

Hence, for Jessop, the state is a "polymorphormous institutional ensemble": if the notion of 'the state' implies the existence of a single thing with its own agency (as implied by unqualified claims such as "the state intervened to" or "the state's role is"), then the state does not exist in reality. What does exist is a 'state project' – a political vision and programme – which presupposes the improbability of a unified state system. In the language of complexity theory, Jessop argues,

> A state project denotes the political imaginaries, projects, and practices that (1) define and regulate the boundaries of the state system vis-à-vis the wider society and (2) seek to provide the state apparatus thus demarcated with sufficient substantive internal operational unity for it to be able to perform its inherited or redefined "socially accepted" tasks. . . . The state apparatus, considered as an assemblage, does not exist as a fully constituted, internally coherent, organizationally pure, and operationally closed system. It is an emergent, contradictory, hybrid, and relatively open system.
>
> *(Jessop, 2016:84)*

From this perspective of his SRA (which uses terms remarkably similar to non-equilibrium economics), Jessop identifies five dynamics that since the 1970s have contributed to the rise of relational governance (Jessop, 2016:174–175). The first is the "de-hierarchization of the state" which refers to increasing dependence on 'partnerships' with the private sector and civil society at the national/local levels and increasing "pooling or sharing of sovereignty" at the international level (e.g. European Union, regional trade agreements). The second is the "*recalibration of state power* as government makes more extensive use of networks and other modes of governance as a way of maintaining its political efficacy in the face of growing societal complexity". Thirdly, there is "destatization of the polity" as the focus shifts from the state apparatus itself to an expansion of the "general organization" of a "decentred" polity. This decentred polity is characterized by a set of (often state-coordinated) networks rather than by the traditional hierarchies of state apparatuses. Fourthly, when governance networks emerge "beyond the polity", the result is the "depolitization of power" as market-based solutions emerge to mobilize the resources of non-state actors for ostensibly public purposes, or the state becomes a weak 'actor' in – or 'member' of – networks that are driven by non-state actors (especially powerful corporates) who have the resources to act in strategically meaningful ways. Finally, there is what Foucaldians refer to as the emergence of powerful forms of neoliberal "governmentality" which refers to ways that states develop new techniques for mobilizing and disciplining civil society in order to "govern social relations at a distance rather than through direct command and control". These so-called "dispositifs" get "organized around various discursively constituted problems" in order to build support for market-based solutions that would otherwise remain unattainable (at least for those who can pay the required price).

There is, however, a counter-trend to the five governance dynamics referred to earlier that has great significance for our understanding of governance for non-equilibrium economies. This is what Jessop refers to as "collibration". The five dynamics that since the 1970s have shifted the centre of polities from government to governance are really "meta-governance" innovations that respond to the failure of traditional forms of state-centric governance under conditions of increasing complexity and ideological adherence to neoliberal dogma. These innovations were, of course, framed by this dogma. The upshot has been the weakening of the directional role of the state. This, in turn, is highly problematic because to catalyse transformation processes over the long-term to achieve particular ends, it will be necessary to strengthen rather than weaken the directional role of state institutions. This weakening of the directional capabilities of states was, after all, the intended outcome of neoclassical economics.

One option is to reverse governance with a view to recapturing a state-centric form of political power. Given rising complexity, this would probably only be possible if a significant measure of coercion is deployed to forcibly reduce complexity to clear the way for this kind of recapture. This is, of course, happening in some places. However, the alternative is emerging where some governments have started to address the need for the "governance of governance", which is what Jessop refers to as "collibration". This entails specific modes of intervention to harness the potential of governance for a particular political/state project adopted by a political leadership. It usually entails establishing a new generation of public agencies with high degrees of autonomy to facilitate governance arrangements. This would entail mobilizing the forces opposed to the weakening of the state in order to harness and strengthen a new conception of state institutions and their role in transformation processes. It is worth quoting Jessop on the modalities and dynamics of collibration at length here:

> Specifically, governments [engaged in collibration] provide the ground rules for governance and the regulatory order through which governance partners can pursue their aims; they ensure the compatibility or coherence of differ-ent governance mechanisms and regimes; they create forums for dialogue or act as primary organizers of the dialogue among policy communities; they deploy a relative monopoly of organizational intelligence and information in order to shape cognitive expectations; they serve as courts of appeal for disputes arising within and over governance; they seek to rebalance power differentials and strategic bias in regimes by strengthening weaker forces or systems in the interest of system integration and social cohesion; they try to modify the self-understanding of identities, strategic capacities, and interests of individual and collective actors in different strategic contexts, and hence they alter the implications of this self-understanding for preferred strategies and tactics; they organize redundancies and duplication in order to sustain resilience through a requisite variety, in response to unexpected problems; they take material and symbolic flanking and supporting measures to stabilize forms of coordination deemed valuable but prone to collapse; they subsidize the production of public goods; they organize side payments for those who make sacrifices for the sake of facilitating effective coordination; they con-tribute to the meshing of short-, medium- and long-term time horizons and temporal rhythms across different sites, scales, and actors, in part to prevent opportunistic exit and entry into governance arrangements; and they also assume political responsibility as addressees of last resort in the event of gov-ernance failure in domains that go beyond the state.
>
> *(Jessop, 2016:172–173)*

In short, collibratory governance re-establishes a central place for state agencies in the polity, but without returning to the statist uni-centrism of the pre-1970s era. Instead of reasserting its sovereignty (backed by overt references to a monopoly over coercion), it becomes a "primus inter pares in a complex, heterogeneous, and multilevel network of social relations. . . . an interconnected, reinforcing series of symbolic and material state capacities" (Jessop, 2016:173). Derived from the SRA, this is a conception of governance that enabled the *"respecification of structure and agency in relational terms"* (Jessop, 2016:55 – emphasis added). Political leadership in this context becomes a new kind of relational statecraft.

In practice, however, relational statecraft seems conceptually far clearer than the real-world dynamics of actual governance practices. This will be explored further in Chapter 7 where the interface between developmental states and sustainability transitions will be discussed. Furthermore, there is a dark side of this idealized conception which is only fleetingly acknowledged by Jessop:

I refer here to the way the relational shift towards governance within polities can morph into neo-patrimonial modes of governance as political leaders fuse long-term developmental commitments with rent-seeking. This theme will be further explored in Chapter 9, drawing on the South African experience of 'state capture' during the 2010s. Notwithstanding these qualifications (discussed further in Chapters 7 and 9), collibratory governance is an extremely useful concept that helps to make sense of the types of governance configurations that are appropriate for intervening in non-equilibrium economies under conditions of increasing complexity.

To link this back to the previous section's discussion of non-equilibrium economics, we need to ask why this profoundly relational notion of collibratory governance is appropriate for governing non-equilibrium economies. If it is accepted that economies tend towards disequilibrium, it follows – contra-neoliberal assumptions – that interventions by state institutions become a necessity. However, this does not mean reverting back to the classical Keynesian modes of intervention that emerged after the 1929 financial crash, maturing over the subsequent four to five decades. Increasing complexity – reinforced by the profound impact of the information and communication revolution since the 1970s (Castells, 2009) – makes a return to statist uni-centrism unrealistic. Sure, this can conceivably happen, but it will probably entail the use of coercion to reverse complexity with negative economic consequences as the spaces for innovation get shut down. This reach for certainty in an uncertain world may be what is happening in the increasing number of neo-patrimonial regimes fostered by both right-wing populism (e.g. the Trump White House) and left-leaning rent-seeking developmentalists (e.g. South Africa under Jacob Zuma until end of 2017) (see Chapter 9).

Instead, the notion of collibratory governance seems appropriate for increasingly complex, information-rich non-equilibrium economies. The heterodox economists (e.g. Mazzucato) call for state-leadership in R&D and risk reduction during the early phases of the innovation cycle but pay very little attention to the kinds of institutional configurations appropriate for this task (Mazzucato, 2011). To address this lacunae, I propose to explore further this notion of collibratory governance.

Indeed, pushing this argument further, the rapid insertion of algorithmically controlled transactions to manage everything from daily life via 'smartphones' to urban infrastructures, global trade and national energy grids shows how innovators have responded to the demand to manage complexity as they hunt down any transactions they can 'alogorithmize', wrap up in an app and sell them into the market. This is not the world that can get folded back into a uni-centric Weberian golden age. This is particularly true when it comes to the global renewable energy revolution which would be inconceivable without algorithmically controlled energy grids connected to hundreds of thousands of small and large energy generators (see Chapter 8). Collibratory governance is exactly the framework needed for designing real-world governance solutions appropriate for increasingly complex non-equilibrium economies that will become progressively more reliant on decentralized renewable energy systems. Animated by imaginaries of the future and the anticipated evolutionary potential of the present, collibratory governance, therefore, may well be what is needed for guiding a just transition.

This has brought us to the heart of Jessop's contribution to the line of argument developed thus far. Jessop's SRA and his conception of collibratory governance connect contemporary state theory with non-equilibrium economics. It is an approach that brings politics into the centre of the discussion about a just transition to a more sustainable world. His emphasis is not on some abstract modernist notion of structure as the primary constraint on change but rather – as with complexity theory – on the "importance of the strategic context of action and the transformative power of actions" (Jessop, 2016:55). Two things, therefore, really matter here: firstly, the SRA's *relational approach* to structure and agency brings into focus the kind of collibratory governance that is appropriate for governing political economies embedded within increasingly complex social ecologies that together operate far from equilibrium. Secondly, the approach creates a bridge into a particular theory of change – namely radical incrementalism (see Chapter 5). However, what Jessop does not adequately recognize is the importance of alternative modes of ownership that collibratory governance should foster as part of a wider just transition. This is where the literature on the 'commons' is useful.

Towards a relational post-capitalism

Whereas market fundamentalism since the 1980s is clearly responsible for justifying the policies that led to the current global polycrisis, state-centric alternatives have become less attractive in light of increasing complexity. As a result, the 'middle way' between market fundamentalism and state centrism remains somewhat opaque. For many of those searching for solutions in this conceptual 'in-between' space, the broad and loosely defined notion of 'post-capitalism' has become increasingly attractive (Gibson-Graham, 2006; Mason, 2015; Rifkin, 2015; Srnicek and Williams, 2015). According to Zack Walsh,

> Though there is no uniform agreement among its proponents, post-capitalism generally describes building alternatives to capitalism within the existing system using technologies, business models, and forms of social organization focused on prefiguring the Great Transition. Whereas anti-capitalist politics generally follows an oppositional logic of resistance, *post-capitalist politics redeploys existing infrastructure for activist causes.*
>
> *(Walsh, 2018:48 – emphasis added)*

In short, post-capitalism refers to an ever-widening set of spaces created by radical incrementalists (see Chapter 5) working to design, construct and operate real-world pre-figurative alternatives within the constraints of the existing systems. These alternatives could, however, coalesce into a quantum shift – a transition to a mode of production not dominated by the dominant capital-state matrix but without suppressing markets and public hierarchies (Mason, 2015). These range from small community-based initiatives to very large-scale social enterprises, to cooperatives and social movements. They mesh together into formations that some are referring to as "the commons".

Commons-oriented post-capitalism is the polar opposite of the state-centric optimism shared by many who advocate post-neoliberal alternatives ranging from *Social Democracy 2.0* to democratic socialism (Mazzucato, 2016; Mitchell and Fazi, 2017). While most of those who talk about post-capitalism avoid being specific about practical institutional arrangements/structures, the literature on the 'commons' does address this challenge, including the provision of a conception of transition.

The commons, of course, has deep historical roots in pre-capitalist communities and early industrial societies (e.g. cooperatives). Indeed, consistent with *Ukama*, many (and by no means all) pre-colonial African societies were structured around some form of commons, and in some areas these social forms still exist (albeit not in their original form). The conservative wing of the movement is often associated with the institutional economics of Nobel Prize winner Eleanor Ostrom. Whereas Ostrom was interested in the way communities collaborate to govern common resources (e.g. land, water, forests) (Ostrom, 1990, 1999), the more radical wing of the commons movement is interested in the governance of two types of converging spaces: the social commons (new modes of collaboration) and the knowledge commons (essentially codes, design and interconnected IT infrastructures). Their interest lies in the various formations within these spaces that converge in ways that prefigure more radical post-capitalist alternatives (Bollier and Helfrich, 1978; Bauwens and Ramos, 2018). Using the new information and communication technologies (ICT) for direct 'many-to-many' communications without transacting via a regulator or a market operator, the most sophisticated frameworks propose a radical vision of the commons as a shared, cooperatively governed ecological, social and knowledge space. State/hierarchies and commodification/markets have a place in this vision but subordinated to the logics of P2P and the commons. Indeed, regulation may well become key to ensuring the sustainability of P2P over the long run (Cumbers, 2015).

For Bauwens et al., the peer-to-peer mode of production becomes the basis for what they call a "commons-centric society". Driven by dynamics that emerge from within the old capital-state dominated system, ICT-enabled P2P could go to scale and eventually transcend the old system under certain conditions. Bauwens et al. connect four aspects of this emergent alternative:

- "P2P is a type of *social relations* in human networks, where participants have maximum freedom to connect": permissionless entry into flat algorithmically managed networks creates the basis for humans to associate, learn, innovate and produce across local–global scales in new ways;
- "P2P is also a *technological infrastructure* that makes the generalization and scaling up of such relations possible": unlike in the pre-digital age when markets and hierarchies were needed to take an isolated invention/innovation to scale, ICT

networks make many-to-many mutual coordination possible on a global scale without relying on hierarchies or markets;

• "P2P thus enables a new *mode of production and property"*: these new modes of production are based on a mix of collaborative voluntary construction of shared know-how in the commons and commercialization via socially and ecologically responsible entrepreneurial businesses within socially embedded markets;

• "P2P creates the potential for a *transition* to an economy that can be generative towards people and nature": once commons-based peer production starts attracting more financial flows than traditional for-profit businesses (via, for example, investments by state-owned Development Finance Institutions), a transition becomes possible – but political re-alignments within the polity will be necessary to reorient states to support this transition (Bauwens, Kostakis and Pazaitis, 2019:1).

As Castells argued, the ICT revolution introduced the 'network' as an alternative to market and hierarchical modes of organization, and ICTs also made possible 'self-managed mass communication' as a new mode of communication that was never before technically feasible (Castells, 2009). Over the past two decades, these two modes of organization and communication have fused together resulting in massive self-organization on a global scale. This has transformed and reinforced global capitalism (Castells, 1997). But "[i]t also allows", Bauwens et al. argue, "for the creation of a new mode of production and new types of social relations outside the state-market nexus" (Bauwens, Kostakis and Pazaitis, 2019:4).

The new mode of P2P production has already been co-opted by for-profit organizations into profoundly capitalist enterprises – Facebook, Uber, Bitcoin and AirBnB being the classic examples. However, there are equally significant examples of non-profit social enterprises that deploy the same P2P methods – examples include Wikipedia, Enspiral, Farm Hack, Wikihouse, Linux, Apache HTTP servers, Mozilla Firefox, Wordpress, blockchain-based trading systems and currencies, open source software platforms and the 'maker commons' (where designers load designs into open source environments so that others can improve on them, thus accelerating learning/innovation).

As discussed in Chapter 4, during the first four major pre-digital techno-industrial epochs since the industrial revolution, emerging new technologies necessitated new modes of hierarchical and/or market organization to take the new technologies to scale. The collaborative small group P2P dynamics that catalysed the innovations in the first place ('niche innovations') were too costly and institutionally impractical to replicate on scale. New (usually state) hierarchies and (profit-oriented) markets were needed (in different combinations, depending on the prevailing economic theory of the time) to reorganize societies to absorb the new technologies, sometimes with a considerable degree of force (think enclosure movements, slave labour, apartheid, 'Kulakization', cultural revolution). Those societies that had the financial and human resources to adapt the fastest and most effectively to the new modes of organization during transition periods ended up outcompeting their rivals during the deployment period because they could take the new technologies to scale. Of course, the fifth techno-industrial transition – the ICT revolution – catalysed something new: the network and self-managed mass communication. For sure, this has enabled new for-profit global corporations, but it also enables the opposite:

> Today . . . it is also possible to scale projects through new coordination mechanisms, which can allow small group dynamics to apply at the global scale. It is, thus, possible to combine "flatter" structures and still operate efficiently on a planetary scale. This has never been the case before.
>
> *(Bauwens, Kostakis and Pazaitis, 2019:4)*

In short, at exactly the historic moment when we face the distinct possibility of extinction as a species because of our capacity to unequally accumulate wealth and over-exploit nature, we also have at our disposal the capacity to create a completely new more inclusive and equitable mode of production through planetary-scale many-to-many self-managed mass communications embedded within a vast multiplicity of local economies.

P2P learning, innovation and production both reinforces the current capital-state dominated system and creates the basis for transcending it. Bauwens et al. capture this duality by referring to the "immanent" and "transcendent" potential of P2P (Bauwens, Kostakis and Pazaitis, 2019:4). The ICT-enabled P2P mutual coordination mechanisms have already been rapidly absorbed by capitalism. Indeed, capital investment in network-based systems helped to rejuvenate capital accumulation during the period leading up to the double crisis of 2001/2009 (see Chapter 4). This is the 'immanent' aspect of P2P. But, Bauwens et al. argue,

> such mechanisms can [also] become the vehicle of new configurations of production and allocation, no longer dominated by capital and state. This is the "transcendent" aspect of peer production, as it creates a new overall system that can subsume the other forms. In the first scenario, capital and state subsume the commons under their direction and domination, leading to a new type of commons-centric capitalism. In the second scenario, the commons, its communities, and institutions become dominant and, thus, may adapt state and market modalities to their interests.
>
> *(Bauwens, Kostakis and Pazaitis, 2019:4–5)*

A 'commons-centric society' is a profoundly *relational society*, thus expressing in practice the thrust of the arguments presented thus far in Chapters 1 and 2 with respect to the post-human relational self, the relationality of Metatheory 2.0 and the relational orientation of the SRA and non-equilibrium economics. The primary organizing principle of such a society is ICT-enabled mutual coordination that effectively hardwires relationality as a mode of economic production. As Bauwens et al. put it, "What market pricing is to capitalism and planning is to state-based production, mutual coordination is to peer production" (Bauwens, Kostakis and Pazaitis, 2019:5). Enacting the ethos of relationality as organized mutual coordination can then become the way resources are allocated without having to coordinate redistribution via hierarchical systems:

> As a result, the emergence and scaling of these P2P dynamics point to a potential transition in the main modality by which humanity allocates resources: from a market-state system that uses hierarchical decision-making (in firms and the state) and pricing (amongst companies and consumers), towards a system that uses various mechanisms of mutual coordination. The market and the state will not disappear, but the configuration of different modalities – and the balance between them – will be radically reconfigured.
> *(Bauwens, Kostakis and Pazaitis, 2019:5)*

The commons-based peer production system massively reduces dependence on the corporate and state systems that have proven incapable of absorbing labour or resolving the current global economic crisis. In 2018, 90% of all financial assets generated negative returns! State and market systems have no solution for this massive global deployment of unproductive ungenerative assets. In a world where millions cannot make a decent living and where access to ICTs is increasingly affordable, a commons-based peer production system is exactly what is needed: it creates a new cycle of accumulation based on shared value through collaboration, algorithmically regulated participatory governance, rapid learning and innovation and shared outputs with limited extraction of surplus value. This is a new logic of socialized accumulation that rivals the classic capitalist logic of capital accumulation. In societies plagued by poverty, unemployment and inequality, it is the only alternative to violent revolution, war or implosion – the three ways wealth has historically been destroyed as a prelude to new modes of production. This is not to say that redistribution is not required; rather it is the underlying logic for redistribution and how it could work in practice that changes.

This may sound idealistic but, as Bauwens et al. show, major commons-based peer production (CBPP) systems already exist, with characteristic features (Table 2.2). These include the following:

TABLE 2.2 Commons-based peer production (CBPP) systems

Productive community	Linux	Mozilla	GNU	Wikipedia	Wordpress
Entrepreneurial coalition	For example, Linux Professional Institute, Canonical	For example, Mozilla Corporation	For example, Red Hat, Endless, SUSE	For example, Wikia company	For example, Automatic company
For-benefit association	Linux Foundation	Mozilla Foundation	Free Software Foundation	Wikimedia Foundation	Wordpress Foundation

Source: Bauwens, Kostakis and Pazaitis, 2019:16) Note: GNU = General Public License.

Bauwens et al. show that each of these major global initiatives have three exemplary features: a 'productive community' of people who voluntarily beaver away at creating new and improving existing know-how in the commons; an 'entrepreneurial coalition' that is licensed to exploit the know-how in the commons in the wider market but with controls over the distribution of surplus; a 'for-benefit association' that gets supported from the revenues generated to reinvest in the capabilities of the productive community and wider environment. These three become the organizational template for building up from below the CBPP. To make it happen, so-called 'transvestments' from the traditional capitalist sector into this CBPP will be required until such time that the CBPP sector has its own autonomous capital base.

Conclusion

Relationality or relatedness has become the central organizing principle of large swathes of contemporary scholarship and practice. Rooted in the deep paradigmatic shift within Western science and philosophy from the iconic Vitruvian Man to the relational self discussed in Chapter 1, I have shown in this chapter how relationality is at the centre of the convergences that have resulted in Metatheory 2.0, non-equilibrium economics, the 'strategic-relational approach' to governance and, of course, the alternative economic vision of commons-based peer production. I have also shown that the rise of relationality/relatedness in Western scholarship aligns with the much longer Sub-Saharan African traditions associated with the notion of *Ukama*.

To conclude, this chapter has elaborated a metatheoretical conception of a relational subjectivity and reality that provides a foundation for the discussions that follow in subsequent chapters. As will be argued in chapter 4, a deep transition may be a necessary condition for the just transition anticipated in the Preamble to the SDGs, but it is by no means a sufficient condition. For a deep and just transition to occur, we will need to make far more explicit than has hitherto been the case that a fundamental change in mindset will be required. This, in turn, will not happen in an intellectual vacuum: to harness and focus a vast amalgam of diverse but related intellectual trends that converge in a conception of relationality, disciplined intellectual work will be required to consolidate a core body of knowledge that makes sense of what is emerging. However, this must be propelled forward as a transformative force that transcends the limitations of both modernist and postmodernist thinking.

The synthesis of influential metatheoretical trends into what Bhaskar et al. call Metatheory 2.0 serves this purpose well. However, to translate this into the real world, it will be necessary to reconstitute the epistemological and ontological foundations of mainstream economics. Although computing power has allowed neoliberal economics to build increasingly complex models, this tradition still assumes there is a natural tendency towards equilibrium. As a contribution to this intellectual project, this chapter has explicitly counterposed neoliberal economics and the laws of thermodynamics. The result is a conception of non-equilibrium economics that is consistent with Metatheory 2.0 and underpins a particular theory of collibratory governance. Indeed, collibratory governance of non-equilibrium economies will need to be at the very centre of the discussion about the dynamics of deep and just transitions. Most significantly of all, by coupling together a theory of non-equilibrium economics and a theory of collibratory governance, a conceptual and strategic space is cleared for imagining the highly complex non-equilibrium dynamics of CBPP.

Non-equilibrium economics leads to two seemingly contradictory conclusions: that because economies tend towards disequilibrium, state intervention will be necessary; and that economies become more complex over time. From the perspective of traditional state theory, greater intervention will mean reducing complexity because the aim of intervention is about gaining greater control for the sake of directionality. But from the SRA perspective, the opposite becomes possible: instead of reducing complexity to implement a traditional conception of 'state intervention', a new generation of collibratory institutions must take responsibility for the 'governance of governance' of increasingly complex transitional dynamics. This, in turn, will be imperative if the CBPP is to evolve within the current system, but then transcend it as the transition deepens. Without the emergence of an appropriate set of collibratory governance institutions led by new political coalitions within the polity to support the emergence and consolidation of the CBPP, the CBPP will forever be limited to a subsector of the global capitalist system. And Metatheory 2.0 will remain a marginalized imaginary.

Notes

1 I was part of a small group that participated in a day-long engagement in Paris between my Stellenbosch University colleague Paul Cilliers, Edgar Morin and Basarab Nicolescu which significantly influenced my way of thinking.
2 It is impossible within the confines of this introductory chapter to do justice to these underlying frameworks. Only the most basic essence of each framework is extracted that are relevant for the overall argument. Furthermore, there are those who contributed to the volume edited by Bhaskar, Esbjorn-Hargens, Hedlund and Hartwig (2016) who question the value of a synthesis of the three metatheories. They have a point if they assume the synthesis trumps the three underlying metatheories – that is not the point.
3 Georgescu-Roegen sacrificed a promising career as a competent conventional economist when he changed his views after engaging with non-equilibrium thermodynamics to become one of the founders of non-equilibrium economics – see Georgescu-Roegen (1971).
4 Chapter 3 is a detailed review of a particular body of knowledge that documents this socio-metabolic flow of resources through the global economy.
5 Although the EROI approach uses the term 'energy', as Ayres shows they are actually primarily talking about exergy, i.e. the usable part of energy.

References

Ahmed, N. M. (2017) *Failing States, Collapsing Systems: Biophysical Triggers of Political Violence.* Cham, Switzerland: Springer.

Arthur, B. (2010) 'Complexity, the Santa Fe approach, and non-equilibrium economics', *History of Economic Ideas*, 18(2), pp. 149–166.

Arthur, W. B. (1999) 'Complexity and the economy', *Science*, 284(April), pp. 107–110.

Arthur, W. B. (2015) *Complexity and Economy*. Oxford: Oxford University Press.

Ayres, B. (2016) *Energy, Complexity and Wealth Maximization*. Berlin: Springer.

Bauwens, M., Kostakis, V. and Pazaitis, A. (2019) *Peer To Peer: The Commons Manifesto*. London: University of Westminster Press.

Bauwens, M. and Ramos, J. (2018) 'Re-imagining the left through an ecology of the commons: Towards a post-capitalist commons transition', *Global Discourse*, pp. 1–19.

Behrens, K. (2014) 'An African relational environmentalism and moral considerability', *Environmental Ethics*, 36(1), pp. 63–82.

Beinhocker, E. (2006) *The Origin of Wealth: Evolution, Complexity and the Radical Remaking of Economics*. Cambridge, MA: Harvard Business Review Press.

Bhaskar, R., Esbjorn-Hargens, S., Hedlund, N. and Hartwig, M. (eds.) (2016) *Metatheory for the Twenty-First Century: Critical Realism and Integral Theory in Dialogue*. New York: Routledge.

Bollier, D. and Helfich, S. (eds.) (1978) *Patterns of Commoning*. Amherst, MA: Off the Common Books.

Bujo, B. (1997) *The Ethical Dimension of Community: The African Model and the Dialogue Between North and South*. Nairobi: Paulines Publications.

Capra, F. (1996) *The Web of Life*. New York: Anchor Books.

Castells, M. (1997) *The Information Age Volumes 1, 2 and 3*. Oxford: Blackwell.

Castells, M. (2009) *Communication Power*. Oxford: Oxford University Press.

Cilliers, P. (1998) *Complexity and Postmodernism: Understanding Complex Systems*. London: Routledge.

Coetezee, P. and Roux, A. (eds.) (1998) *Philosophy from Africa: A Text with Readings*. Jacana: Johannesburg.

Coetzee, A. (2017) *African Feminism as a Decolonising Force: A Philosophical Exploration of the Work of Oyeronke Oyewumi*. Stellenbosch: Stellenbosch University.

Crutzen, P. J. (2002) 'The anthropocene: Geology and mankind', *Nature*, 415, p. 23.

Cumbers, A. (2015) 'Constructing a global commons in, against and beyond the state', *Space and Polity*, 19(1), pp. 62–75. doi: 10.1080/13562576.2014.995465.

Ehrlich, P. (2000) *Human Natures: Genes, Cultures and the Human Prospect*. Washington, DC: Island Press.

Esbjorn-Hargens, S. (2016) 'Developing a complex integral realism for global response: Three meta-frameworks for knowledge integration and coordinated action', in Bhaskar, R., Esbjorn-Hargens, S., Hedlund, N. and Hartwig, M. (eds.) *Metatheory for the Twenty-First Century*. New York: Routledge, pp. 99–139.

Eze, M. (2009) 'What is African communitarianism? Against consensus as a regulative ideal', *South African Journal of Philosophy*, 27(4), pp. 386–399.

Fischer-Kowalski, M., Rovenskaya, E., Krausmann, F., Pallua, I. and Mc Neill, J.R. (2019). 'Energy transitions and social revolutions', *Technological Forecasting and Social Change*, 138, pp. 69–77. doi: 10.1016/j.techfore.2018.08.010.

Georgescu-Roegen, N. (1971) *The Entropy Law and the Economic Process*. Cambridge, MA: Harvard University Press.

Giampietro, M., Mayumi, K. and Sorman, A. H. (2012) *Metabolic Pattern of Societies: Where Economists Fall Short*. Abingdon and New York: Routledge.

Gibson-Graham, J. (2006) *A Postcapitalist Politics*. Minneapolis, MN: University of Minnesota Press.

Hajer, M. (2010) *The Energetic Society: In Search of a Governance Philosophy for a Clean Economy*. The Hague: PBL. Available at: www.pbl.nl/sites/default/files/cms/publicaties/Energetic_society_WEB.pdf.

Hajer, M., Nilsson, M., Raworth, K., Bakker, P., Berkhout, F., et al. (2015) 'Beyond cockpitism: Four insights to enhance the transformative potential of the sustainable development goals', *Sustainability*, 7, pp. 1651–1660.

Hall, S., Held, D., Hubert, D. and Thompson, K. (1996) *Modernity: An Introduction to Modern Societies*. Oxford: Blackwell.

Harari, Y. N. (2011) *Sapiens: A Brief History of Humankind*. London: Vintage.

Hedland, N., Esbjorn-Hargens, S., Hartwith, M. and Bhaskar, R. (2016) 'Introduction: On the deep need for integrative metatheory in the twenty-first century', in *Metatheory for the Twenty-First Century*. New York: Routledge, pp. 1–35.

International Resource Panel. (2019) *Mineral Resource Governance in the 21st Century: Gearing Extractive Industries Towards Sustainable Development*. Nairobi: United Nations Environment Programme.

Jessop, B. (2016) *The State: Past Present Future*. Cambridge: Polity Press.

Kagame, A. (1976) *Contemporary Bantu Philosophy*. Paris: Presence Africaine.

Keane, J. (2009) *The Life and Death of Democracy*. London: Simon & Schuster.

Lavoie, M. (2006) *Introduction to Post-Keynesian Economics*. doi: 10.1111/j.1813-6982.1961.tb02406.x.

Mama, A. (2001) 'Talking about feminism in Africa', *Agenda*, 50, pp. 58–63.

Marshall, P. (2016) 'Towards a complex integral realism', in Bhaskar, R., Esbjorn-Hargens, S., Hedlund, N. and Hartwig, M. (eds.) *Metatheory for the Twenty-First Century*. New York: Routledge, pp. 140–182.

Masolo, D. (1994) *African Philosophy and in Search of Identity*. Indiana: Indiana University Press.

Mason, P. (2015) *Postcapitalism: A Guide to Our Future*. London: Penguin.

Mazzucato, M. (2011) *The Entrepreneurial State*. London: Demos.

Mazzucato, M. (2016) *Rethinking Capitalism: Economics and Policy for Sustainable and Inclusive Growth*. West Sussex: John Wiley & Sons.

Mbembe, A. (2002) 'African modes of self-writing', *Public Culture*, 14, pp. 239–273.

Mishra, P. (2018) *Age of Anger: A History of the Present*. London: Penguin Random House.

Mitchell, W. and Fazi, T. (2017) *Reclaiming the State: A Progressive Vision of Sovereignty for a Post-Neoliberal World*. London: Pluto Press.

Morin, E. (1999) *Homeland Earth*. Cresskill, NJ: Hampton Press.

Murove, M. (ed.) (2009a) *African Ethics: An Anthology of Comparative and Applied Ethics*. Durban: University of KwaZulu-Natal Press.

Murove, M. (2009b) 'Beyond the savage evidence ethic: A vindication of African ethics', in Murove, M. F. (ed.) *African Ethics: An Anthology of Comparative and Applied Ethics*. Scottsville, South Africa: University of KwaZulu-Natal Press, pp. 14–32.

Murray, J. and King, D. (2012) 'Oil's tipping point has passed', *Nature*, 481, pp. 433–435.

Ostrom, E. (1990) *Governing the Commons*. Cambridge: Cambridge University Press.

Ostrom, E. (1999) 'Revisiting the commons: Local lessons, global challenges', *Science*, 284, pp. 278–282.

Oyewumi, O. (1997) *The Invention of Women: Making an African Sense of Western Gender Discourses*. Minnesota: University of Minnesota Press.

Peat, D. (2002) *From Certainty to Uncertainty: The Story of Science and Ideas in the Twentieth Century*. Washington, DC: Joseph Henry Press.

Picketty, T. (2014) *Capital in the Twenty-First Century*. Boston, MA: Belknap Press.

Preiser, R., Biggs, O., de Vos, A. and Folke, C. (2018) 'A complexity-based paradigm for studying social-ecological systems', *Ecology and Society*, 23(4), p. 46. doi: 10.5751/ES-10558-230446.

Prigogine, I. (1997) *End of Certainty*. Kinston, NC: The Free Press.

Rifkin, J. (2015) *The Zero Marginal Cost Society: The Internet of Things, Collaborative Commons, and the Eclipse of Capitalism*. New York: St. Martin's Griffin.

Sekera, J. (2017) *Missing from the Mainstream: The Biophysical Basis of Production and the Public Economy*. Working Paper No. 17–02. Medford, MA: Global Development and Environment Institute, Tufts University.

Simone, A. (2004) *For the City Yet to Come: Changing African Life in Four Cities*. Durham, NC and London: Duke University Press.

Srnicek, N. and Williams, A. (2015) *Inventing the Future: Postcapitalism and a World Without Work*. New York: Verso.

Swilling, M. and Annecke, E. (2012) *Just Transitions: Explorations of Sustainability in an Unfair World*. Tokyo: United Nations University Press.

Walsh, Z. (2018) 'Navigating the great transition via post-capitalism and contemplative social sciences', in Giorgino, V. and Wash, Z. (eds.) *Co-Designing Economies in Transition: Radical Approaches in Dialogue With Contemplative Social Sciences*. Cham, Switzerland: Palgrave MacMillan, pp. 43–62.

Whitehead, A. (1929) *Process and Reality*. London: Rider and Company.

Wiredu, K. (1995) *Conceptual Decolonization in African Philosophy: Four Essays by Kwasi Wiredu*. Ibadan: Hope Publications.

Wiredu, K. (ed.) (2004) *A Companion to African Philosophy*. Oxford: Blackwell Publishing.

PART II
Rethinking global transitions

3

UNDERSTANDING OUR FINITE WORLD

Resource flows of late modernity[1]

Introduction

This chapter will analytically review the contribution made by the *International Resource Panel* (IRP)[2] to our understanding of the dynamics of the deep transition discussed in Chapter 4. The South African Government nominated me to be a member of this body in 2007, the year it was founded. UNEP (as it was then called) decided to establish the IRP in the wake of the Fourth Assessment Report of the Intergovernmental Panel on Climate Change (IPCC) that won the Nobel Prize in 2007. This report argued that the decarbonization of the global economy would only be possible if it was transformed. However, climate science cannot provide the framework for how to do this.

After serving three four-year terms, my participation will end in 2019. I have been involved from the start in shaping the IRP's intellectual project, in particular through its founding report on decoupling of which I was co-lead author. The IRP can be understood as a collaborative effort by a diverse group of researchers to document the socio-metabolic case for why the industrial epoch has effectively reached the end of its 250-year historical cycle. Although this documentary evidence suggests that the necessary conditions are in place for a socio-metabolic transition to a more sustainable epoch (as part of a wider deep transition), this by no means implies that the IRP has developed a view on whether sufficient conditions exist for such a transition to happen. Now that the Sustainable Development Goals (SDGs) have been approved, this may provide the context for such a task. The IRP has yet to pay attention to the key factors that will determine the nature of such a transition, namely the social actors, their networks and the highly complex dynamics of the institutions that make up the polities of each nation-state.

The IRP was established by UNEP – now known as United Nations Environment (UNE)[3] – in 2007. By 2017 it had 24 members from 26 countries. It is not constituted like the IPCC as an *intergovernmental* expert panel. Instead, it is a panel of experts funded by governments and UNE. It has a Steering Committee comprised of government representatives who consider the scientific reports of the Panel members but without the requirement that reports must first be approved by the Steering Committee before they are published. The Steering Committee, however, does have the power to approve the initiation of reports. The Panel members come from a wide range of scientific disciplines and intellectual traditions, with some closely allied to their respective governments while others are thoroughly independent and even oppositional within their domestic policy environments.

The original objectives of the IRP were to:

- provide independent, coherent and authoritative scientific assessments of policy relevance on the sustainable use of natural resources and their environmental impacts over the full life cycle;
- contribute to a better understanding of how to decouple economic growth rates from the rate of resource use and environmental degradation.

Energy, resources and human civilization

There is growing acceptance across a wide range of audiences that 'modern society' is currently facing historically unprecedented challenges. The advent of the 'Anthropocene' comes with an all-pervasive sense that landscape pressures like climate change, resource depletion and ecosystem breakdown threaten the conditions of existence of human life as we know it (Crutzen, 2002). The result of the converging techno-economic, socio-technical and socio-metabolic crises discussed in Chapter 4 is an interregnum Edgar Morin has usefully called a 'polycrisis' (Morin, 1999:73).

This chapter aims to deepen our understanding of the complex interactions between two primary complex adaptive systems (with their own interdependent myriad of subsystems): the socio-economic systems that comprise industrial modernity (or what Ahmed calls "human civilization" (2017)) and the biophysical systems that these socio-economic systems depend on for energy and resources. These interactions are understood here from a socio-metabolic transition perspective. This means I am interested in what Giampietro et al. refer to as the "metabolic pattern of society" (Giampietro, Mayumi and Sorman, 2012), namely the flow of exergy and resources *through* the global social-ecological system from natural systems into the economies that make up the global economy and back out into natural systems in context-specific ways. To increase capacities to extract, retain and deploy energy and resources over time in the real world, increasingly complex adaptive systems – with increasingly sophisticated institutional/regulatory capacities and technological capabilities – get assembled and extended for managing these dissipative structures (see Chapter 2) (Giampietro, Mayumi and Sorman, 2012; Ahmed, 2017). Following the argument in Chapter 2, they operate thermodynamically far from equilibrium because up to a certain point there is an ever-increasing flow of exergy and materials through them. However, there comes a point where no matter the capacity to further extend the complexity of a given set of socio-economic systems (e.g. via IT systems to increase efficiencies), there is no escaping the biophysical thermodynamic limits to the energy and materials that have been over-exploited over time. This is made worse by the fact that the dissipation of exergy and resources in the natural environment is taking place exactly when there is accelerated rising demand for energy and resources in the new emerging industrializing nations, some of whom are large developing countries. This is the empirical reality addressed in this chapter, namely the biophysical conditions that make the reconstitution of socio-economic systems an urgent necessity. The evidence suggests that biophysical conditions of existence of these socio-economic systems over the longer term can no longer be taken for granted.

How socio-economic systems – or 'human civilization'– adapt to these conditions, however, is dependent on the political power dynamics of ownership of the land, resources and technology that enables the production of particular kinds of exergy and resources under capitalist conditions. This path-dependent pattern is locked in by particular configurations of institutionalized political power. Political coalitions within the polity committed to fostering deep transitions (just or not) may well start taking over governments in their respective countries. How they reconfigure the polities they take over and then deploy state institutions in partnership with non-state actors with the capacity for initiating sustainability-oriented innovations is what will make all the difference. Without such a shift in the balance of power, we are likely to see a rapid rise in the extent and frequency of resource conflicts (Swilling and Annecke, 2012:Chapter 7; Ahmed, 2017), on the one hand, and a spreading and deepening of incrementalist solutions with potential to coalesce into new regimes on the other. The first can, of course, catalyse the latter under certain conditions. Before these political dynamics are addressed in forthcoming chapters (see Chapters 4 and 5), the summary overview of the biophysical limits to industrial modernity must be discussed in detail.

Contextualizing the work of IRP

Three conditions make this particular deep transition unique, of which only one is given sufficient emphasis in most reports. The first – which *is* generally recognized – is the fact that it is probably going to depend on the collective intent of specific constellations of actors who will need to collaborate at global, national and local levels. It is for this reason that the GACGC Report argues as follows:

> The imminent transition must gain momentum on the basis of the scientific findings and knowledge regarding the risks of continuation along the resource intensive development path based on fossil fuels, and shaped

by policy-making to avoid the historical norm of a change in direction in response to crises and shocks.

(German Advisory Council on Climate
Change, 2011:84)

This statement clearly defines the historic role of anticipatory science as key driver of the next great transformation (Poli, 2014). This is why the work of the IRP, the IPCC, Intergovernmental Science-Policy Platform on Biodiversity and Ecosystem Services (IPBES), Future Earth and many other global scientific initiatives is significant. If they can contribute to the translation of anticipatory science into an anticipatory culture, then accumulated evidence about the risks we face and the potentials that can be exploited might just tilt the balance in favour of human survival (Poli, 2014). This, however, is a big "if". Wider cultural and political changes will be required before this can really happen.

The co-evolutionary dynamics of anticipatory science (Poli, 2018), the network mode of organization (Castells, 2009), the ICT-enabled CBPP learning (in their capitalist and post-capitalist forms) and the reconfiguration of spaces of agglomeration caused by accelerated urbanization (Swilling, Hajer, Baynes, Bergesen, Labbe, et al., 2018) create conditions that make it possible to consider *how* the four dimensions of transition discussed in Chapter 4 could converge into a deep transition. This provides the context for understanding the enormous significance of the rapidly expanding body of work that has been generated by the IRP since 2007.

At the most simplest level, the IRP is providing the documented evidence across a range of fields that it is no longer possible to conceive of a future for modern society that rests on the assumption that there are unlimited resources available for ensuring the well-being of over 9 billion people on a finite planet by 2050. In other words, the IRP is documenting the end of the industrial socio-metabolic epoch and by implication anticipates a deep transition to a more sustainable socio-metabolic order. However, the IRP has also put in place within the global policy community a way of thinking that is different to the two other mainstream bodies of sustainability science, namely climate science and ecosystem science (International Resource Panel, 2019b).

By thinking of socio-technical and techno-economic systems as socio-metabolic systems that consume, transform and dispose of resources extracted from natural systems, the IRP has put in place a key conceptual framework for imagining the dimensions and modalities of the deep transition. The notion that we need to decouple economic development and well-being from the rising rate of resource consumption is potentially a very radical idea, especially if this implies massive reductions in resource use per capita for people living in rich countries and a redefinition of development for those policymakers in poorer countries committed to poverty eradication. It is a notion, however, that has been robustly criticized for legitimizing 'economic growth' (Jackson, 2009; Naess and Hoyer, 2009; Ward, Sutton, Werner, Costanza, Mohr, et al., 2016). Used to imply that the current economic system can be 'greened', this criticism is valid. But a sustainable future based on ever-rising extraction of natural resources is inconceivable. How we build a more equitable world of over 9 billion people by 2050 without destroying the planet will not only depend on the mainstreaming of an appropriate political economy to replace neoliberalism (Picketty, 2014; Mason, 2015; Mazzucato, 2016) but will also mean imagining a deep transition premised on fundamentally reconfiguring the flow of non-renewable and renewable resources through our socio-technical and techno-economic systems. The research assessments generated by the IRP since 2007 provide a significant starting point and partial foundation for imagining this deep transition.

Overview of the work of the IRP

Unlike the IPCC, up until 2019 when the Global Resources Outlook was produced, the IRP has not produced an integrated report at specific points in time. Instead, the IRP publishes reports as and when they have been produced by one or more members of the IRP and their respective research teams. This means there is no integrated synthesis of the IRP's body of knowledge. For the purposes of this chapter, the work of the panel has been divided into the following categories[4]:

- *global resource perspectives*, with special reference to decoupling rates of economic growth from rates of resource use by focusing on the importance of resource productivity (Decoupling 1 and Decoupling 2), the environmental impacts of products and materials and the beginnings of scenario thinking;
- *nexus themes*, including cities, food, trade and GHG mitigation technologies;
- *specific resource challenges* with respect to two clusters of issues, namely metals (both stocks-in-use and recycling) and ecosystem services (including water and land use/soils)[5]; and
- *governance* with respect to mineral resources, SDGs and cities.

The global resource perspectives define the IRP's commitment to focus on the resource inputs into the global economy and, therefore, on how future economic trajectories (whether growth-oriented or not) can be decoupled from the prevailing rising level of resource use over time. Without this kind of decoupling, a deep transition will be unlikely. Nexus themes are about specific spheres of action constituted by highly complex socio-technical systems where the potential for decoupling exists. Specific resource challenges are about resource regimes that are under threat from, for example, rising demand and prices and can also be potential threats to larger systems that are dependent on them. The recent emergence of reports on governance represents the start of the IRP's shift into thinking about transition. At its 24th meeting in early 2019 (which took place in Nairobi), it approved a Terms of Reference submitted by myself to prepare a 'think piece' on transition.

Global resource perspectives

The environmental science of pollution, climate science and ecosystem science have traditionally been the three underlying bodies of science that have supported the claims of the environmental movement. In recent years, material flow analysis has emerged as the fourth body of science, with roots in industrial ecology, resource economics and political economy (Fischer-Kowalski, 1998, 1999). Major historical reinterpretations of agricultural and industrial economic transitions have now been written that are clearly extremely useful for anticipating the dynamics of future transitions (Fischer-Kowalski and Haberl, 2007; Giampietro, Mayumi and Sorman, 2012; Smil, 2014). The focus has shifted from the negative environmental impacts of the outputs of industrial processes to the material inputs into a global economy that depends on a finite set of material resources. This is the discursive framework within which the work of the IRP should be located.

One of the first reports produced by the IRP (generally referred to as 'Decoupling 1') entitled *Decoupling Natural Resource Use and Environmental Impacts from Economic Growth* presented evidence on the use of four categories of resources: biomass (everything from agricultural products to clothing material like cotton to forest products), fossil fuels (oil, coal and gas), construction minerals (essentially cement, building sand, etc.) and ores and industrial minerals (Fischer-Kowalski and Swilling, 2011). The Decoupling 1 Report showed that by the start of the twenty-first century, the global economy consumed between 47 and 59 billion metric tons of resources per annum (which is equal to half of what is physically extracted from the crust of the earth). During 1900–2005, total material extraction increased by a factor of 8 and annual GDP increased by a factor of 23. The result is relative decoupling between rates of resource use and global growth rates (Fischer-Kowalski and Swilling, 2011).

As the Decoupling 1 Report shows, rising global resource use during the course of the twentieth century (including the socio-metabolic shift that took place from mid-century onwards as non-renewables grew and dependence on renewable biomass declined in relative terms) corresponded with declining real resource prices – a trend that came to an end in 2000–2002. Since 2000–2002, the macro trend in real resource prices has been upwards (notwithstanding dips along the way).

The data on global resource flows in the IRP's 2011 Decoupling Report was updated in the IRP's 2016 report entitled *Global Material Flows and Resource Productivity* (Schandl, Fischer-Kowalski, West, Giljum, Dittrich, et al., 2016). According to this report, "annual global extraction grew from 22 billion metric tons in 1970 to around 70 billion metric tons in 2010". Unsurprisingly, given the extent of the second urbanization wave, the resources with the highest growth rate were the non-metallic minerals used in construction (mainly building sand and cement).

Equally unsurprising is the fact that the growth in domestic material extraction has grown faster in the Asia/Pacific compared to other regions. If Africa had addressed its development challenges as successfully as the Asian nations, domestic extraction (DE) in Africa also would have been far higher.

The McKinsey Global Institute report (which was published after the IRP report) generally confirms the trends identified by the Decoupling 1 Report. This report also demonstrates that resource prices increased by 147% over the decade starting in 2000. As a result investments in resource productivity over the long-term can generate returns of 10%, more if the $1.1 trillion "resource subsidies" are removed (McKinsey Global Institute, 2011).

When it comes to assessing the significance of material flows from a just transition perspective, much depends on the measurements used. The aforementioned figures respectively reflect global extraction and DE, that is, the total quantity of resources extracted globally and then the total quantity extracted per country aggregated into world regions. However, DE does not equal what is consumed because there are nations that export a significant proportion of their extracted resources and there are nations that import a significant proportion of their extracted resources. Domestic Material Consumption (DMC) refers to what is actually consumed within a country, which includes DE minus exports plus imports. However, what this calculation masks is the quantity of material resources that goes into the production of exports that are consumed in other countries.

Using DMC as its main indicator, Decoupling 1 effectively employed a producer perspective that allocated the 'ecological rucksack' (i.e. materials used to produce exports) of imported goods to the exporting country. If, however, the ecological rucksack is attributed to the importing country, apparent decoupling by burden shifting is no longer possible (Wiedman, Schandl, Lenzen, Moran, Suh, et al., 2013). Indeed, Wiedman, Schandl, Lenzen, et al. calculated that 40% of domestically extracted resources were used to enable the exports of goods and services to other countries (Ibid). Figure 3.1 reflects the material footprint of nations (in tons/cap)

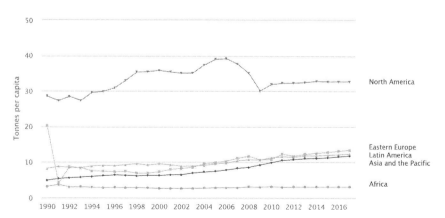

FIGURE 3.1 The material footprint of nations.

Source: Schandl, Fischer-Kowalski, West, Giljum, Dittrich, et al., 2016

where ecological rucksacks are attributed to the consumer and not to the producer country.

This is problematic when looking at a region like Europe. From a DMC perspective, Europe appears to be decoupling its growth rates from rates of increase in DMC. In reality, however, this is because it is importing more finished products produced elsewhere. Imports into Europe, therefore, can be understood to have a resource 'rucksack', that is, the resources used to produce the imports that are attributed to the country of origin. To remedy this problem, Wiedman et al. developed the notion of a 'material footprint' which attributes the resources used to produce imports to the importing nation (Wiedman, Schandl, Lenzen, Moran, Suh, et al., 2013). The result of this calculation is that there is no evidence of decoupling in developed nations that depend on imports, and major developing nations that are large exporters (like China) look far more sustainable (in terms of resource use per capita or per unit of economic output) than would otherwise be the case. Figure 3.1 reflects the global material footprints of all nations. Unsurprisingly, this map correlates with what a global map of material wealth levels would look like.

A key conclusion of the Decoupling 1 Report is that a transition to a more sustainable global economy will depend on "absolute resource reduction" in the developed world and some developing countries and relative decoupling of economic growth rates from rates of resource use in those parts of the developing world where economic growth is still required. Significantly, we did not use the notion of absolute decoupling in this report, despite repeated references to this concept in subsequent IRP reports. This is because we agree with the critique of the notion of 'absolute decoupling' as a concept that legitimizes GDP growth (Naess and Hoyer, 2009; Ward, Sutton, Werner, Costanza, Mohr, et al., 2016).

The Decoupling Report argues that without absolute resource reduction and (where necessary) relative decoupling, the result may well be an increase in total resource use from 60 billion tons in 2005 to 140 billion tons by 2050 if all 9 billion living on the planet by then consume the equivalent of the average European (i.e. 16 tons per annum per capita, which is half of what the average American consumes). However, if the convergence point is 8 tons/cap, the total material requirement would be 70 billion tons by 2050 on a planet of 9 billion people. The Decoupling 1 Report suggests that the material equivalent of living in ways that will result in the emission of 2 tons of CO_2 per annum per capita by 2050 on a planet of 9 billion people (as recommended by the IPCC) may well be 60 billion tons or 6 tons/cap for everyone. This would only be achieved by massive improvements in resource efficiency, reductions in material consumption for those who overconsume (about 1 billion people), a transition to renewable energy (to reduce fossil fuel consumption over the long term) and a drastic reduction in the use of cement (most probably by increasing the use of wood). Although these conclusions are the logical consequence of the science of resource flow analysis and the climate science of the IPCC that all countries have approved, it will require a socio-metabolic transition that will drive a wider deep transition equal in significance to the transitions that resulted in the Neolithic and Industrial Revolutions.

Reinforcing the argument of the Decoupling 1 Report, another early IRP report entitled *Assessing the Environmental Impacts of Consumption and Production: Priority Products and Materials* (referred to as the Priority Materials Report) addressed key questions of relevance to this discussion of socio-metabolic transition. Only three of these questions are addressed here: Which industries are the most responsible for contributing to environmental and resource pressures? What products and services have the greatest environmental impacts? Which materials have the greatest environmental impacts across their respective life cycles? (UNEP, 2010).

As indicated by the Priority Materials Report, the energy industry (26%) followed by industry (19%) and forestry (through deforestation) (17%) are the greatest contributors to climate change, abiotic resource depletion and sometimes eutrophication, acidification and toxicity.

As far as consumption is concerned, the Priority Materials Report shows that transport, housing and food are responsible for 60% of all impacts (UNEP, 2010). Given that these are overwhelmingly configured by urban systems, this prioritization reinforces the argument of the City-Level Decoupling Report (discussed later) that interventions that address these priorities will have to take into account their spatial contextuality.

However, even more important is the unsurprising fact that as incomes go up, so do the environmental impacts (a correlation that was, of course, confirmed in 2016 by the report on *Global Materials and Resource Productivity*). As income increases, so does the environmental footprint for construction, shelter, food, clothing, manufactured products, mobility, service and trade (UNEP, 2010). Clearly, there is no decoupling when it comes to rising incomes and related environmental impacts.

As far as the environmental impacts of materials are concerned, the Priority Materials Report shows that animal products, fossil fuels and key metals (iron, steel and aluminium) have the greatest impacts. However, only integrated data for Europe exists.

The Priority Materials Report concludes that future economic growth and development on a business-as-usual basis will exacerbate these trends. The impact of fossil fuels and agricultural activities are seen as the top two priorities that must be addressed if a transition to a more sustainable order is to be achieved.

In a follow-up to the Decoupling 1 Report and the Priority Materials Report, the IRP report entitled *Decoupling 2: Technologies Opportunities and Policy Options* (launched at Green Week, Brussels, in June 2014) argued that there are three types of decoupling (UNEP, 2014):

- decoupling through maturation: found mainly in developed countries, this is a natural process caused by saturated demand, levelling off or even decline of populations, minimal new construction and a shift towards services;
- decoupling through burden shifting to other countries: by offshoring the resource extraction and related impacts to other countries and then excluding this reality from material use calculations, it is possible for many countries to create the appearance of decoupling – in reality, as recent research has shown,

if the ecological rucksacks are attributed to the consumer and not producer, this apparent decoupling disappears (Wiedman, Schandl, Lenzen, Moran, Suh, et al., 2013);

• decoupling by intentionally improving resource productivity: as a "paradigm shift", this type of decoupling "requires technological and institutional innovations, resource-efficient infrastructure, low-material-intensity manufacturing, public awareness and appropriate attitudes and behaviours" (UNEP, 2014:5).

The Decoupling 2 Report demonstrated that since 2000, metal prices have risen by 176%, rubber by 350%, energy by 260% and food by 22.4% (with some projecting an increase for food of 120–180% by 2030). Demand for water by 2030 is expected to have risen by 40%, exceeding existing capacity by 60%. Possibly even more important than price increases is price volatility and related supply shocks (UNEP, 2014). The Decoupling 2 Report documents a wide range of emerging alternatives that are made possible by these price increases and argues the case for replicating radical improvements in resource productivity on a global scale. Many examples are provided, including the potential to reduce energy and water demand in developed economies by 50–80% using existing energy and water efficiency technologies; how developing countries investing in new energy infrastructure could reduce energy demand by half over the next 12 years if energy efficiency and renewable energy technologies were adopted now rather than later; and that decoupling technologies could result in resource savings equal to US$2.9–3.7 trillion each year until 2030 if the policy, regulatory and technological innovations are put in place (UNEP, 2014).

The most significant contribution of the Decoupling 2 Report is the suggestion that radical resource productivity can be achieved by introducing a resource tax system that is used to gradually and incrementally increase real resource prices over the long term. This tax could be used counter-cyclically to ameliorate rising prices when these occur and to counteract declining resource prices when these occur, thus providing the market with a level of certainty over the long term. This is crucial for counteracting what is inevitably going to happen if nothing of this kind is done, namely increasing price volatility that will tend to reinforce short-term investment perspectives with limited investment in innovation. Long-term innovation-driven investments will not thrive if prices remain volatile.

Nexus themes

Each nexus theme can be defined as a complex of interrelated resource use and environmental impact issues that can be analysed by reference to a particular cross-cutting process. Although all cities are different, cities concentrate the resource use and environmental impact issues. Food systems are globally, regionally and locally constituted in ways that connect incredibly complex flows of nutrients, energy, water, wastes and materials. Trade is about the global flows of resources and their associated ecological rucksacks that can be attributed to the producing or consuming countries with drastically differing results. And GHG mitigation technologies are massive composites that require energy and resource inputs that are intended to produce lower carbon and more resource-efficient outputs.

City-level decoupling[6]

The main aim of the Cities Working Group of the IRP (Swilling, Robinson, Marvin and Hodson, 2013; Swilling, Hajer, Baynes, Bergesen, Labbe, et al., 2018)[7] is to apply the insights generated by the new literature on urban metabolism (Costa, Marchettini and Facchini, 2004; Heynen, Kaika and Swyngedouw, 2006; Barles, 2009, 2010; Weisz and Steinberger, 2010; Kennedy, Pincetl and Bunje, 2011; Ramaswami, Weible, Main, Heikkila, Siddiki, et al., 2012; Farrao and Fernandez, 2013; Robinson, Musango, Swilling, Joss and Mentz-Lagrange, 2013; Swilling, Robinson, Marvin and Hodson, 2013) to the challenge of designing, building and operating more sustainable urban infrastructures.

The first urbanization wave took place between 1750 and 1950 and resulted in the urbanization of about 400 million people in what is now the developed world. The second urbanization wave between 1950 and 2030 is expected to result in the urbanization of close to 4 billion people in the developing world in less than a century. By 2007 just over 50% of the global population lived in cities. Hence, we should be talking about the 'urban Anthropocene'.

According to the 2014 revision of the World Urbanization Project report (United Nations Population Division, 2015), between 2015 and 2050 an additional 2.4 billion people will be added to the global urban population. This will bring the total urban population up to nearly 6 billion by 2050, which is 1 billion less than the size of the total global population in 2010. The percentage of the global population living in urban settlements is expected to rise from 54% in 2015 to 60% by 2030 and to 66% by 2050. Nearly 37% of the projected urban population growth (i.e. 37% of 2.4 billion) to 2050 is expected to come from just three countries: China, India and Nigeria. They will respectively contribute 404 million, 292 million and 212 million new urban dwellers to the global urban population by 2050. Overall, nearly 90% of the global population increase is set to occur in Africa and Asia. Their urbanization levels in 2015 were 40% and 48%, respectively. Even though the number of megacities (10 million plus) are expected to increase from 28 in 2015 to 41 by 2030, the fastest growth is expected to occur in the villages, towns and small- to medium-sized cities in Africa and Asia.

By 2010 the global process of urbanization that began in earnest in 1800 had resulted in the urbanization of just over 50% of households that are expected to live in cities by 2050.

Furthermore, using the ratio derived from the UN Habitat report *Challenge of Slums* (United Nations Centre for Human Settlements 2003) that 1 in 3 urbanites live in slums, this means of the 3.5 billion living in urban settlements by 2010, over 1 billion lived in slums. In other words, 210 years of urbanization had created a decent quality of life for only two-thirds of all urban dwellers. Resolving this problem must, therefore, be seen as integral to a just urban transition by 2050.

It follows, therefore, that just under 50% of the urban fabric that is expected to exist by 2050 must still happen over the four decades to 2050. The significant proportion of the additional urban population of nearly 4 billion people will end up in developing country cities, in particular Asian and African cities. If we include the 1 billion people who live in slums, then it follows that material infrastructures of one kind or another will need to be assembled for an additional 3.4 billion new urban dwellers by 2050.

This raises obvious questions: what will the resource requirements of future urbanization be if business-as-usual socio-technical systems are deployed to assemble built environments? What are the resource implications of more sustainable socio-technical systems? These questions were addressed in the IRP's two reports on cities.[8]

Launched in 2018, the IRP's *Weight of Cities* report (Swilling, Hajer, Baynes, Bergesen, Labbe, et al., 2018) built on the first IRP report on cities entitled *City-Level Decoupling: Urban Resource Flows and the Governance of Infrastructure Transitions* (Swilling, Robinson, Marvin and Hodson, 2013). The latter made the case for paying attention to the fact that the quantity of resources flowing through urban systems is related to how urban infrastructures are configured. However, very little data existed then to demonstrate the case. Another five years of collaborative research work resolved that problem. *Weight of Cities* was the first-ever empirical analysis of total resource flows through urban systems, projected forward to 2050. The report revealed that if the global urban population almost doubles during the four decades to 2050 and if urban development continued to be planned and managed on a business-as-usual basis, the annual resource requirements of the world's urban settlements would increase from 40 billion tons in 2010 to 90 billion tons by 2050. Furthermore, if the long-term historic trend of de-densification of urban settlements by minus 2% per annum continued, urban land use would increase from 1 million km^2 to over 2.5 million km^2 by 2050. Notably, this expansion would be into the most productive farmland in the world (with greatest negative impacts in Asia and Africa), thus threatening food supplies (see Kelley, 2003; for best overview of this trend, see D'Amour, Reitsma, Baiocchi, Barthel, Guneralp, et al., 2017).

The *Weight of Cities* report also explored alternatives. Overall, if the target of 6 tons per capita were achieved, this would reduce resource consumption in urban settlements by half by 2050. Some would argue that this is not enough, because it is more or less equivalent to what was consumed in 2010 but with an extra 3.5 billion consumers. That said, based on life cycle assessments of only three urban systems – district energy systems, green buildings and mass transit – it was possible to show that resource efficiencies of between 36% and 54% of current use could be achieved within each of these sectors (Swilling, Hajer, Baynes, Bergesen, Labbe, et al., 2018). If this is true for these sectors, it is assumed that it is more than likely valid for other sectors. Nevertheless, the 50% reduction in total resource use if the

6 tons/cap target is achieved seems to correlate with the efficiencies achieved using a Life Cycle Assessment of alternatives across 84 cities.

The *Weight of Cities* concludes with an analysis of alternative modes of urban governance. The argument is that a resource-efficient urbanism will only be achieved if a new mode of governance is introduced.

GHG mitigation technologies

Given that the energy transition is going to be the most important driver of the next deep transition, it follows that more needs to be known about the environmental implications of the suite of renewable energy technologies that are regarded as the cornerstone of this transition. In a highly detailed report entitled Green Energy Choices: The Benefits, Risks and Trade-offs of Low-Carbon Technologies for Electricity Production (referred to as the Green Energy Report), the following technologies were assessed using Life Cycle Assessment: wind power, hydropower, *photovoltaics* (PV), concentrated solar power (CSP), geothermal power, natural *gas combined cycle power* (GCCP) with and without CO_2 *capture and storage* (CCS) and coal-fired power with and without CCS (Hertwich, Gibon, Arveson, Bayer, Bouman, et al., 2015).[9] Bioenergy, nuclear energy, oil-fired power plants and ocean energy were not assessed.

The Green Energy Report found that wind, PV, CSP, hydro and geothermal power generate GHG emissions over the life cycle of less than $50gCO_2e/kWh$. This compares favourably to coal-fired power plants that generate $800–1,000gCO_2e/kWh$ over the life cycle and GCCP (without CCS) that generate $500–600gCO_2e/kWh$ over the life cycle. CCS can reduce emissions of fossil power plants by only 200–300g/kWh. As far as pollution and related health impacts are concerned, renewables reduce impacts by 70–90%. Similarly, impacts of renewables on ecosystems are a factor of 3–10 lower than fossil power plants (Hertwich, Gibon, Arveson, Bayer, Bouman, et al., 2015).

By contrast, the Report shows, a global transition to renewables (with some GCCP for peak loading and some coal power plants) would require an increased use of steel, cement and copper in comparison to the continuation of the business-as-usual fossil fuel–based system. Furthermore, renewables depend on various rare earth metals like indium and tellurium as well as silver (Hertwich, Gibon, Arveson, Bayer, Bouman, et al., 2015). There is ongoing debate in the literature on the security of supply over the long-term of these materials. However, their concentration in China is well known.

In short, from a purely technical perspective (that of course ignores institutional change, financing and learning), the environmental impacts of renewables are substantially reduced compared to fossil fuel–based energy supply. However, the resource inputs with respect to steel, cement and copper may be greater if alternative technologies for these aspects of the renewable energy infrastructure are not found. Increased requirements of bulk materials such as steel, cement and copper can be met with current production rates, especially if over time these production processes are themselves decarbonized.

Food systems

As argued by the *Priority Materials* Report, the food system is a major user of resources and a major contributor to negative environmental impacts. Food systems are highly complex global-local systems that are currently in deep crisis as several long-term megatrends accumulate into a perfect storm. Breaking from the dominant tendency to see food insecurity as a problem of production, the Food Working Group of the IRP adopted a food system perspective that, in turn, makes it possible to see food insecurity as a direct and persistent symptom of a flawed global food system (Hajer, Westhoek, Ozay, Ingram and van Berkum, 2015). Food security is defined as a situation where all people, at all times, have physical, social and economic access to sufficient, safe and nutritious food which meets their dietary needs and food preferences for an active and healthy life (Food and Agriculture Organization, 1996).

Considered in terms of the distribution of dietary energy supply, 868 million people around the world were considered chronically undernourished in 2013 (FAO, WFP and IFAD, 2012: ix). In addition, a further 2 billion people experienced the negative health consequences of micronutrient deficiencies (Ibid:4). About 850 million out of the 868 million people estimated to be undernourished lived in developing countries (Ibid: 8). Food insecurity is one of the key indicators of a system incapable of responding to the pressures that it faces. The capacity of the food system to ensure food availability and thus food security is shaped by a wide variety of factors, but the increase in population, urbanization and improved welfare are the most important (Ibid:ix) drivers of food system change. The conceptual framework captured in Figure 3.2 represents this complex set of actors and networks.

This diagram captures several dynamics: the food system as a highly complex globalized system, the underlying drivers of change and how they manifest in context-specific ways, the impacts on natural systems and how these, in turn, contribute to a range of environmental impacts.

The IRP's report on the food system essentially mounts the following argument (Hajer, Westhoek, Ozay, Ingram and van Berkum, 2015): the global food system is now dominated by large-scale modern systems that have replaced localized family farm–based food economies with large-scale globalized processing and retail activities, long value-chains, regulatory standards and transnational companies. One result of neoliberalization since the 1980s has been the shift in food governance from largely localized upstream governance systems to the big global downstream players, in particular the food processors and retailers. The result is that the food system is now primarily configured for short-term profit rather than the long-term continuity of farming systems and the ecosystems they depend on. Global and national governance systems tend to reinforce this orientation because it is perceived to be more 'efficient'. As a result, concentration in the off-farm sectors of the food value chain is high and rising: the three largest seed companies control 50% of the market; the top 10 agro-processors have 28% of market share and top 10 food retailers control 10% of the market. It is this shift in power that is contested by the agro-ecological

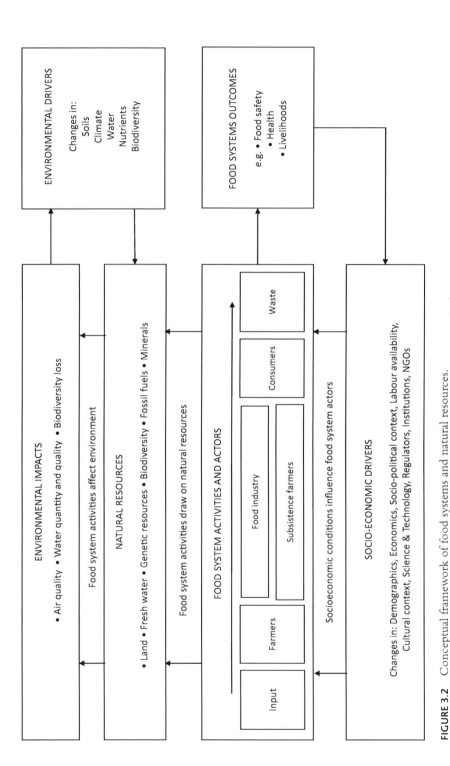

FIGURE 3.2 Conceptual framework of food systems and natural resources.

Source: Adapted from Hajer, Westhoek, Ozay, Ingram and van Berkum, 2015:12

movements who want a return to local food bio-economies where a commitment to sufficiency ensures the long-term sustainability of the underlying ecosystems.

The Food System Report goes on to argue that population growth, urbanization and improved welfare imply a 10% increase in food demand by 2025, with the fastest growth in demand taking place where logistical infrastructures are weakest. Given that urbanization and economic growth in developing countries imply an expanding middle class, a nutrition transition is underway from calorie-rich diets (cereals) to energy-rich diets (meat, vegetable oils and sugars). Energy-rich food requires far greater natural resource inputs, including the fact that instead of being consumed directly by humans, grains are used as inputs for livestock production. This, in turn, increases the demand for land for cereal production and grazing. Furthermore, now that supermarkets have become the dominant food delivery systems in all regions where middle-class consumers are significant, energy-rich food is transported over longer distances, requires more packaging and depends on vast globally structured networks of interconnected specialized companies and value chains. The combined impact of these processes includes soil degradation as land is over-exploited, depletion of aquifers and fish stocks, eutrophication due to nutrient losses (rising by 20% over next 40 years) and diminished biodiversity. Furthermore, climate change is expected to reduce crop production in key regions of the world (Hajer, Westhoek, Ozay, Ingram and van Berkum, 2015).

The Food System Report concludes that there are significant opportunities to increase resource use efficiency in the food system, while simultaneously reducing environmental impacts. On the supply side, important options include increasing yields in certain low-yield regions with higher potential using a more balanced mix of natural resources (including agro-ecological systems and higher input of minerals) leading to an increase in the output per unit of land, water and human labour; increased nutrient use efficiency in the food chain and consequent reduction of nutrient losses to the environment; development of resource-efficient aquaculture systems and sustainable land and water management using agro-ecological approaches. On the demand side, the two key strategies would be reduction of food losses and wastes and a shift to less resource-intensive diets, especially in regions with 'Western' diets by lowering the consumption of meat, dairy and eggs (Hajer, Westhoek, Ozay, Ingram and van Berkum, 2015).

Trade

The core question of the *International Trade in Resources* report (referred to as the Trade Report) (Fischer-Kowalski, Dittrich, Eisenmenger, Ekins, Fulton, et al., 2015) is whether or not the global trading system contributes to greater resource efficiency and diminished environmental impacts.

The Trade Report clearly shows that although trade in volume increased by a factor of 2.5 between 1980 and 2010, trade measured in monetary terms increased dramatically to 28% of global GDP in 2010. Fifteen per cent of all extracted materials were traded internationally in 2010. The Report argues that while incentivizing increased extraction, one key result of trade was increased financial revenues for poor resource-rich countries that rapidly became major exporters. In theory, this should have positive developmental consequences that would need to be weighed up against the environmental costs. However, as Collier has argued, in reality the more dependent on resource rents economies become, the less likely they will have the governance mechanisms to translate resource rents into developmental benefits – a dynamic known as the resource curse (Collier, 2010).

The Trade Report goes on to argue that a closer look at physical trade reveals that the total volume of materials traded between 1980 and 2010 more than doubled, with fossil fuels making up around 50% of total volume traded. However, reflecting the levelling off of oil production and trade globally since 2005 (Murray, 2012), the growth rates in trade in oil have declined since 2005.

The Trade Report describes who the largest exporters and importers were in 2010. Representing a shift from twentieth-century trends, by 2010 only 30% of all countries were net material suppliers while 70% of all countries had become net importers. Significantly, this distribution of importing-exporting nations was not neatly split along north-south lines. South American countries, Scandinavian, West and Central Asian countries, as well as Canada, Australia and the Southeast Asian Islands have become the largest suppliers of materials. The largest importers were the United States, Japan and Western Europe.

The Trade Report argues that the twentieth-century international division of labour was characterized by declining resource prices that in general made it possible for northern industrialized countries to act as importers of primary resources and exporters of manufactured goods, with southern countries as the exporters of primary resources and importers of manufactured goods. The Trade Report confirms that this picture is rapidly changing in the twenty-first century. In a context of rising resource prices, some fast industrializers in the global South have become both importers of primary resources and exporters of manufactured goods, and some industrialized countries like Canada and Australia have become increasingly important exporters of primary resources. In general, there are an increasing number of countries dependent on resource imports and a declining number of countries that are providing an ever-greater proportion of resource exports (Schandl, Fischer-Kowalski, West, Giljum, Dittrich, et al., 2016). Trade makes physical burden-shifting possible, a phenomenon that becomes clear if ecological rucksacks are attributed to the consumer and not to the producer.

The Trade Report ends by saying that a conclusive answer to the core question about the role of trade in resource use and environmental impacts is not possible at this stage, especially if a balanced view of environmental and developmental factors is taken into account. This, however, is not the question that guides the primary concerns of this chapter – this chapter is interested in the socio-metabolic dynamics of transition. From this perspective, the declining number of countries providing more and more primary resources within the context of a long-term super-cycle of rising resource prices is clearly the most important limiting factor. The rise of 'resource nationalism' in Africa (together with rising labour costs in China which makes manufacturing through beneficiation increasingly viable in Africa) (Swilling, 2013) suggests that rising resource prices over a longer-term super-cycle are unlikely to be reversed in the near future. The adoption by the European Union of a Resource Efficiency strategy suggests that rich resource-importing countries will start to find ways of reducing their dependence on resource imports (European Commission, 2011). If resource efficiency results in reduced demand that is not balanced out by increased demand within resource-rich countries, then this could have negative economic implications on resource-rich developing economies. Both trends signal new directions of change with potentially transformative implications.

Specific resource challenges

The series of IRP reports that deal with specific resource challenges have addressed metals (four reports), water (two reports), land use and soils, and forests.[10] They all recognize that these resources will in one way or another be required by society irrespective of whether there is a sustainability-oriented transition or not. It therefore follows that it is necessary to understand the complex dynamics that will shape the availability of these resources over time and what actions will be required to ensure that these resources are managed and used in more sustainable ways as part of a wider deep transition.

Metals

The Metals Working Group has published four peer-reviewed reports and one working paper:[11]

- Report 1: Metal Stocks in Society (2010)
- Report 2a: Recycling Rates of Metals (2011)
- Report 2b: Metal Recycling: Opportunities, Limits, Infrastructure
- Report 3: Environmental Risks and Challenges of Anthropogenic Metals Flows and Cycles (2013)
- Working Paper: Estimating Long-Term Geological Stocks of Metals

All economies, no matter their level of development, depend on metals of various kinds. The rise of the Information Age and related increased demand for hi-tech electronic goods has resulted in rapid increases in demand for specialty (or rare earth) metals like lithium and indium. Simultaneously, the accelerated growth and rapid urbanization in the BRICs Plus countries – especially China – has resulted in massive increases in demand for base metals. In combination, these two driving forces of demand have deepened the criticality of a wide range of metals. For example, as the Metals e-Book makes clear, the future demand for zinc, copper, nickel and aluminium just for the expansion of the global energy system is in each case several magnitudes greater than current demand (e.g. demand for aluminium is expected to grow from 500 Gg/y to over 5,500 Gg/y by 2050 just for non–fossil fuel infrastructure).

The increasing complexity of electronic goods is a major driver of demand for a wide range of metals to produce the compounds required by these goods. In the 1980s, 12 elements were required to make a microchip. This increased to more than 60 required elements in the 2000s.[12]

Although a lack of information prevents high-confidence estimations about resource depletion (Smil, 2014), what is clear from the work of the Metals Working Group is that there are also other factors that increase supply risks. These include, according to reports of this Working Group, challenging technological conditions (depth, composition of ore as ore grades decline), economic variables (adequacy of infrastructure, size of deposit), environmental constraints (natural habitats, ecosystem services) and geopolitical dynamics (trade barriers, political instability, weak states) (International Resource Panel Working Group on Global Metal Flows, 2013).

Global metals production is a major contributor to environmental pollution and energy demand. The Working Group's reports shows that no less than 7–8% of global energy use and therefore energy-related GHG emissions can be attributed to metals production. Whereas 20 MJ of energy is needed to make 1 kg of steel, 200 000 MJ is needed to make 1 kg of platinum (International Resource Panel Working Group on Global Metal Flows, 2013). A major driver of increased future energy demand of metals production is the declining ore quality – three times more material must be moved today to extract 1 kg of ore compared to a century ago.

Report 1 estimated the quantity of metals being used by society for the period 2000–2006. The average for aluminium was 80 kg/cap, with a range of 350–500 kg/cap for developed economies and 35 kg/cap for the least developed economies. Similarly, for copper, 35–55 kg/cap is the global average, ranging from 140–300 kg/cap and 30–40 kg/cap; and for iron, 2,200 kg/cap, ranging from 7,000–14,000 kg/cap to 2,000 kg/cap. Obviously, the same pattern replicates itself for each metal (International Resource Panel Working Group on Global Metal Flows, 2013). The implication is that global development targets aimed at eradicating poverty and achieving greater equity will entail significant increases in metals consumption in developing countries.

To diminish the environmental impact and energy requirements of metal production, it will be necessary to increase the recycling rates of metals. Report 2a demonstrated that the End of Life-Recycling Rates (EoLRR) for metals are very low: EoLRR of above 50% can be found for only 18 metals. Forty-two metals have an EoLRR of below 50%, 34 of which have an EoLRR of below 1% (International Resource Panel Working Group on Global Metal Flows, 2013).

The Report shows that part of the explanation for low EoLRR is rising demand and the long in-use life of metals. However, another more important explanation is that the design of products has not hitherto taken into account the need for end-of-life recovery and reuse. Disassembly and metals recovery are not what designers have been required to do. To increase EoLRR to 50% or more for all metals as part of a wider sustainability-oriented transition, it will be necessary – the Report argues – to radically change the way products are designed (i.e. design for disassembly) and substantial investments in a new set of resource efficient of infrastructures will be necessary (International Resource Panel Working Group on Global Metal Flows, 2013). As resource prices continue to rise as demand continues to grow driven mainly by the requirements of the Information Age and urbanization, the financial viability of design for disassembly will more than likely improve (International Resource Panel, 2018). This will be a crucial driver of a more fundamental transformation.

Land use and soils

A century of steadily declining food prices came to an end at the turn of the millennium. Since then, food prices have been rising and so has the number of large-scale land transactions (including so-called land grabbing). Whereas food prices rose steadily during the first decade of the twenty-first century, there has been no previous decade since 1900 where there is evidence of steadily rising food prices across the entire ten-year span. This pattern of rising food prices is expected to continue with major implications for land use and food security (Swilling and Annecke, 2012: Chapter 6).

Following Scherr, the total ice-free land surface of the Earth is 13 billion hectares (Bha) of which 1.5 Bha is unused 'wasteland' and an additional 2.8 Bha is unused and inaccessible. This leaves 8.7 Bha which humans can choose to 'use' for a wide variety of purposes, including pasture, forests and cropland. Of this, 3.2 Bha are potentially arable, the rest being marginal land from a cultivation perspective and covered by forest, grassland and permanent vegetation.[13] Of the potentially arable land, only 1.3 Bha is deemed to be moderate to highly productive. Nearly half of the 3.2 Bha of potentially arable land (1.47 Bha) is cultivated as cropland. This means that just over 10% of the ice-free land surface of the Earth is the resource on which humans depend for the bulk of their food. This 1.47 Bha of cropland and approximately 3.2 Bha of permanent pasture and 4 Bha of permanent forest and woodland are what makes up the 8.7 Bha of 'usable' land (Scherr, 1999).

Half of the developing world's arable and perennial cropland is in just five countries – Brazil, China, India (with 22%), Indonesia and Nigeria. It is noticeable that the African countries with very extensive or moderately extensive arable land resources are Nigeria, Ethiopia, South Africa and Sudan. The majority of African countries have limited arable land resources with high population pressures and it is estimated that 65% of Africa's agricultural land is degraded (Scherr, 1999). Yet African countries are earmarked by all the models of the future for substantial yield increases – it is also where most of the land grabs are taking place (Cotula, Vermeulen, Leonard and Keeley, 2009).

Global land use, rising food prices, soil degradation and accelerated land transactions (as countries scramble to secure food supplies) provide the context for the IRP report entitled *Assessing Global Land Use: Balancing Consumption with Sustainable Supply* (generally referred to as the *Land and Soils Report*) (Bringezu, Schutz,

Pengue, O'Brien, Garcia, et al., 2014). Launched at the World Economic Forum (WEF) in January 2014, the report raises very serious questions about the sustainability of expanding agricultural production in a world dominated by a resource-inefficient food system that does not cater for the needs of the nearly billion or so people who are undernourished.

Since 1961, inputs (nitrogen, phosphorus, potassium, tractors) have tended to rise at a faster rate than crop yields. While the irrigation area and cropland doubled between 1961 and 2003, the use of nitrogen as fertilizer experienced a sixfold increase while phosphorus experienced a threefold increase (Bringezu, Schutz, Pengue, O'Brien, Garcia, et al., 2014).

At the same time, soil degradation continues, with 23% of soils degraded by 1990. Approximately 2–5 Mha are degraded per annum.

The international division of agricultural output is clearly reflected in Figure 3.3 that reveals the gap between yields in Europe and North America where high external input–intensive industrial farming is prevalent and the yields in developing countries where soil degradation levels are high, infrastructures are poor and farming is still dominated by 400 million small farmers (only 40% of whom use chemical inputs).

Expanding agricultural land use is driven in part by rising demand for food and non-food biomass which cannot be compensated by higher yields. This net expansion occurs at the expense of grasslands, savannahs and forests. However, expansion is also driven by the need to compensate for expanding cities and soil degradation. This plus net expansion results in gross expansion. Based on an assessment of a wide range of studies (for sources of data cited, see original diagram in *Land and Soils Report*), the dimensions of the net and gross expansion of agricultural land are represented in Table 3.1.

Table 3.1 from the Land and Soils Report shows the future land requirements to meet food supply (after exhausting yield growth potential) are estimated to be

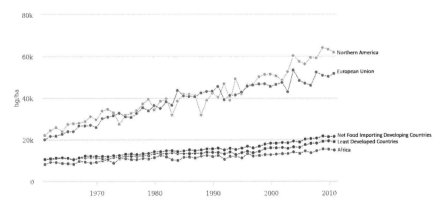

FIGURE 3.3   Cereal yields by selected world regions, 1961–2011.

Source: FAO data cited in Bringezu, Schutz, Pengue, O'Brien, Garcia, et al., 2014

TABLE 3.1 High and low estimates of net and gross expansion of agricultural land, 2005–2015

Business-as-usual expansion	Low estimate (Mha)	High estimate (Mha)
Food supply	71	300
Biofuel supply	48	80
Biomaterial supply	4	115
Net expansion	123	495
Compensation for built environment	107	129
Compensation for soil degradation	90	225
Gross expansion	320	849

Source: Bringezu, Schutz, Pengue, O'Brien, Garcia, et al., 2014 – note: for detailed references to the sources of data cited here, see original diagram in this report

between 71 Mha and 300 Mha. The rapidly expanding demand for land to grow biofuel crops is estimated to be 48–80 Mha, and the requirement for additional biomaterials (wood, textile crops, etc.) is estimated to be 4–115 Mha. To compensate for the expanding built environment (that tends to destroy the most valuable agricultural land), 107–129 Mha may be needed. Assuming that a significant proportion of degraded soils cannot be restored,[14] estimates of the requirements to compensate for degradation range from 90 Mha to 225 Mha. This means that the estimates for gross additional agricultural land requirements to meet growing needs range between 320 Mha and 849 Mha. This suggests that the needs are much greater than FAO's estimate of 120 Mha (Bringezu, Schutz, Pengue, O'Brien, Garcia, et al., 2014).

It needs to be recognized that land use change in favour of agriculture is one of the primary drivers of rising CO_2 emissions and biodiversity loss. We, therefore, need to accept that there are absolute limits to the quantity of global land that can be used for agriculture. Taking into account various factors, the *Land and Soils Report* proposes that the expansion of global cropland should be halted by 2020 at which point it is estimated that global cropland will have expanded from about 1.5 Bha to 1.64 Bha. In other words, although an additional 140 Mha of cropland is bound to have very negative environmental effects, the *Land and Soils Report* nevertheless estimates that it may be possible to remain within this 'safe operating space' and thus avoid the far more negative consequences of an expansion in the 320–849 Mha range that the sum of existing research tends to suggest. To achieve this reduction in future requirements, the *Land and Soils Report* recommends the following (Bringezu, Schutz, Pengue, O'Brien, Garcia, et al., 2014):

- massively increase the existing land potential by restoring degraded soils and using existing soils optimally – how to do this was the focus of a subsequent IRP report;
- ensure that national governments have the capacity to control expansion of agricultural land in order to avoid uncontrolled destruction of biodiversity, forests and pastures;

- limit meat/dairy consumption and foster changes in the way the food system works – again, as mentioned earlier in this chapter, this is the focus of a forthcoming IRP report.

The Report recommends the following specific sets of interventions (Bringezu, Schutz, Pengue, O'Brien, Garcia, et al., 2014):

- reducing the demand for meat/dairy products and reducing the levels of food waste could save 96–135 Mha;
- halving the global biofuel targets could save 24–40 Mha;
- controlling the demand for biomaterials could save up to 57 Mha;
- limiting the expansion of cities into productive agricultural land by just 10% of the expected impact could save 11–13 Mha;
- restoring a third of degraded soils could save 30–74 Mha.

In short, a mix of strategies and measures to reduce overconsumption of certain foods, reduce food waste and limit the consumption of non-food biomass products while at the same time improving land management could save 160–320 Mha by 2050. Cropland area would still expand to meet, in particular, the demand for increased food production to meet the needs of those who do not have enough but not as much.

Water

According to the Water Decoupling Report (Urama, 2015), integrated water resources management faces two closely interlinked obstacles – one on the supply side of unpolluted freshwater resources for a growing world population and the other on the demand side of water for increased agricultural output, water-intensive industries and domestic use. The problems associated with the supply and demand for water, such as significant increases in water pollution and freshwater withdrawals, are driven by population increase, urbanization, rising living standards, unsustainable water governance (which includes inefficient supply and demand management), agricultural land uses (specifically irrigation), industrial production, ecosystem degradation and climate change.

The *Water Decoupling Report* addressed the challenge of water availability and use in light of mounting global challenges to security of supply (Urama, 2015). Water withdrawals on a global scale have increased at a rate almost double the human population growth rate, from 600 billion m^3 in 1900 to 4,500 billion m^3 in 2010. This could grow to 6,350–6,900 billion m^3 by 2030 if an average economic growth scenario and efficiency gains are assumed. This represents a 40% demand gap above currently accessible water resources, including return flows.

Table 3.2 illustrates the expected increases in water withdrawal demand for human activities by 2030. The highest incremental demand is expected to occur in sub-Saharan Africa at 283%, while the lowest is expected in North America at 43%.

TABLE 3.2 Increases in annual water demand, 2005–2030

Region	Projected change from 2005
China	61%
India	58%
Rest of Asia	54%
Sub-Saharan Africa	283%
North America	43%
Europe	50%
South America	95%
Oceania	109%

Source: McKinsey (2009), cited in Urama, 2015

In terms of global freshwater use to support human activities, currently 70% is used in agriculture (estimated to increase by another 65% by 2030), of which 15–35% is considered unsustainable, especially in cases where groundwater is extracted faster than it can be recharged. An additional 22% of freshwater is used in industry (estimated to grow by an additional 25% by 2030), but this can range from as high as 60% in industrialized countries to as low as 10% in some developing countries. Lastly, 8–11% is for domestic use (estimated to grow by another 10% by 2030), at an average of about 50 l per person per day, although also with great variability (International Water Management Institute, 2007 and Gleick, 2006 both cited in Urama, 2015).

On the supply side, it is estimated that over the next 20 years water supply would need to increase by about 140% to meet increasing demand and ensure accessible, reliable and sustainable provision of existing water supplies (Urama, 2015). The obstacles to achieving this are as follows. Readily available sources of freshwater are under significant stress already, with the shrinking of many freshwater lakes, the drying up of rivers that subsequently never reach the ocean and the overuse of groundwater resources that is already occurring in many regions. Further limiting these water resources, the Water Report argues, are increasing rates of pollution with over 405 dead zones globally currently on record in coastal waters. Lastly, water is lost due to inefficient technologies and related infrastructures. The most pertinent example is the loss of drinking water from municipal distribution systems before it even reaches the consumer, where on average 30% (and in extreme cases reaching up to 80%) of water is lost. This is equal to over $18 billion worth of water per year that does not generate revenue, indicating the need for efficiency and productivity gains (Urama, 2015).

Especially in the BRICS countries, the estimated number of people living in water-stressed areas between 2005 and 2030 is rising rapidly. In 2005, approximately 3 billion people experienced severe water stress and this is estimated to increase by another 1 billion people in 2030 (UNEP, 2011 cited in Urama, 2015). The OECD estimated that nearly 3.9 billion people will experience severe water stress by 2030.

Water pollution is the greatest threat to water supply for human activities. The most significant pollution sources are mining activities, agriculture, landfills and industrial and urban wastewater effluents. Main pollutants from agriculture, for example, include pesticides, organic compounds and nutrients from fertilizer that end up in water bodies, causing eutrophication and ultimately leading to "dead zones"(Urama, 2015). Furthermore, pollution results from industrial activities – 70% of untreated industrial wastes are estimated to be dumped into water bodies (UN-Water, 2009 cited in Urama, 2015).

In many developing countries, sanitation and wastewater treatment cause major water pollution, and scenarios have been found where as much as 85–95% of sewage is discharged directly into rivers, lakes and coastal areas, causing large amounts of revenue to be spent on dealing with waterborne diseases instead of generating new wealth (Tropp, 2010 cited in Urama, 2015).

Lastly, the number of people vulnerable to water-related disasters, particularly flooding, as a result of climate change, deforestation, population growth, rising sea levels and human settlement in flood-prone lands, may reach 2 billion by 2050 (Urama, 2015).

All these obstacles make a strong case for water decoupling. Water decoupling means using fewer units of water resources per unit of economic output, while also reducing other adverse socio-economic and environmental downstream impacts.

Achieving sustainable decoupling in the water sector will require innovative structural transformations in economic pathways. Integrated water management policy and practices at local, national, river basin and global scales will be required, in addition to substantive investments in improved technologies and innovations for improving water efficiency and productivity at the appropriate temporal and spatial scales. Improving technical and allocative efficiency and resource productivity in the key water use sectors could offset up to 60% of the anticipated growth in demand for water by 2030 (Urama, 2015).

Governance

Governance is a relatively new theme for the IRP. Two reports have raised the question of governance in anticipation of transition to a more sustainable world:

- the penultimate chapter of the *Weight of Cities* report; and the
- Mineral Resources Governance in the Twenty-First Century.

Weight of Cities concludes that building resource-efficient cities will require a fundamental rethink of urban governance. Given the diverse nature of cities, a one-size-fits-all approach will not work. Instead, the proliferation of experimentation across all world regions needs to be encouraged and supported to go to scale (discussed further in Chapter 5). This will require a new mode of "entrepreneurial governance" - or what was called 'collibratory governance' in Chapter 2 - that shifts the constituents of urban political coalitions from the property developer-banking-estate agent-services coalition of the neoliberal period leading up to the 2007/2009 crisis to a new coalition of innovators, knowledge workers, civil society organizations, entrepreneurs, venture capital and DFI actors with a sustainable development agenda (Swilling, Hajer, Baynes, Bergesen, Labbe, et al., 2018). Inspired by Mazzucato's conception of the "entrepreneurial state" (Mazzucato, 2011), urban entrepreneurial governance defines an active role for the state but primarily to provide long-term strategic direction via interventions in land use, infrastructure investments and strategic coordination of innovation-oriented coalitions rooted in the proliferation of experimentation.

The *Mineral Resources Governance* report addresses the challenge the adoption of the SDGs in 2015 presents for the extractive industries (International Resource Panel, 2019a). Whereas prior to 2015 extractive industries needed a 'social license to operate', since 2015 they require a 'sustainable development license to operate'. Because there is essentially no understanding of what this means in practice, this report provides an analysis of the state of the mineral resources sector, and what a 'sustainable development license to operate' could be in practice. Over 80 frameworks for managing the behaviour of extractive industries were reviewed, with none leading to successful benefit sharing within robust local economies. The report recommends reform at both international and national levels. At national level, it suggests that countries adopt a Strategic Plan for the mining sector and other sectors impacted by it. The Plan should be set in the context of sustainable development and could include a mining law that enshrines the principles of consultation, transparency and reporting, as well as explicitly recognizes the rights of local populations. The Plan should also facilitate the creation of three core public institutions to promote and regulate the development of mines and metals industries – an Environmental Directorate in charge of developing environmental policies, laws and regulations; a Mining Directorate in charge of mines and metals-related policies; a Geological Survey in charge of acquiring, conserving, managing, modelling and disseminating geological, geophysical, geochemical and other data. At international level, the report proposes an International Minerals Agency, or an international agreement, to, among others, coordinate and share data on economic geology and mineral demand needs and promote transparency on impacts and benefits. Ultimately, what is required, is a governance arrangement that facilitates the transition to post-extractivism.

Discussion: implications of the work of the IRP for global transition thinking

The work of the IRP documents the limits of the socio-metabolism of industrial modernity. Although the notion of decoupling is contested on the grounds that it implies that fundamental structural change can be avoided while greening consumption (Jackson, 2009), the global resource perspectives provided in the Decoupling 1, Decoupling 2 and Environmental Impacts reports all confirm that unless the global systems of production and consumption are in fact radically transformed, it will not be possible to build a world without poverty where average consumption (of around 6 tons/cap) is consistent with what available planetary systems can provide on a sustainable basis. This message goes way beyond the carbon-centred argument of the IPCC that has succeeded in establishing the notion of a low-carbon transition within the global policy community. From the perspective of the IRP, this will not be sufficient. A low-carbon destruction of planetary resources is not an appropriate future trajectory.

The conclusions of the work on the various nexus themes confirm that superficial modifications to the socio-technical systems that we depend on will not suffice. To double the extent of the world's urban settlements, it will be necessary to fundamentally change the way we design, build and operate cities. Many aspects of urban living that are taken for granted will clearly have to be replaced with information-rich alternatives that embed cities in sustainable technological and ecological cycles.

The crisis of the current highly complex tightly coupled global food system poses a major risk to the survival of the global population, in particular the poor. Although this might be the most difficult socio-technical system to change, fundamental changes to this system might well be driven by the social and health consequences of deepening food insecurity.

The global trade in material resources is already changing rapidly as fewer and fewer countries become increasingly important exporters of primary resources to an increasing number of industrializing and industrialized resource-importing countries. The rise of resource nationalism in many resource-rich developing countries and the resource efficiency movement in many resource-importing developed countries suggests future trajectories that will have major implications for global trade in material resources.

Finally, although the transition to renewable energy technologies will clearly have beneficial environmental impacts compared to business-as-usual, it would be naïve to assume that they are a panacea for an environmental utopia. Like everything we humans do, resources are required that we derive from the crust of the earth in one way or another. A future world of 9 billion people where we consume 6 tons/cap and emit 2 tons of CO_2 is still a world that will require the extraction of resources on a scale equal to levels of extraction that pertained in 2010. This is early warning that the renewable energy revolution is not a simplistic break that miraculously heals the planet.

The IRP work on the specific resource challenges in the metals and ecosystems sectors clearly shows that a deep transition will depend on extraordinary efforts to change the way we use the three most basic ingredients of contemporary modern living: metals (in particular for the global electronic infrastructure), water and land. No matter what we do, there is no way we can do without these three key resources. Indeed, it is clear from the evidence that we will need more of them and that this will have to be done in a way that ensures that those who currently live in poverty gain greater access to these resources. The challenge, however, is to trigger new consumption and production systems that create new economic opportunities out of the need, for example, to 'design for disassembly' when it comes to metals; or design and build decentralized urban water and sanitation systems that can use water more efficiently and recycle all wastewater; or replicate on a massive scale the agro-ecological farming methods that have proven to be able to increase yields by restoring the soils and ecosystems.

This discussion raises the question about whether the IRP should go beyond its current mandate and become the global body that addresses in an integrated way the challenge of making the sustainability-oriented transition happen. The recent attention being paid to urban and mineral resource governance is a step in this direction. However, to take transition more seriously, it will be necessary to embed the research done to date on resource flows within an analysis of the multiple transition dynamics discussed in Chapter 4.

Conclusion

It may be appropriate to use the long-wave theories of transition discussed in Chapter 4 to understand the contribution of the IRP to the wider field of anticipatory science. What is anticipated by those who use this perspective is that in some way the socio-metabolic, techno-economic and socio-technical regimes of the industrial epoch will be replaced over time by an alternative more sustainable epoch characterized by more sustainable socio-metabolic flows. Although the IRP's work does not directly address transition per se, when read together the various strands of thought and evidence in the completed and current work do suggest that it is highly unlikely that the industrial epoch can continue into the medium- and long-term future if it depends on the continuous increase in consumption of natural resources. There are elements, of course, across the reports that could be woven into a more robust and systematic conception of transition: the types of decoupling envisaged in the Decoupling 2 Report, the recommended dietary and land-use changes in the Land and Soils Report, the key role of cities in the City-Decoupling Report and the unintended consequences of a transition to clean energy, to cite only some examples.

Three broad conclusions flow from this analysis. Firstly, when collected together, the totality of evidence mobilized by the IRP supports the notion that future well-being and development (whether growth-based or not) will have to be decoupled from rising rates of resource use. Relative decoupling is not sufficient. Absolute reductions in resource use will be necessary. To implement this idea, however, a fundamental restructuring of prevailing modes of production and consumption will be necessary. Decoupling is not simply sophisticated greenwash. It will mean significant changes for consumers in developed economies and in developing countries committed to poverty eradication It will be necessary to replace resource-intensive development pathways with resource-efficient development pathways that end up delivering to more people a fairer deal resulting in less inequality and therefore more long-term democratic stability.

Secondly, the IRP's work reveals the futility of naïve assumptions about what will be attainable in a sustainable world populated by over 9 billion people, most of whom will be living in cities. All past human activity has depended on the exploitation of natural resources in one way or another. Humans currently have technical and institutional capabilities to exploit these resources on an unprecedented scale. Anticipatory science is needed to show that this cannot continue. However, if the results from IRP research are anything to go by, massive reductions in resource use are possible but they cannot be eliminated or reduced to insignificant levels. A world of over 9 billion people without poverty may well need what was extracted from the earth in 2000. The finding that more of certain materials might be needed, as suggested in the Green Energy Choices Report, is highly significant. The only question is how this will be done. Will these socio-technical processes become part of closed-loop techno-industrial and ecological cycles or not? That will become the key longer-term question, not simply a zero–sum calculation based on how much less can be consumed. This, in turn, might make it possible to make a shift from only focusing on 'resource limits' to focusing more on 'resource potential'. This shift is already underway in a number of reports.

Thirdly, the IRP work on resource limits and potentials needs to get integrated into a wider holistic theory of economic development that is not GDP-centred. The gradual dismantling of neoliberalism is already underway as the intolerance of poverty and inequality reaches new heights in the wake of the global economic crisis and developmental states in the developing world disassociate themselves from the hegemonies of Western thinking. This is clearly a positive movement. However, if an alternative gets reconstructed that anticipates once again that there is an unlimited supply of natural resources, then a major opportunity will have been missed. The economic theories informing the developmental states run the risk of making this mistake. However, this is unlikely to become a mainstream habit because the century-long decline in resource prices ended in 2002. If those who have predicted a long-term super-cycle of 40–60 years of rising resource prices prove to be correct, then we can safely anticipate that the economic theory that replaces the reductionist simplicities of neoliberalism will indeed need to come to terms with the expanding body of work produced by the IRP. This will surely justify the efforts by those who have made this work possible.

Notes

1 Aspects of this chapter were first published in Swilling, M. (2016) 'Preparing for global transition: Implications of the work of the International Resource Panel', in Brauch, H.G., Spring, U.O., Grin, J. and Sheffran, J. (eds.) *Handbook on Sustainability Transition and Sustainable Peace.* Hexagon Series on Human and Environmental Security and Peace, Vol. 10. Zurich: Springer, pp. 391–418.

2 The arguments presented in this chapter have been developed exclusively by the author and do not in any way reflect the views of the IRP, UNE or individual members of the IRP.

3 For more details on UNE, see www.UNE.org/resourcepanel/

4 All completed reports referred to below are available on the IRP website: www.UNE. org/resourcepanel. I have been co-lead author of three reports: Decoupling Economic Growth and Resource Use, City-Level Decoupling and Weight of Cities. I was also involved in the reports on water, land use and food.

5 The IRP has also produced reports on biofuels and forests, but these have not been included in this analysis.

6 Since the writing of this chapter, I was co-lead author of a new report on cities published in 2018 by the IRP – see Swilling, M., Hajer, M., Baynes, T., Bergesen, J., Labbe, F., et al. (2018) *The Weight of Cities: Resource Requirements of Future Urbanization*. Paris: International Resource Panel. Available at: www.internationalresourcepanel.org. Chapter 5 draws on this report and is not included in the assessment of the IRP's work in this chapter.

7 I was lead author for both reports.

8 I was lead author for both these reports.

9 Note that at the time of writing this report had not gone through the UNE peer-review process.

10 The reports on biofuels and forests have not been included, partly because the implications of biofuels is incorporated into the land and soil group, and the forests report was compiled in a way that does not quite fit into the overall orientation of the IRP.

11 A summary is contained in an e-book available on the IRP website – www.UNE.org/resourcepanel/; unless alternative sources are specified, the data referred to in this section is taken from this e-book (International Resource Panel Working Group on Global Metal Flows, 2013).

12 Quoted in a PowerPoint presentation by Tom Graedel, November 2013, Stellenbosch University.

13 Lambin and Meyfroidt estimate that there is approximately 4 Bha available for 'rain-fed agriculture' (Lambin and Meyfroidt, 2011:3466).

14 Although seriously degraded soils are difficult and costly to restore, there is still about 300 Mha of lightly degraded soils that can be restored mainly by changing management practices.

References

Ahmed, N. M. (2017) *Failing States, Collapsing Systems: Biophysical Triggers of Political Violence*. Cham, Switzerland: Springer.

Barles, S. (2009) 'Urban metabolism of Paris and its region', *Journal of Industrial Ecology*, 13(6), pp. 898–913.

Barles, S. (2010) 'Society, energy and materials: The contribution of urban metabolism studies to sustainable urban development issues', *Journal of Environmental Planning and Management*, 53(4), pp. 439–455.

Bringezu, S., Schutz, H., Pengue, W., O'Brien, M., Garcia, F., et al. (2014) *Assessing Global Land Use: Balancing Consumption with Sustainable Supply*. Nairobi: United Nations Environment Programme.

Castells, M. (2009) *Communication Power*. Oxford: Oxford University Press.

Collier, P. (2010) 'The political economy of natural resources', *Social Research*, 77(4), pp. 1105–1132.

Costa, A., Marchettini, N. and Facchini, A. (2004) 'Developing the urban metabolism approach into a new urban metabolic model', in Marchettini, N., et al. (eds.) *The Sustainable City III*. London: WIT Press, pp. 31–40.

Cotula, L., Vermeulen, S., Leonard, R. and Keeley, J. (2009) *Land Grab or Development Opportunity? Agricultural Investment and International Land Deals in Africa*. London: International Instituted for Environment and Development; FAO; International Fund for Agricultural Development.

Crutzen, P. J. (2002) 'The anthropocene: Geology and mankind', *Nature*, 415, p. 23.

D'Amour, C., Reitsma, F., Baiocchi, G., Barthel, S., Guneralp, B., et al. (2017) 'Future urban land expansion and implications for global croplands', *Proceedings of the National Academy of Science*, 114(34), pp. 8939–8944.

European Commission. (2011) *Life and Resource Efficiency: Decoupling Growth from Resource Use*. Luxembourg: European Commission.

FAO, WFP and IFAD. (2012) *The of Food Insecurity in the World 2012: Economic Growth Is Necessary but Not Sufficient to Accelerate Reduction of Hunger and Malnutrition*. Rome: Food and Agriculture Organisation.

Farrao, P. and Fernandez, J. E. (2013) *Sustainable Urban Metabolism*. Cambridge, MA: MIT Press.

Fischer-Kowalski, M. (1998) 'Society's metabolism: The intellectual history of materials flow analysis, Part I, 1860–1970', *Journal of Industrial Ecology*, 2(1), pp. 61–78.

Fischer-Kowalski, M. (1999) 'Society's metabolism: The intellectual history of materials flow analysis, Part II, 1970–1998', *Journal of Industrial Ecology*, 2(4), pp. 107–136.

Fischer-Kowalski, M. and Haberl, H. (2007) *Socioecological Transitions and Global Change: Trajectories of Social Metabolism and Land Use*. Cheltenham, UK: Edward Elgar.

Fischer-Kowalski, M., Dittrich, M., Eisenmenger, N., Ekins, P., Fulton, J., et al. (2015) *International Trade in Resources: A Biophysical Assessment*. Nairobi: UNEP.

Fischer-Kowalski, M. and Swilling, M. (2011) *Decoupling Natural Resource Use and Environmental Impacts from Economic Growth*. Report for the International Resource Panel. Paris: United Nations Environment Programme.

Food and Agriculture Organization. (1996) *World Food Summit: Rome Declaration on World Food Security*. Rome: Food and Agriculture Organisation. Available at: www.fao.org/WFS.

German Advisory Council on Climate Change. (2011) *World in Transition: A Social Contract for Sustainability*. Berlin: German Advisory Council on Global Change.

Giampietro, M., Mayumi, K. and Sorman, A. H. (2012) *Metabolic Pattern of Societies: Where Economists Fall Short*. Abingdon and New York: Routledge.

Gleick, P. (2006) *The World's Water (2006–2007): The Biennial Report on Freshwater Resources*. Washington, DC: Island Press.

Hajer, M., Westhoek, H., Ozay, L., Ingram, J. and van Berkum, S. (2015) *Food Systems and Natural Resources*. Paris: UNEP.

Hertwich, E. G., Gibon, T., Arveson, A., Bayer, P., Bouman, E., et al. (2015) *Green Energy Choices: The Benefits, Risks, and Trade-Offs of Low Carbon Technologies for Electricity Production – Technical Summary*. Paris: UNEP.

Heynen, N., Kaika, M. and Swyngedouw, E. (2006) *In the Nature of Cities: Urban Political Ecology and the Politics of Urban Metabolism*. London and New York: Routledge.

International Resource Panel. (2018) *Re-defining Value – The Manufacturing Revolution: Remanufacturing, Refurbishment, Repair and Direct Reuse in the Circular Economy*. Nabil Nasr, Jennifer Russell, Stefan Bringezu, Stefanie Hellweg, Brian Hilton, Cory Kreiss, and Nadia von Gries. Nairobi: United Nations Environment Programme.

International Resource Panel. (2019a) *Mineral Resource Governance in the 21st Century: Gearing Extractive Industries Towards Sustainable Development*. Nairobi: United Nations Environment Programme.

International Resource Panel. (2019b) *Natural Resources for the Future We Want*. Nairobi: United Nations Environment Programme.

International Resource Panel Working Group on Global Metal Flows. (2013) *E-Book: International Resource Panel Work on Global Metal Flows*. Nairobi: UNEP.

Jackson, T. (2009) *Prosperity Without Growth? The Transition to a Sustainable Economy*. London: Sustainable Development Commission.

Kelley, P. (2003) 'Urbanization and the politics of land in the Manila region', *The ANNALs of the American Academy of Political and Social Science*, 590, pp. 170–187.

Kennedy, C., Pincetl, S. and Bunje, P. (2011) 'The study of urban metabolism and its applications to urban planning and design', *Environmental Pollution*, 159, pp. 1965–1973.

Lambin, E. F. and Meyfroidt, P. (2011) 'Global land use change, economic globalization, and the looming land scarcity', *Proceedings of the National Academy of Science*, 108(9), pp. 3465–3472.

Mason, P. (2015) *Postcapitalism: A Guide to Our Future*. London: Penguin.

Mazzucato, M. (2011) *The Entrepreneurial State*. London: Demos.

Mazzucato, M. (2016) *Rethinking Capitalism: Economics and Policy for Sustainable and Inclusive Growth*. West Sussex: John Wiley & Sons.

McKinsey Global Institute. (2011) *Resource Revolution: Meeting the World's Energy, Materials, Food, and Water Needs*. London: McKinsey Global Institute. Available at: www.mckinsey. com/client_service/sustainability.aspx (Accessed: 10 May 2011).

Morin, E. (1999) *Homeland Earth*. Cresskill, NJ: Hampton Press.

Murray, J. and King, K. (2012) 'Oil's tipping point has passed', *Nature*, 481, pp. 433–435.

Naess, P. and Hoyer, K. G. (2009) 'The emperor's green clothes: Growth, decoupling, and capitalism', *Capitalism Nature Socialism*, 20(3), pp. 74–95.

Picketty, T. (2014) *Capital in the Twenty-First Century*. Boston, MA: Belknap Press.

Poli, R. (2014) 'Anticipation: A new thread for the human and social sciences?', *CADMUS*, 2(3), pp. 23–36.

Poli, R. (ed.) (2018) *Handbook for Anticipation*. New York: Springer.

Ramaswami, A., Weible, C., Main, D., Heikkila, T., Siddiki, S., et al. (2012) 'A social-ecological-infrastructure systems framework for interdisciplinary study of sustainable city systems', *Journal of Industrial Ecology*, 16(6), pp. 801–813.

Robinson, B., Musango, J., Swilling, M., Joss, S. and Mentz-Lagrange, S. (2013) *Urban Metabolism Assessment Tools for Resource Efficient Cities*. Stellenbosch: Sustainability Institute.

Schandl, H., Fischer-Kowalski, M., West, J., Giljum, S., Dittrich, M., et al. (2016) *Global Material Flows and Resource Productivity*. Paris: United Nations Environment Programme.

Scherr, S. (1999) *Soil Degradation: A Threat to Developing-Country Food Security by 2020?* Food, Agriculture and the Environment Discussion Paper. Washington, DC: International Food Policy Research Institute. Available at: www.ifpri.org.

Smil, V. (2014) *Making the Modern World: Materials & Dematerialization*. Chichester, UK: Wiley.

Swilling, M. (2013) 'Economic crisis, long waves and the sustainability transition: An African perspective', *Environmental Innovation and Societal Transitions*, 6. doi: 10.1016/j. eist.2012.11.001.

Swilling, M. and Annecke, E. (2012) *Just Transitions: Explorations of Sustainability in an Unfair World*. Tokyo: United Nations University Press.

Swilling, M., Hajer, M., Baynes, T., Bergesen, J., Labbe, F., et al. (2018) *The Weight of Cities: Resource Requirements of Future Urbanization*. Paris: International Resource Panel. Available at: www.internationalresourcepanel.org.

Swilling, M., Robinson, B., Marvin, S. and Hodson, M. (2013) *City-Level Decoupling: Urban Resource Flows and the Governance of Infrastructure Transitions*. Available at: www.unep.org/ resourcepanel/Publications/AreasofAssessment/Cities/tabid/106447/Default.aspx.

UNEP. (2010) *Assessing the Environmental Impacts of Consumption and Production: Priority Products and Materials*. A Report of the Working Group on the Environmental Impacts of Products and Materials to the International Panel for Sustainable Resource Management. Paris: UNEP.

UNEP. (2014) *Decoupling 2: Technologies Opportunities and Policy Options*. Paris: UNEP.

United Nations Centre for Human Settlements. (2003) *The Challenge of Slums: Global Report on Human Settlements*. London: Earthscan.

United Nations Population Division. (2015) *World Urbanization Prospects: The 2014 Revision.* New York: United Nations.

Urama, K., et al. (2015) *Water Decoupling Report.* Paris: UNEP.

Ward, J., Sutton, P., Werner, A., Costanza, R., Mohr, S., et al. (2016) 'Is decoupling GDP growth from environmental impact possible?', *PLoS One*, pp. 1–14. doi: 10.1371/journal. pone.0164733.

Weisz, H. and Steinberger, J. K. (2010) 'Reducing energy and materials flows in cities', *Current Opinion in Environmental Sustainability*, 2, pp. 185–192.

Wiedman, T. O., Schandl, H., Lenzen, M., Moran, D., Suh, S., et al. (2013) 'The material footprint of nations', *Proceedings of the National Academy of Science.* Available at: www.pnas. org/cgi/doi/10.1073/pnas.1220362110.

4

GLOBAL CRISIS AND TRANSITION: A LONG-WAVE PERSPECTIVE

Introduction

The Preamble to the SDGs refers to the need for a 'transformed world'. However, it says nothing about how this may come about or the historical trends that might coalesce to create the conditions for a transition to a transformed world. Following a tradition of thinking about "great transformations" pioneered by Polanyi (Polanyi, 1946) and extended into the contemporary era by Berry (Berry, 1999) and Perez (Perez, 2002), this chapter will aim to address this challenge by proposing a framework for understanding the complex dynamics of transition that are already underway. A growing body of popular and academic literature has turned to long-wave theory to contextualize the crisis and comprehend the dynamics of possible future trajectories of transition. The problem with this literature is that contributors tend to search for single long-waves of transition that are then used to better understand the wider dynamics of change. In reality, there are multiple historic dynamics unfolding at any one time that intersect in complex and unpredictable ways. This chapter breaks away from most long-wave traditions of scholarship by establishing a framework for making sense of the complexity of our current transition as a multiplicity of related, asynchronic and intersecting long-wave cycles.

It will be argued in this chapter that we need to understand the dynamics of the current global polycrisis as the emergent outcome of intersections between four dimensions of transition: socio-metabolic transition, techno-economic transition, socio-technical transition and long-term global development cycles. When understood as multiple cycles that intersect concurrently and asynchronistically across these four dimensions, the emergent outcome can be understood as a deep transition.[1] However, a deep transition is a quantum shift that must also – from a normative perspective – be a just transition. While an unjust transition is conceivable (decarbonized enclaves for rich elites and middle classes who want to be, or are compelled/incentivized to be, 'green'), this will not provide a stable political basis for sustaining a deep transition over the long term. This normative perspective imposes additional challenges that are frequently avoided in most analyses of transition.

These four dimensions of transition can combine into a deep transition in different ways depending on the dynamics and impacts of policy decisions and social struggles for more or less just outcomes. This goes to the *politics* of the deep transition. While there is agreement that we are in some kind of interregnum and that, therefore, a transition of some sort is almost inevitable, there is little agreement on how this transition should be understood. For some, exemplified by Carlota Perez (Perez, 2002, 2009, 2013), the transition amounts to another phase in the evolution of industrial modernity with a reformed capitalism remaining as the primary mode of accumulation (for a similar position, see Mazzucato, 2016). For others, we should anticipate a deep and fundamental transition to a post-industrial epoch of some kind (Schot and Kanger, 2018). This position bifurcates into two, with some arguing this can only happen in a post-capitalist future of some kind (Eisenstein, 2013; Streeck, 2014; Escobar, 2015; Mason, 2015; Ahmed, 2017), while others assume that a post-industrial era is more or less synonymous with new more sustainable modes of production and consumption that remain essentially capitalist but in a very different form (Korten, 2007; Grin, Rotmans, Schot, Geels and Loorbach, 2010; Geels, 2013; Schot and Kanger, 2018). If capitalism means private control of the large bulk of capital by a tiny fraction of the population, this will be inconsistent with the requirements of a just transition. A deep and just transition creates conditions for post-capitalist alternatives, in particular commons-based peer production systems (see Chapter 2) with special reference to the renewable energy (RE) sector (see Chapter 8).

In "Rethinking the polycrisis from a long-wave perspective" section, the current global economic crisis is defined as a 'polycrisis' that can, in turn, be usefully understood from the perspective of long waves of historical development across different temporal scales. The section that follows, "Socio-metabolic transitions", describes the primary socio-metabolic transitions – the agricultural and industrial revolutions – in order to propose a template for thinking about what may turn out to be the next 'deep transition'. Section "Techno-economic surges of development" describes the dynamics and modalities of techno-economic transitions and argues that we may be moving into a new phase of global development that is driven by both the deployment phase of the Information Age and the installation phase of the green-tech revolution. The section on "Socio-technical transitions" takes this argument further by suggesting that the post–World War II period ending in the economic contraction of 2009 can be seen as a long-term global development cycle that has now come to an end. The next cycle will be shaped not only by the usual financial and economic drivers (capital, labour and knowledge) but also now by the drivers of the next socio-metabolic transition. The final concluding section poses some key questions for future research.

Rethinking the polycrisis from a long-wave perspective

The global economic crisis has generated a new literature that draws on long-wave theory to reimagine present and future landscapes. The writers in this neo-Polanyian tradition articulate what Geels would refer to as clusters of discursive and cultural ontologies of probable futures (Geels, 2010). These include consultant's advisories and popular literature aimed at business audiences (Allianz Global Investors, 2010; Bradfield-Moody and Nogrady, 2010; Rifkin, 2011); the policy-oriented research-based literature generated from a variety of academic, UN, advisory and consulting agencies (Von Weizsacker, Hargroves, Smith, Desha and Stasinopoulos, 2009; McKinsey Global Institute, 2011b; United Nations Environment Programme, 2011b), the theory-laden academic literature (Drucker, 1993; Perez, 2009, 2010; Gore, 2010; Smith, Voss and Grin, 2010; Pearson and Foxon, 2012; Swilling and Annecke, 2012; Mason, 2015) and the post-developmental "transition discourses" (Escobar, 2015). These texts have to a greater or lesser extent drawn on a tradition (originating in the works of Kondratieff and Schumpeter[2]) that depicts economic history in terms of a succession of long-term waves or cycles of economic development lasting between 40 and 60 years (for useful overviews of some of the main schools of long-wave – or what Foxon calls "co-evolutionary" – thinking, see Foxon, 2011; Köhler, 2012).[3]

Table 4.1 summarizes the state of the literature on long-wave thinking about the historic periods that characterize the industrial epoch.

Instead of seeing the crisis as an accident of history, long-wave theory provides a set of heuristic conceptual framings that make it possible to depict the crisis as a particular moment in a much wider and deeper set of historical trajectories that have not only occurred before but can be expected to unfold rather unpredictably in contextually specific ways in future.

But before proceeding to elaborate this framework, it is necessary to recognize the critiques of long-wave theory (Rosenberg and Frischtak, 1983; Fagerberg, 2003; Verspagen, 2005; Broadberry, 2007) and the relationship between long-wave theory and the multi-level perspective (MLP) (Köhler, 2012). Whether referring to the more classical Kondratieff cycles used by development economists (Gore, 2010) or the S-curves at the centre of the MLP (Grin, Rotmans, Schot, Geels and Loorbach, 2010) or the structural evolutionary approaches (Freeman and Louca, 2001; Perez, 2002), the obvious danger is that they are prone to techno–economic determinism: technological innovations do the 'acting' and socio-political institutions do the 'reacting'. One solution to this problem offered by the "co-evolutionary framework" is to analyse the co-evolution of socio-economic, institutional and ecological systems and their causal interactions (Pearson and Foxon, 2012). Foxon and Pearson argue that long-wave theorists

> are keen to stress that these attributes of technologies do not "determine" wider socio-economic change, but they enable co-evolutionary changes in institutions and practices that, together with technology changes, give rise to significant macroeconomic impacts.
>
> *(Pearson and Foxon, 2012:121)*

TABLE 4.1 Long-wave thinking

	Techno-economic transition – Carlota Perez[4]	Kondratieff as per Mason (2015)	History of industrial capitalism – Peter Drucker (1993: 25–42)	Overlapping periods according to other authors	Schwab – Fourth Industrial Revolution	Industrialization of knowledge – Yochia Benkler (2003, 2006)	Major trends (Swilling and Annecke, 2012:56; Mason, 2015:115)
From 1771	First: *The "Industrial Revolution"*	1790–1848: steam, canals, factory system (1st)	*The "Industrial Revolution"*	Start of the first machine age (Brynjolfsson and McAfee, 2014:7)		Industrial economy	Small family-owned enterprises
From 1829	Second: *Age of Steam and Railways*	1848–90s: railways, the telegraph, ocean-going steamers, machine-produced machinery (2nd)	*The "Mechanical Revolution"*		First Industrial Revolution		
From 1875	Third: *Age of Steel, Electricity and Heavy Engineering*	1890s–1945: heavy industry, telephone, mass production, electrical engineering (3rd)	*The "Productivity Revolution"*		Second Industrial Revolution	Industrial information economy	Emergence of modern stock-holding corporations – separation of ownership; Scientific management
From 1908	Fourth: *Age of Oil, the Automobile and Mass Production*	Late 1940s–2008: transistor, synthetic materials, factory automation, nuclear power, computerization (4th)	*The "Management Revolution"*				
From 1971	Fifth: *Age of Information and Tele-communications*		*The "Information Revolution"*		Schwab (2017:6–7), the 3rd Industrial Revolution – 1960–2000: resulted in mainframe computing (1960s), personal computing (1970s, 1980s) and the Internet (1990s). Rifkin (2011) refers to this as version 3.0		Consumerism, Globalization, Financialization, Shortermism
From 1990s		From late 1990s: network technology; mobile communications, globalization, information goods (5th)					
From 2000s				2000s–Present Second Machine Age (Brynjolfsson and McAfee)	Fourth Industrial Revolution[5]	Networked information economy	Digitization, Automation, Decentralization

Source: Compiled from Drucker, 1993; Benkler, 2003, 2006; Perez, 2009; Rifkin, 2011; Swilling and Annecke, 2012; Brynjolfsson and McAfee, 2014; Mason, 2015; Schwab, 2017

It is problematic to assume that there is a grand wave of economic development that somehow takes hold simultaneously everywhere and – using language from neoliberal ideology – 'lifts all boats'. Instead, innovations originate in particular countries for quite specific well-documented reasons related to institutions, culture, labour markets and economic dynamics (Pearson and Foxon, 2012). They then radiate outwards absorbing others into mutually reinforcing new economic and financial circuits, while still others get excluded from innovations and investments in human capital and institutional reform or subordinated to providers of primary materials (for the uneven development impact of the information revolution, see Castells, 1997). As radical geographers have argued for decades, it needs to be accepted that uneven development has been intrinsic to all the different phases of capitalist development since the start of the industrial era (Smith, 2008).

A framework will be proposed here that differs from existing approaches because it deals with four interactively asynchronous long-wave dynamics that operate at different temporal scales and with reference to four *dimensions of transition* (all explained in detail later):

- 'socio-metabolic transitions' that focus on the flow of materials and energy through socio-ecological systems across the pre-industrial, industrial and (potentially more sustainable) post-industrial epochs (Fischer-Kowalski and Haberl, 2007; Fischer-Kowalski and Swilling, 2011);
- 'techno-economic transitions' – or what Perez prefers to call "great surges of development" – comprising the evolution of the five main clusters of 'general-purpose technologies' (Lipsey, Carlaw and Bekar, 2005) that have partially driven and shaped the fundamental changes in production and consumption during the industrial era (Perez, 2009);
- 'socio-technical transitions' – this refers to the Multi-Lateral Perspective (MLP) on the dynamics of change as 'landscapes' interact with 'regimes' and 'niches' within particular socio-technical sectors such as energy or transport (Geels, 2011);
- 'long-term global development cycles' is the Schumpeterian focus on cycles of economic growth, prices, crises and 'creative destruction' (Gore, 2010).

Although there are huge bodies of literature that relate to each of these themes, leading contributions have been selected that effectively articulate the different but linked long-wave dynamics that operate at these different scales. Drawing on – but going much further than – Swilling and Annecke (Swilling and Annecke, 2012), these are then synthesized not for the purpose of constructing a new 'grand theory' of transitions but with the much more limited aim of assessing where we are in the global polycrisis and what the possible dynamics of transition might be at different levels of analysis.

Polanyi was interested in a 'double movement': the fragmentary nature of laissez-faire capitalism that emerged during the second half of the nineteenth century and shaped during the first half of the twentieth century and the parallel emergence of integrative dynamics of micro-level pacts and associations. Reading this 'double movement', he anticipated the post–World War II 'grand transformation'. Social democracy after World War II seemed to realize this vision. Today's conditions exhibit the same double movement – crisis, inequality, division and potential collapse versus the power of global grassroots movements expressing real liveable alternatives (see Chapter 6).

There is today increasing interest in the possibility of some sort of epochal shift, leading to a post-industrial world that is more or less sustainable – the "transformed world" referred to in the Preamble to the Sustainable Development Goals (SDGs). Schot and Kanger refer to this as the "second deep transition" (Schot and Kanger, 2018), while the German Advisory Council on Climate Change (explicitly invoking Polanyi) refer to another "great transformation" similar in significance to the agricultural and industrial revolutions 13,000 and 250 years ago, respectively (German Advisory Council on Climate Change, 2011). For Mason we are transitioning to a "post-capitalist" epoch, while Nafeez Ahmed envisages a "civilizational transition" that will of necessity transcend capitalism (Ahmed, 2017). For Perez and Mazzucato, capitalism can be reformed around new developmental and environmentally sustainable imperatives (Mazzucato, 2016; Perez, 2016).

None of these perspectives deal with all four of the aforementioned dimensions of transition: Schot and Kanger synthesize Perez's 'great surges of development' and the MLP 'socio-technical transitions'; the GACCC only really considers the 'socio-metabolic transitions' literature (discussed further in subsequent pages); Ahmed and Mason both deploy a complex adaptive systems perspective in a way that is suggestive of a post-capitalist future but with Mason incorporating a reinterpretation of a Perezian/Kondratieff perspective.

It is worth considering Schot and Kanger's definition of a 'deep transition':

> A Deep Transition is formally defined as a series of connected and sustained fundamental transformations of a wide range of socio-technical systems in a *similar direction*. Examples of this directionality include (the post WWII) move towards increased labour productivity, mechanization, reliance on fossil fuels, resource-intensity, energy-intensity, and reliance on global value chains. Our assumption is that this process of building connections between change processes in multiple systems takes on wave-type properties, unfolds through centuries, and is implicated in broader transformations of societies and economies. In this conceptualization each wave is broadening and deepened in the Deep Transition, but should not be seen as a Deep Transition in itself. The Deep Transition refers to the overall change process, and is thus comparable to what Polanyi (2001 [1944]) called the Great Transformation.
>
> *(Schot and Kanger, 2018:1 – emphasis added)*

While agreeing with this definition of a 'deep transition' (which is remarkably similar to my conception of 'epochal transition' developed in *Just Transitions*), it is only ontologically coherent when all four dimensions of transition are integrated: socio-metabolic transition (between historical epochs), techno-economic transition (between industrial periods), socio-technical transition (between socio-technical systems in particular sectors) and long-term development cycles (reflected in price and growth cycles that occur over historical periods during the industrial epoch). I have, therefore, adopted Schot and Kanger's elegant notion of a 'deep transition' but broadened out its meaning to include the four dimensions of transition underpinned by an overarching normative commitment to a just transition.[6]

It may be more useful to understand the deep transition to a more sustainable world as an emergent outcome (as per complex adaptive systems theory) of the four dimensions of transition. When integrated in this way, the following proposition becomes possible: the 'deep transition' from industrial modernity to the 'transformed world' referred to in the Preamble to the SDGs is not merely about the extended survival of industrial modernity, but rather it implies a deep (socio-metabolic) transition to a new sustainable epoch whose directionality and pace will depend on the three other dimensions of transition that emerge from within the contradictions of industrial modernity (techno-economic transitions, socio-technical transitions and long-term development cycles). Depending on actually existing social struggles for change and how, in particular, the energy transition pans out (see Chapter 8), the outcome will be more or less *just. A just transition* may well need to be an information-based hybrid of capitalist and post-capitalist commons-based peer-to-peer (CBPP) systems. The former would entail a socially embedded market, subordination of finance to the 'real economy' and continuation of aspects of private ownership and private investment; while the latter entails a significantly expanded 'commons' where ownership is neither state nor private, socially and/or publicly owned financial institutions with major investment resources, expanded non-market transactions, a burgeoning social entrepreneurship sector and a non-exploitative non-extractive relationship with natural systems. How exactly this pans out will more than likely be very different to what can be imagined from 'this side of history' (Frase, 2016). Either way – more or less just – the actual empirical drivers of change are incrementalist in nature, and this is dealt with in subsequent chapters (see Chapters 5 and 6). We need a balance between incrementalist approaches to the evolutionary potentials of the present and a sense of the unfolding logics within all four dimensions of change.

Socio-metabolic transitions

There is an increasingly common trend within academic and non-academic analyses of the crisis to identify purely economic causes of the crisis (of various kinds), followed by a set of conclusions about remedies that then add on at the end suggestions that the next phase of global growth will more than likely also be 'green', 'low carbon' or even 'sustainable'. This move amounts to an afterthought that recognizes the negative economic consequences of 'externalities', but these externalities are left out of the analysis of the initial causes.[7] But, as Fischer-Kowalski points out, it is only possible to refer to the unsustainability of a system relative to another system (Fischer-Kowalski, 2011). In order to do this, she argues, the unit of analysis needs to be the socio-metabolic flow of materials and energy through different configurations of coupled natural and social systems. The materials assessed usually include biomass, fossil fuels, construction minerals (mainly cement and building sand) and metals. These materials are measured in tons (see Chapter 3).

As demonstrated in Chapter 3, by 2010 the global economy depended on the use of about 70 billion tons of stuff per annum. The reproduction of our current industrial civilization depends on a constant increase in the use of resources plus the production of about 500 Exajoules of energy that generate most of the GHG emissions that drive climate change. It is also an industrial civilization that is highly unequal, with at most 20% of the population consuming over 80% of the worlds' resources and energy (United Nations Development Programme, 1998).

Understanding social-ecological systems in terms of the through-flow of materials from eco-systems through socio-economic systems and back into eco-systems (as 'waste') helps explain the deep transitions from one socio-metabolic regime to another. Fischer-Kowalski and Haberl use this framework to reinterpret economic history since the last ice age. This includes deep transitions from hunter-gatherers to the agricultural socio-metabolic epoch some 13,000 years ago as soils, seeds and land became usable resources. Followed by the agricultural epoch to the industrial socio-metabolic epoch over 250 years ago as fossil fuels, metals and minerals were added to the resource pool. The "inevitable but improbable" (Fischer-Kowalski, 2011:153) deep transition to a sustainable socio-metabolic epoch would when it is no longer possible to depend on large quantities of non-renewable materials and cheap fossil fuels (Fischer-Kowalski and Haberl, 2007). By rooting the analysis of the polycrisis within the endogenous thermodynamics of material and energy flows, it becomes possible to anticipate futures where natural resources (and not just carbon) are used more sustainably as a necessary condition for the emergence of a future potentially sustainable long wave of ecologically sustainable and inclusive economic development.

This perspective has been operationalized within the contemporary global policy space by the International Resource Panel (IRP) that was established in 2007 by (what is now called) United Nations Environment to deal with global material flows, resource depletion and decoupling growth rates from rates of resource use[8] (see Chapter 3). The IRP distinguishes between four categories of resources: biomass, fossil fuels, construction materials and metals.[9] In one of the IRP's first reports (discussed in detail in chapter 3), the IRP advocated the controversial notion that well-being and economic growth rates could be decoupled from rising rates of resource use (Fischer-Kowalski and Swilling, 2011). Although this notion of decoupling is heavily criticized in the political ecology community (Jackson, 2009), what is not recognized by the critics is that this report made a distinction between "relative decoupling" and "absolute resource reduction" irrespective of the prevailing growth rates and improvements in well-being.

The distinction between "relative decoupling" (economic growth plus *increases* in resource use but at a slower rate than economic growth) and "absolute resource reduction" is helpful from a transition perspective. The latter is only possible if three conditions can be met: more resource efficiency (doing more with less); greater sufficiency (more for most who do not have enough, much less for the over-consumers) and more renewables which means using a lot more of previously unused materials (e.g. what is required to generate solar energy).

As the IRP reports discussed in the previous chapter show, rising global resource use during the course of the twentieth century corresponded with declining real resource prices – a trend that came to an end in 2000–2002. Since 2000–2002, the macro trend in real resource prices has been upwards (notwithstanding dips in 2008/2009 and in 2012).

McKinsey argues that if resource subsidies are reduced, a carbon price of at least \$30/ton is introduced and an additional \$1 trillion per annum is invested in resource-efficient production systems to meet growing demand, the result will be the creation of a whole new set of "productivity opportunities" with an internal rate of return of at least 10% at current prices (McKinsey Global Institute, 2011a). However, these are unlikely to become the focus for investments to drive economic recovery if resource subsidies continue to be defended by the institutionalized politically powerful interests of the dominant regimes of the mineral-energy complex that vigorously defend "carbon lock-in" (Pierson, 2000).

As argued in Chapter 3, the IRP's work essentially documents the resource limits of the industrial epoch and establishes empirically the rationale for a deep transition to a more sustainable epoch. However, like the bulk of similar research in industrial ecology, ecosystem services and climate change, a convincing case is made for *why* a deep transition is needed (mainly by accumulating vast amounts of quantitative analysis) but without any conception of *how* this will happen and *what* needs to be achieved incrementally over the short, medium and long term to ensure a sustainability-oriented deep transition actually takes place. In short, the IRP's work establishes the necessary conditions for a transition but not the sufficient conditions. For this it is necessary to focus on the complex interactive dynamics of accumulation, institutional power and technological change within the other three dimensions of transition. Governance is discussed further in Chapters 7 and 9.

Techno-economic surges of development

The substantial body of work by Venezuelan economist Carlota Perez has deeply influenced those who write about techno-economic cycles. Perez identified five techno-economic transitions or what she preferred to call "great surges of development", each associated with specific technological revolutions that emerged at particular historic moments since the onset of the industrial revolution in the 1770s (Table 4.2). Each followed the familiar S-curve with an installation and a deployment phase bifurcated by a financial crisis (Perez, 2002, 2007).

During the installation phase, financial capital (i.e. those who pursue capital gains by buying and selling shares in companies) tends to gravitate towards a particular cluster of promising socio-technical innovations. Because investors have a

TABLE 4.2 The industries, infrastructures and paradigms of the five technological revolutions

Technological revolution	New technologies and new or redefined industries	New or redefined infrastructures	Techno-economic paradigm 'Common-sense' innovation principles
First: from 1771 The 'Industrial Revolution' Britain	Mechanized cotton industry Wrought iron Machinery	Canals and waterways Turnpike roads Water power (highly improved water wheels)	Factory production; division of labour; efficiency Mechanization Productivity/ time keeping and time saving Fluidity of movement (as ideal for machines with water-power, for transport through canals and other waterways and for human work on products from task to task) Local networks
Second: from 1829 Age of Steam and Railways In Britain and spreading to continent and the United States	Steam engines and machinery (made in iron; fuelled by coal) Iron and coal mining (now playing a central role in growth)* Railway construction Rolling stock production Steam power for many industries (including textiles)	Railways (use of steam engine) Universal postal service Telegraph (mainly nationally along railway lines) Great ports, great depots and worldwide sailing ships City gas	Economies of agglomeration/Industrial cities/ National markets Power centres with national networks: decentralized centralization Scale as progress Standard parts/ machine-made machines Energy where needed (steam) Interdependent movement (of machines and of means of transport) Free markets as ideal context
Third: from 1875 Age of Steel, Electricity and Heavy Engineering The United States and Germany overtaking Britain	Cheap steel (especially Bessemer) Full development of steam engine for steel ships Heavy chemistry and civil engineering Electrical equipment industry Copper and cables Canned and bottled food Paper and packaging	Worldwide shipping in rapid steel steamships (use of Suez Canal) Worldwide railways (use of cheap steel rails and bolts in standard sizes) Great bridges and tunnels Worldwide Telegraph Telephone (mainly nationally) Electrical networks (for illumination and industrial use)	Giant structures (steel) Economies of scale of plant/vertical integration Distributed power for industry (electricity) Science as a productive force Worldwide networks and empires (including cartels) Universal standardization Cost accounting for control and efficiency Great scale for world market power/ 'small' is successful, if local

(Continued)

TABLE 4.2 (Continued)

Technological revolution	New technologies and new or redefined industries	New or redefined infrastructures	Techno-economic paradigm 'Common-sense' innovation principles
Fourth: from 1908 *Age of Oil, the Automobile and Mass Production* In the United States and spreading to Europe	Mass-produced automobiles Cheap oil and oil fuels Petrochemicals (synthetics) Internal combustion engine for automobiles, transport, tractors, airplanes, war tanks and electricity Home electrical appliances Refrigerated and frozen foods	Networks of roads, highways, ports and airports Networks of oil ducts Universal electricity (industry and homes) Worldwide analog telecommunications (telephone, telex and cablegram) wire and wireless	Mass production/mass markets Economies of scale (product and market volume)/ horizontal integration Standardization of products Energy intensity (oil based) Synthetic materials Functional specialization/ hierarchical pyramids Centralization/metropolitan centres – suburbanization National powers, world agreements and confrontations
Fifth: from 1971 *Age of Information and Telecommunications* In the United States, spreading to Europe and Asia	The information revolution: cheap microelectronics Computers, software Telecommunications Control instruments Computer-aided biotechnology and new materials Artificial intelligence, Robotics, Internet of Things	World digital telecommunications (cable, fibre optics, radio and satellite) Internet/ Electronic mail and other e-services Multiple source, flexible use, electricity networks High-speed physical transport links (by land, air and water)	Information-intensity (microelectronics-based ICT) Decentralized integration/ network structures Platforms Knowledge as capital/intangible value added Data as raw material Heterogeneity, diversity, adaptability Segmentation of markets/proliferation of niches Economies of scope and specialization combined with scale Globalization/interaction between the global and the local Inward and outward cooperation/clusters Instant contact and action/instant global communications

Source: Perez (2002) Ch. 2, Tables 2.2 and 2.3 (updated by the author)

* These traditional industries acquire a new role and a new dynamism when serving as the material and the fuel of the world of railways and machinery.

'herd mentality', large amounts of money get concentrated in a selected inter-connected set of innovations that, in turn, get catapulted into wider markets and aggressively promoted via combinations of supply- and demand-side measures. This over-excited concentration of investments in innovations that must still be gener-ally accepted within the wider economy and society leads to over-investment, share price rises incommensurate with underlying value and, eventually, a financial bub-ble emerges. When this bubble bursts, a financial crisis occurs. This amounts to an across-the-board devaluation of listed wealth that can, in turn, have wider economic repercussions. The crisis can trigger corrective state interventions to restore macro-economic stability. This usually entails creating the conditions for the absorption of the new disruptive technologies. The result is a shift in power from finance capi-tal to productive capital. The latter is more long-termist and patient because it is dividend-seeking rather than capital gains–oriented. After the 1929 crash (and some initial mistaken responses that exacerbated the crisis), these state interventions were essentially inspired by Keynesian economic thinking,[10] establishing the basis for the growth of the post–World War II welfare states. After the onset of recessionary conditions after the 1973 oil crisis, from the late 1970s onwards the interventions were essentially inspired by neoliberal economic thinking establishing, in turn, the basis for neoliberal governance, financialization and globalization during the 1980s and 1990s. Contrary to the rhetoric of neoliberalism, state intervention – and the increasingly significant role of the World Bank, IMF and financial regulators (who were, in turn, obsessed with deregulation) – was decisive in the global restructuring that occurred during this period (Boldeman, 2007; Stiglitz, 2010a; Turner, 2016).

A key problem with Perez's argument is that state interventions only really become significant in her analysis at times of crisis and she underestimates the dynamics of capital accumulation. There is now a widely accepted body of scholar-ship that accepts that states play a crucial role during the early phases of the inno-vation cycle with respect to two kinds of interventions (Mazzucato, 2015, 2016): investments in R&D and investments and mechanisms (e.g. sovereign guarantees) to reduce risk during the early phases of the innovation cycle. These two make sense because in the case of the former, private investors are often reluctant to invest in R&D because the benefits accrue to society as a whole, not the specific investor; while in the latter case risks tend to be too high for the private sector during the early phases of the innovation cycle. Thus, the 'crowding in' of private investment depends on both knowledge construction and risk reduction (via direct invest-ments, or the provision of guarantees, or discounted loan finance). Mazzucato has applied this framework to explain both the rise of the internet in the US context and (as discussed in Chapter 8) the rise of renewables in the German and Chinese context (Mazzucato, 2011, 2015).

For Perez, there have thus been five major techno-economic transitions, each corresponding with the emergence of a particular set of interdependent general-purpose socio-technical innovations with transformative implications for regula-tory regimes and institutional configurations. These are reflected in Table 4.2. Each techno-economic transition, however, goes through the S-curve that starts off with a finance-driven installation period and ends off with a productive capital-driven deployment phase, with the state playing key roles during the early phases of the innovation cycle and during the crisis-driven transition from finance to productive capital.

Following Perez (Perez, 2002), the dominant techno-economic paradigm today is what Castells referred to as the "Information Age" (Castells, 1997). Its origins lie in the United States in the early 1970s as innovators (with state support) assembled the basics of what later became the micro-computing revolution that built on, but also transcended, the parameters of the fourth (fossil fuel based) techno-economic transition. However, Mason's reworking of this periodization makes more sense. Arguing against Perez's argument that the third transition lasted over 70 years (from 1908 to early 1970s), he prefers to date the fourth techno-economic transition from 1945 to 2008 (i.e. corresponding with Gore's conception of the post–World War II long-term development cycle - see below). For Mason, the fourth techno-economic transition is about the maturation of the fossil fuel era. It follows that for Mason the fifth techno-economic transition originates in the late 1990s rather than the early 1970s, reflected in the mainstreaming of ICT into many aspects of economic and daily life. For him and Perez, the crisis of 2007/2008 marked the mid-point of the fifth and the ending of the fourth techno-economic transition. The oil crisis of the early 1970s marks the mid-point crisis of the fourth techno-economic transition, not the start of the fifth as suggested by Perez (Mason, 2015). While Mason's schema is more convincing, he does not recognize the significance of the rise of 'green-tech' and the RE revolution in the context of a gradual shift in power from finance to productive capital. 'Greening' is not just an extension of the fifth techno-economic transition. Instead, overlapping with the fifth techno-economic transition (no matter whether one prefers Perez's or Mason's periodization) is the start of the sixth techno-economic paradigm in 2004, reinforced by the onset of the global financial crisis in 2007/2008 and reflected in the brief appearance of 'green Keynesianism' during 2009 (Geels, 2013). As Chapter 8 shows, massive increases in public and private investment began in 2004 and accelerated after 2007/2009. Thus, breaking with both Perez and Mason, it can be argued that a sixth techno-economic transition commenced in the early 2000s that is probably best described as the 'green tech' revolution, with RE as the lead sector (see Figure 4.2).

Perez has argued that the global economic crisis of the "Information Age" (the fifth techno-economic transition) has, in fact, experienced a 'double bubble' – the so-called 'dot.com' bubble of 1997–2000, followed by the financial bubble of 2004–2007. The former was triggered by over-investments in so-called 'dot.com' shares during the years leading up to the 'dot.com' crash in 2000 and the latter by over-investments in financial instruments. Perez has argued that these "two bubbles of the turn of the century are two stages of the same phenomenon" (Perez, 2009:780), that is, together they mark the mid-point crisis of the fifth techno-economic transition (whether or not it is seen as originating in the 1970s or the 1990s). She argues against the Keynesian argument that explains the financial crisis as a 'Minsky moment' in terms of which debt markets have an in-built tendency towards financial instability, which can only be mitigated by increased state spending after the crisis-driven devaluation hits (Krugman, 2012). Instead, she argues that the most significant crises are triggered by the financial opportunities created by new technologies that result in 'major technology bubbles' that eventually burst. This is what the 'internet mania' of 1997–2000 was all about. However, instead of triggering an economic recession that would have necessitated extensive state intervention to prepare the way for productive capital to take over from financial capital after the bubble burst in 2000/2001, the post-crisis recession was mitigated by the rapid financialization of the global economy (enabled by deepening deregulation of the banks), the rise of China as the manufacturer and de facto funder of the world economy, and the accelerated expansion of information-based global trade (with flexible specialization and just-in-time systems as key operating procedures).

Cheap Chinese exports (achieved in part by 'artificially' keeping the value of the Chinese yuan down) not only brought down the cost of mass consumer goods (which effectively raised the value of wages in the West at a time when wages were flat or in decline) but also made it possible for China to become one of the world's largest providers of debt to developed world consumers via the purchase of massive quantities of government bonds (which were then on-loaned to the private banks). Indeed, the preference for liquid assets and quick operations within the paper economy that Chinese surpluses made possible generated skyrocketing capital gains for finance capital in the lead-up to the dot.com crisis between 1996 and 2000, while profits in the real economy (i.e. sectors outside of IT) remained flat or even negative. After the 'dot.com' crash, instead of interventions to restrain financial capital, the opposite happened as various interventions by the Federal Reserve and neoliberal governments around the world effectively allowed the paper economy to mushroom into a gigantic unregulated global casino (Gowan, 2009; Turner, 2016) – what former UK Prime Minister Gordon Brown liked to call 'light touch' regulation. The resulting bubble was a Ponzi-type "easy liquidity bubble" driven by massive concentrations of investments in what eventually became worthless paper assets (Perez, 2009). The 'nationalization' of private debt, constraints on short-term capital flows and 'quantitative easing' became the instruments that were then used to salvage the resulting wreckage (Turner, 2016).

For Perez, a key condition for a successful transition is the disciplining of capital gain-seeking finance capital to make way for dividend-seeking productive capital to drive the deployment phase. However, in her recent writing (Perez, 2013), she has started to factor in environmental externalities by emphasizing the role that innovations for greening the economy will play in the deployment phase of the fifth cycle. But it is unclear what will drive these innovations – finance or productive capital? Nor does Perez define a new historic mission for finance capital after it has been disciplined to make way for productive capital. Indeed, there is little evidence that finance capital is in fact being effectively disciplined (see discussion in subsequent pages).

The alternative lies in accepting that a sixth cycle – the 'green-tech revolution' – is emerging that is increasingly driven by finance capital supported by state interventions to generate R&D and reduce risk for investors (see Chapter 8). Surely, this is the new historic mission for finance capital, especially during a period of falling RE prices and awareness of the potential threats of climate change. And would this not create a growth-catalysing installation phase of an emergent sixth cycle to complement the deployment phase of the fifth cycle? Perez is reluctant to accept this line of argumentation, as is Mason. Yet this may well be what is underway.

Socio-technical transitions

Although the literature on socio-technical transitions is rooted in the broader literature on systems innovation, evolutionary economics and the sociology of technology (Rip and Kemp, 1998), for the purpose of this book I use the Multi-Level Perspective (MLP) on socio-technical transitions (Geels, 2005; Smith, Voss and Grin, 2010; for a critique from a political ecology perspective, see Lawhon and Murphy, 2011). According to the MLP, socio-technical transitions result in 'deep structural changes' over long time periods within particular sectors (e.g. transport, energy, water, sanitation, waste communications) that involve fundamental reconfigurations of technologies, markets, institutions, knowledge, consumption practices and cultural norms (Geels, 2011:24). They are explained in terms of complex non-deterministic interrelations between three levels of reality: landscape pressures (macro), regime structures (meso) and niche innovations (micro). This framework is then used to address the challenge of the complex transition(s) to a more sustainable world which is defined as 'human well-being in the face of real bio-physical limits' (Meadowcroft, 2011:71) and 'an open-ended orientation for change' (Grin, Rotmans, Schot, Geels and Loorbach, 2010:2). However, as Hausknost and Haas argue, there is insufficient emphasis in this literature on state intervention to promote a sustainability-oriented directionality (Hausknost and Haas, 2019).

Major socio-technical regimes comprise a core set of technologies that co-evolve with social functions, social interests, market dynamics, policy frameworks, knowledge infrastructures and institutional regulations. These socio-technical regimes are shaped by a broad constituency of technologists, engineers, policymakers, business interests, NGOs, consumers and so on. The interrelationships of these interests through regulations, policy priorities, consumption patterns and investment decisions, among other things, hold together to stabilize socio-technical regimes and their existing trajectories. Regimes set the parameters for what is possible:

> reconfiguration processes do not occur easily, because the elements in a socio-technical configuration are linked and aligned to each other. Radically new technologies have a hard time to break through, because regulations, infrastructure, user practices, maintenance networks are aligned to the existing technology.

(Geels, 2002:1258)

The concept of 'landscape' is important in the MLP in seeking to understand the broader 'conditions', 'environment' and 'pressures' for transitions. The landscape operates at the macro level beyond the immediate efficacy of human agency, focusing on issues such as political cultures, economic growth, macro-economic trends, land use, utility infrastructures and so on. The landscape applies pressures on existing socio-technical regimes creating opportunities for responses, for example, to climate change and the need for the expansion of RE. Landscapes are characterized as being 'external' pressures that have the potential to impinge upon – but do not determine – the constitution of regimes and niches: they are an external context "that sustains action and makes some actions easier than others. These external landscape developments do not mechanically impact niches and regimes, but need to be perceived and translated by actors to exert influence" (Geels and Schot, 2007:404).

The idea of socio-technical niches, which operate at a micro level, is one of 'protected' spaces, usually encompassing small networks of actors learning about new and novel technologies and their uses. These networks agitate to get new technologies onto 'the agenda' and promote innovations by trying to keep alive novel technological developments. The constitution of networks and the expectations of a technology they present are important in the creation of niches.

The MLP is particularly useful when it comes to explaining why particular sectors introduce sustainability-oriented alternatives, from green packaging (retail sector), lighter materials (car industry), recycling (waste industry), mass transit (mobility sector), organic foods (food system), and RE (energy sector). In each case, it is relatively easy to demonstrate how a pre-existing regime became unviable as a result of the impact of landscape pressures (climate change, prices of renewables, water scarcities, urbanization, etc.) and how regimes react to the combination of these pressures and expanding niche innovations (Smith, Stirling and Berkhout, 2005). However, for change to occur, niche innovations need to mature to a point where they can become an alternative regime or else get absorbed by incumbent regimes that decide to survive by transitioning (e.g. some of the large European energy companies that have become major RE players). Once again, this cycle unfolds over time, normally a 40–60-year period.

Global development cycles

To improve our understanding of the linkages between the techno-economic transitions that Perez has identified and the socio-technical transitions that emerge from the MLP (with special reference to the RE revolution) and the long-wave dynamics of global economic development, it is necessary to turn to the work of former UNC-TAD economist Charles Gore (Gore, 2010). He has located the techno-economic cycles described by Perez within what he refers to as the Kondratieff-like "global development cycle" that began in the 1950s and ended with the global economic contraction of 2009 (see Figure 4.1). For Gore, the year 2009 marks a key turning point because it was the only year since World War II that the global economy actually shrank. For Gore, a Kondratieff-type cycle cannot be equated to the techno-economic cycles that Perez has in mind. While techno-economic cycles typically follow the well-known S-curve (found in Perez and to some extent the MLP) of *'irruption-crisis-deployment'*, Gore[11] suggests the global development cycles adhere to very different logics. As reflected Figure 4.1 (A), the global development cycle starts off with *'growth-plus-price-inflation'* during the spring-summer period (1950s/1960s) ending in a stagflation crisis driven in part by over-investment in infrastructures during the growth phases (1970s). These over-investments push up interest rates prior to the benefits of the infrastructure investments working their way through the economy as a whole. This is then followed by *growth-with-limited-inflation* during the autumn-winter period (1980s/1990s) ending in deflationary depression driven in part by diminishing returns on mature technologies, while returns on the new technologies have yet to materialize and the inflationary pressures have not kicked in yet (2007 onwards). Significantly, the first half of the post–World War II long-term development cycle was dominated in the West by a Keynesian 'golden age' of welfarism, inclusion, solidarity and liberation (including decolonization in the peripheries) within national development policy frameworks; while the second half was dominated by neoliberalism, financialization through deregulation, exclusion, structural adjustment, commodification, individualism and rising inequalities in an increasingly globalized world.

Although Perez tried to link technological cycles to economic growth, in her later work she gave up this effort. Although Gore admits there is no evidence to support the notion that growth phases are driven exclusively by techno-economic transitions (Gore, 2010), in Figure 4.1 he has enriched the overall picture by

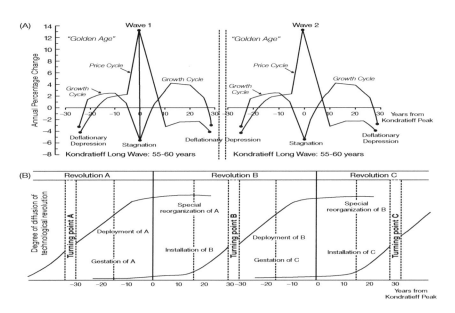

FIGURE 4.1 The synchronisation of growth cycles, price cycles and the lifecycles of technological revolutions. A, Growth cycles and price cycles in the Kondratiev long wave (based on USA). B, The lifecycle of technological revolution

Source: Gore 2010: 718

showing the asynchronous correlation of the Kondratieff cycles with the techno-economic cycles derived from Perez's work. Read together these are very rough approximations of actual growth cycles without in any way suggesting that the actual complex drivers of economic growth at any moment in time are reducible to techno-economic dynamics. His key insight seems to confirm Köhler's argument that S-curves do not run consecutively (as represented by both the MLP and, to some extent, by Perez), but instead they tend to overlap with the deployment phase of a previous cycle and the installation of the new cycle acting as co-drivers of growth-oriented processes (Köhler, 2012). These overlapping deployment and installation phases interact asynchronistically with the Kondratieff cycles in ways that help explain the long-term post-WWII development processes.

Specifically, Gore argues that the post-1970s growth phase was driven by *two* key drivers: the *first* was the economics of the deployment phase of the fourth techno-economic transition (investments in fossil fuels, mass production, inclusive infrastructures, global trade) ('Revolution A' in Figure 4.1) and the *second* was the installation phase of the fifth techno-economic transition (information and communication technologies and their associated applications) ('Revolution B' in Figure 4.1). The mid-point crisis of the fifth techno-economic transition also marks the *end* of the post–World War II global development cycle that ended in 2009. Like the interregnum between 1929 and the 1940s, an interregnum between 2007/2009 and the take-off of the next long-term development cycle can be expected. However, the next global long-term development cycle could only emerge if radical institutional reconfigurations not only displace finance capital to unleash productive capital (following Perez's script) but also displace the powerful and highly subsidized regimes of the mineral-energy complex that depend on the continuities of "carbon lock-in" (Pierson, 2000).

It is worth noting that like Mason, Gore connected Kondratieff cycles and Perez's techno-economic transitions. However, Gore retained Perez's periodization: the fifth transition originating in the early 1970s and reaching its mid-point in 2009. What Gore refers to as the "post-WWII long-term development cycle" is identical to the periodization of Mason's fourth techno-economic transition (1945–2009). This is unsurprising, because Mason is trying to find a single cycle, whereas Gore has recognized that cycles can be asynchronous. Mason's fifth cycle (the Information Age) originates in the early 1990s rather than the 1970s (as per Perez and Gore), but it still hits a crisis nearly 20 years later in 2007/2009.

Towards a synthesis

There is significant evidence that since 2007/2009, there has been a significant increase in investments in RE (United Nations Environment Programme, 2018), communications (Baller, Dutta and Lanvin, 2016) and mobility (Oxford Economics, 2015). These empirical realities are significant because they provide the required focus for convincingly connecting the four dimensions of transition. Consider the following statements:

- confirming that the commencement of the 'spring period' of a long-term development cycle (or Kondratieff cycle) usually corresponds with significant increases in investment in a new interconnected set of foundational energy, communications and mobility infrastructures, there is evidence that this is happening (World Bank, 2017): there is significant evidence of upticks in investments in RE (mainly finance capital, with state subsidies) (see United Nations Environment Programme, 2018), mass transit (productive capital, with a big role for DFIs) (Oxford Economics, 2015) and Web 3.0 (productive capital, mainly the newly consolidated giant IT companies) (Baller, Dutta and Lanvin, 2016), with the last providing the operating system for the first two (as argued by Rifkin, 2011);
- the commencement of a Perezian techno-economic surge corresponds with the increases in investments by finance capital in new techno-economic innovations: the move by finance capital into 'green-tech' (especially RE) suggests that the installation phase of a sixth techno-economic transition has begun (United Nations Environment Programme, 2018; see also Chapter 8);
- a regime-shift within a particular sector that results in a socio-technical transition (as per the MLP) corresponds with the coalescing of niche innovations into a new regime that is more appropriately aligned with landscape pressures: the RE revolution follows the ideal-typical dynamic suggested by the MLP's sectoral focus (discussed in detail in Chapter 8) with niches forming into regimes that emerge in parallel to the old fossil fuel energy regimes, but there is also evidence of old regimes absorbing niches as in the case of large fossil fuel–based energy companies who change their business models to include – or even prioritize – RE (e.g. Enel, Italy's largest energy company) (Aklin and Urpelainen, 2018);
- a socio-metabolic transition is underway when there is a fundamental shift in resource use patterns, often taking many years to occur: the drastic decline in the EROI ratio and the rise of renewables seems to suggest that such a "civilizational transition" (Ahmed, 2017) may already be starting, and if what is unfolding is a resource price super-cycle (see Chapter 3), this could well accelerate the decarbonizing dynamics of the socio-metabolic transition.

In short, the increased investments in energy, communications and mobility confirm there is evidence of transitional dynamics across the four dimensions of transition. This provides the basis for concluding that a deep transition of some kind has commenced. But under what conditions will this be a just transition? If it is not also a just transition, it can be a deep transition that could well be destabilized and even attenuated by social movements and political backlashes.

A deep transition is now conceivable: as represented in Figure 4.2 the deployment phase of as represented in Figure 4.2 the fifth techno-economic transition,

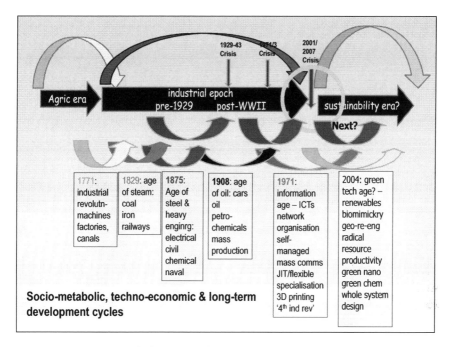

FIGURE 4.2 Socio-metabolic cycles, long-term development cycles and techno-industrial cycles, 1770s to post-2009.

Source: author

the installation phase of the sixth (originating in the crisis of 2007/2009) and the commencement of the next Kondratieff-type long-term global development cycle as conceived by Gore are all conceptualized as being embedded within the wider socio-metabolic dynamics of transition from the industrial to the sustainable epoch as conceptualized by Fischer-Kowalski et al. and the material flow analysis of the IRP.

Nevertheless, an unjust transition is a distinct possibility. A just transition implies that new forms of finance, equitable access to property and productive capital may well be required if the replication of inequality in a decarbonized world is to be avoided. For this, however, new political coalitions will be required which would need to succeed in securing control of governments that reorient their respective polities to achieve three things: (a) definancialization of the fifth techno-economic surge (i.e. the shift to productive capital or, more colloquially, 'real economy' investments); (b) interventions in R&D and risk reduction to support the emergence of the sixth techno-economic surge; and support for the emerging commons-based P2P systems that contribute to a just transition orientation by mobilizing capital and technologies in ways that reinforce inclusive local economies and value chains.

A just transition – as opposed to a mere transition to a decarbonized economy – may well depend on state institutions (e.g. the new generation of Development Finance Institutions (DFIs) and Sovereign Wealth Funds (SWFs)) and large

TABLE 4.3 The global development cycle, 1950s–2030s

	1950s	1960s	1970s	1980s	1990s	2000s	2010s	2020s	2030s
1 Phase of Kondratieff cycle (Gore)	←Spring→		Summer	←Autumn→		Winter	←Spring?→		Summer?
2 Price cycle	Rising prices			Falling prices			Rising prices (driven by rising resource prices, debt devaluation and capital demand/interest rates)		
3 Growth cycle	Growth acceleration		Growth deceleration	Growth acceleration		Growth deceleration		Stagnation	Growth acceleration, beginnings of deceleration from the late 2030s?
4 Fourth industrial transition (Perez)		Deployment phase		Maturity but persistence of Oil Age socio-technical regimes			Decline?		
5 Fifth industrial transition (Perez)			Irruption	Frenzy		Crisis	Synergy		Maturity
6 Sixth industrial transition (Perez)						Irruption			Frenzy, start of crisis by late 2030s
7 Nature of financial crisis			Stagflation crisis			From 2007: deflationary crisis			Start of stagflation crisis from late 2030s?
8 Patterns of economic development		Equalizing (welfarism, Keynesianism, actually existing socialism, decolonization)		Unequalizing (globalization, privatization, deregulation, markets)			Equalizing? (rise of the BRICs, return of Keynesianism, developmental states, etc.)		
9 Resource flows (Fischer-Kowalski and Swilling)	Mainly biomass 10–20 bt/yr	Doubling of non-biomass materials 20–30 bt/yr			Non-biomass materials become dominant, increase to 50 bt/yr		Two-thirds non-biomass, 60 bt/yr, relative decoupling	Relative and absolute resource reduction?	Absolute resource reduction?
10 Resource prices (IRP/McKinsey)		Declining resource prices	Rising resource prices				Rising resource prices	Stable/declining resource prices?	
11 Socio-ecological regime (Fischer-Kowalski)	Industrial socio-ecological regime						(transition to) sustainable socio-ecological regime		

Source: Compiled and adapted from Gore, 2010; Perez, 2010; Fischer-Kowalski and Swilling, 2011; Haberl, Fischer-Kowalski, Krausmann, Martinez-Alier and Winiwarter, 2011; McKinsey Global Institute, 2011a; United Nations Environment Programme, 2011a, 2011b; Swilling and Annecke, 2012

non-profit financial institutions (e.g. Triodos Bank, or the German Church banks) stepping in to create appropriate financial eco-systems for channelling dividend-oriented long-termist investments into the CBPP systems discussed in Chapter 2. Furthermore, there is a growing body of literature on 'energy democracy' (ED) that argues that the material reality of distributed decentralized RE infrastructures favours the building of CBPP systems (Burke and Stephens, 2018). As demonstrated in Chapter 8, the example of Germany stands out: nearly 50% of RE by 2012 were socially or publicly owned. This is the kind of institutional-cum-financial innovation that will make it possible for a new generation of tech-savvy social enterprises to emerge across a wide range of different contexts, from rural electrification in East Africa to the advanced so-called smart grid apps in the developed world that could potentially create the operating system for new modes of ED.

Table 4.3 summarizes the synthesis of these approaches to socio-metabolic transition, techno-economic transitions and global development cycles. Rows 1, 2 and 3 capture the Kondratieff-like price and growth cycles, while rows 7 and 8 depict moments of crisis and periods of inclusion/exclusion. Rows 4 and 5 summarize Perez's techno-economic transitions, revealing their non-alignment with the rhythms of the global development cycle referred to in row 1 (read together with rows 2 and 3). Row 6 factors in the sixth ('green-tech') techno-economic transition and how this is not just a necessary driver for the spring/summer period of the next global development cycle (rows 1 and 3) but is also a necessary condition for the socio-metabolic shift anticipated in rows 9 and 11 (assisted by the dynamics of resource prices – row 10). Obviously, none of the projections here for the 2010s–30s are inevitable. While they are dependent entirely on policy choices by a wide range of actors that must still be made and the associated social struggles that will inevitably emerge, they do reflect patterns of what may be possible. Specifically, the initiation of the next global development cycle, driven by the deployment period of the fifth and the installation of the sixth techno-economic transitions, could provide the conditions required for a fundamental socio-metabolic transition that is also more equalizing and inclusive.

It remains debatable that conditions have matured to a point where the present interregnum can be transcended in a way that could result in a more inclusive and sustainable long-term development cycle. While there is some debate about whether the low carbon and resource efficiency technologies have matured sufficiently (Janicke, 2012; versus Pearson and Foxon, 2012), what is becoming very clear is that the consolidation – through a spate of mergers and acquisitions – within the information and communication sector is preparing the way for the deployment phase of the Information Age. With a strategic focus on 'digitization' and 'integrated value chains', the conditions are in place for productive capital to take the lead (Acker, Grone and Schroder, 2012). However, many analysts admit that this time round, it might not be so easy to discipline financial capital to make way for productive capital (Gore, 2010; Turner, 2016). In this respect the Marxists have a point when they argue that the structural nature of contemporary global capitalism is such that finance capital has managed to establish a hegemonic role for itself that may allow it to resist the transition to productive capital (Altvater, 2009; Gowan, 2009; Harvey, 2009; Blackburn, 2011).

For many observers, by 2018 there seemed little evidence of any fundamental restructuring of the global financial system which may confirm Gore's original argument that the global crisis is a "structurally blocked transition" (Gore, 2010). We have the rivalry between China and the United States about the value of the Chinese currency and trade barriers; the ongoing financial instabilities in the European Union, exacerbated by the multiple sovereign debt crises; the de facto bankruptcy of the United States masked by 'Quantitative Easing' (read: printing money) and the (short-term) fracking bonanza; the relatively unfettered flow of speculative finance through global markets despite the Dodd-Frank regulatory reforms in the United States (which the Trump Administration has now dismantled); the hoarding of cash as investors wait for short-term capital gains opportunities to return, instead of looking for long-term productive investments in the real economy; the continued expansion of derivatives and the power of hedge funds; the uptick in investments in fossil fuels since the commencement of the Trump Presidency; and national governments who, having experienced massive devaluations in the past, continue to build up currency reserves to counteract financial shocks, thus keeping much-needed investment capital away from productive investment. The increased indebtedness of the United States as it cuts taxes and increases spending has most likely triggered a trajectory that could culminate in the next financial crash.

Yet there is no denying the reality of the polycrisis. The world's largest institutional investors seem to be making sustainability commitments and many have withdrawn from the coal sector (Buckley, 2019). According to a 2012 study, of the 283 of the 'very large corporations' that dominate investments in listed stocks, a majority had joined one or more sustainability platform. However, these long-term commitments are undermined by the "short-termism" of equity and financial markets. They found that 36.9% of all shares in the 283 global corporations were owned by investors who had some form of formal commitment to climate change action. These 'climate interested investors' can – and do – make a difference where their shareholdings are large enough (1.5% or above). However, the endemic price volatility of listed shares and the absence of a globally agreed way to price carbon over the long term reinforce what these authors refer to as a low-risk "short-termism" which, in turn, disincentivizes the high-risk investments that will be needed to drive the 'green-tech revolution' out of its niches and into the mainstream. They pursue capital gains but in a financialized world where "short-termism" is regarded as low risk (Peetz and Murray, 2012).

However, there are also counter-movements, including the rise of the DFIs and SWFs as key dividend-oriented long-term investors (Saldinger, 2019), the rise of social enterprises (e.g. in the renewables sector in Germany) (Debor, 2018), the emerging CBPP economy and associated socially responsible financial institutions (Bauwens, Kostakis and Pazaitis, 2019), the shift into 'green-tech' by increasing numbers of investment funds (The Global Commission on the Geopolitics of Energy Transformation, 2019), the creation of new global funds for mitigating climate change (including G20 initiatives in this regard), the rising significance of infrastructure investment funds focused on urban infrastructure and the increasing number of major businesses that are committing to more sustainable practices.

Energy democracy and the just transition

The argument thus far is that a deep transition will be the emergent outcome of dynamics that play out within the four dimensions of transition. However, the directionality of this transition is the subject of social and political contestation. There is, of course, a distinct danger that one direction is towards a decarbonized ecologically sustainable future that leaves existing inequalities intact. This would amount to an unjust transition – the first SDG ('End Poverty') would not be achieved. This raises the question about which dynamics will have the greatest influence on the overall directionality of the deep transition from a just transition perspective, with special reference to the global RE transition which is the leading force of the wider deep transition. We can find an answer to this question in the rapidly expanding new literature on ED (for an overview of this literature, see Burke and Stephens, 2018). In essence, this literature captures new thinking in activist/social movement, progressive academic and certain local governance circles about the potential for more socially just – and therefore more democratic – outcomes of the rapidly expanding number of RE installations emerging across all regions of the world.

When read together with the literature on ICTs and CBPP, the literature on ED explores the political implications of the fusion of renewable and information technologies. Although the informational technologies of the *installation* phase of the fifth techno-economic surge have overlapped with the fossil fuel–based technologies of the *deployment* phase of the fourth techno-economic surge, this has started to change. Since 2007, the informational technologies of the *deployment* phase of the fifth techno-economic surge have begun to overlap with the installation phase of the sixth techno-economic surge. As Rifkin has argued, whenever a new communications technology (in this case the internet/Web 3.0) *conjoins* with a new set of energy technologies (renewables), the result has been a far-reaching industrial revolution (similar in significance to when steam conjoined with the printing press, and later on when the combustion engine conjoined with long-distance telephony and AC electricity). Hence for Rifkin, the conjoining of informational and renewable technologies is bound to trigger what he has called the "third industrial revolution" (Rifkin, 2011). And in many ways this makes sense: only when electricity grids are algorithmically coordinated to manage increasingly complex two-way transactions involving numerous actors (including generators and end-users, or 'prosumers') will it be possible for a multiplicity of 'mini-grids' (to manage embedded generation) and utility-scale RE power plants to emerge within decentralized energy systems across vast geographical territories. Undoubtedly, this will be a key driver of the deep transition.

Whereas Rifkin is a techno-determinist who ignores power relations, the ED literature does not make this mistake. Although the ED literature recognizes the enormous potential of the conjoining of informational technologies (including blockchain-based currencies) and the inherently decentralized nature of actually existing energy technologies, there is an acceptance that the outcome from a just transition perspective is not inevitable. Much will depend on the struggles for – and consolidation of – a particular broadly shared vision of ED and the incrementalist strategies to achieve it. As the case of Germany seems to confirm (see Chapter 8), the decentralized and geographically distributed nature of renewables is such that it creates the potential material base for the emergence of a vast multitude of locally constituted commons-based 'energy democracies'.

The ED literature (explored further in Chapter 8) provides a point of convergence for key arguments made thus far: Mason's celebration of ICT-enabled post-capitalist social relations (Mason, 2015), the literature on CBPP that describes the nascent institutional form of these post-capitalist relations (Bauwens, Kostakis and Pazaitis, 2019) and Jessop's notion of collibratory governance for directionality without sacrificing complexity (Jessop, 2016). In the words of the best synthesizer of the grey and published literature:

> The energy democracy movement represents a contemporary expression of ongoing struggles for social and environmental justice through engagement with technological systems. Energy democracy redefines individual energy consumers as energy citizens, energy commodities as public goods, and energy infrastructure not as a class of assets but rather as public works or common resources. Recognizing an opportunity in the renewable energy transition, the agenda for energy democracy calls for opposing fossil fuels and other centralized energy systems agenda, reclaiming the energy sector within the public sphere, and restructuring energy systems technologies and governance for greater democracy and inclusivity. Above all, energy democracy allows for a *vision of renewable energy transitions as pathways for democratic development*.
>
> *(Burke and Stephens, 2018:90 – emphasis added)*

For ED to become a lead sector for a socially just development pathway rather than a mere niche for progressives, then – following Mazzucato – political coalitions will be needed that put in place appropriate policy and regulatory instruments to support R&D for the sector and reduce risks during the early phases of the innovation cycle. However, unlike her conventional 'green Keynesian' perspective that advocates large-scale capital investment in the green economy (see Mazzucato, 2015), it would be to support more 'post-capitalist' forms of finance, informational commons, institutional collaboration, open source technology development and a regulatory environment that favours CBPP economies. Given the race to prevent runaway global warming, the more rapidly these socio-technical niche innovations coalesce into full-scale regimes the better (see Chapter 8 for an extensive discussion).

Conclusion

Building on the meta-frameworks developed in Chapter 1, this chapter provides a meso-level framework for understanding the highly complex long-wave dynamics of the deep transition to the "transformed" world referred to in the Preamble to the SDGs. Four dimensions of transition were used to develop a stratified understanding of these dynamics. The deep transition, however, will not necessarily be a just transition.

The most significant transitional dynamics that get expressed in different ways across the four dimensions of transition are as follows: the declining EROI which suggests a socio-metabolic transition may be underway; the accelerating RE revolution that signals a socio-technical transition; the commencement of a sixth techno-economic surge reflected in rising investments in 'green-tech'; and increases in the combined investments in new modes of mobility, energy and communications may signal the start of the next long-term development cycle. A deep transition from the industrial to a potentially sustainable epoch may well be the emergent outcome of transitions across all four of these dimensions.

Gore may be right when he argues we may be facing a "structurally blocked transition", no matter how just it may be. Much will depend on whether the grip of global finance on the global economy can be broken (by adopting, for example, the recommendations of the Stiglitz Report or a punitive Tobin tax) (Stiglitz, 2010b). The failure to dislodge the hegemony of finance capital will have three consequences. The first is that despite the considerable consolidations since 2007 in the ICT sector to drive the digitization agenda via greater value-chain integration, productive capital will remain relatively weak which, in turn, could undermine the developmental potential of the deployment phase of the Information Age. The second is that despite the potential of the green-tech revolution, it lacks the magnitude of high-risk investments needed to go beyond niche markets – investments that should be provided by capital gains–seeking finance capital with an appetite for high-risk investments or a new class of socially responsible or even non-profit finance, or a judicious mix of both. Whereas a third of the world's largest investors appreciate the significance of the green-tech agenda, most remain locked into what Peetz and Murray have called low-risk finance-driven "short-termism". And the third is that continued financialization will infect the RE sector resulting in the centralized extractive financing of a decentralized energy infrastructure which would otherwise have huge potential for a more democratic and socially just outcome.

It is clearly valid to conclude from the arguments presented in this chapter that for the deep transition to also be a just transition, much will depend on how the RE sector unfolds (see Chapter 8). It is possible to predict with great certainty that there will be intense social and political struggles over the directionality of the accelerating RE transition. If the balance of forces results in the unfolding of energy democracies at local and regional levels, the overall directionality of the deep transition this may catalyse could well result in a more just transition.

Notes

1 This term is borrowed from Schot and Kanger as discussed further below (Schot and Kanger, 2018).

2 See Kondratief (1935) and Schumpeter (1939).
3 What is left out of this review are long-wave perspectives originating in evolutionary economics that do not include a reference to ecological cycles – a perspective originating in Nelson and Winter (1982) and expressed at a popular level in many references within business circles to super-cycles – see report from global banking firm Standard Chartered (Standard Chartered, 2010).
4 Perez (2009:9,14,18).
5 The Fourth Industrial Revolution according to Schwab (2017:6–7) saw the emergence of the Internet of Things, Big Data and Nanotechnology. Rifkin (2011) does not agree that this constitutes a separate revolution but concedes that we could refer to it as version 3.1.
6 In my review of Schot and Kanger's work, I proposed this broadening out of their concept, but my suggestion was not accepted for reasons that were not explained. The notion of a 'deep transition' is thus used here to refer to what was called a "epochal transition" in previous work (Swilling and Annecke, 2012).
7 Of the literature cited thus far, the works by Allianz (2010) and Perez (2010b – including her contribution to this volume) are representative of this approach.
8 See www.unep.org/resourcepanel/
9 Note that water and land resources are excluded from this categorization of global material flows – for a justification see, Fischer-Kowalski and Swilling (2011:8–9).
10 Economic theories based on the theories of the UK economist John Maynard Keynes.
11 See Gore (2010:718) for the full figures illustrating the asynchronistic alignment of growth cycles, price cycles and life cycles of technological revolutions.

References

Acker, O., Grone, F. and Schroder, G. (2012) 'The global ICT 50: The supply side of digitization', *Strategy and Business*, Autumn, pp. 52–63.

Ahmed, N. M. (2017) *Failing States, Collapsing Systems: Biophysical Triggers of Political Violence*. Cham, Switzerland: Springer.

Aklin, M. and Urpelainen, J. (2018) *Renewables: The Politics of a Global Energy Transition*. Boston, MA: MIT Press.

Allianz Global Investors. (2010) *The Sixth Kondratieff – Long Waves of Prosperity*. Frankfurt: Allianz Global Investors. Available at: www.allianzglobalinvestors.de/capitalmarketanalysis (Accessed: 1 March 2012).

Altvater, E. (2009) 'Postneoliberalism or postcapitalism? The failure of neoliberalism in the financial market crisis', *Development Dialogue*, January.

Baller, S., Dutta, S. and Lanvin, B. (2016) *The Global Information Technology Report 2016: Innovating in the Digital Economy, WEF, INSEAD*. Geneva: World Economic Forum. doi: 10.17349/jmc117310.

Bauwens, M., Kostakis, V. and Pazaitis, A. (2019) *Peer To Peer: The Commons Manifesto*. London: University of Westminster Press.

Benkler, Y. (2003) 'Freedom in the Commons: Towards a Political Economy of Information', *Duke Law Journal*, 52(6), pp. 1245–1276.

Benkler, Y. (2006) *The Wealth of Networks: How Social Production Transforms Markets and Freedom*. New Haven: Yale University Press.

Berry, B. J. L. (1991) *Long-Wave Rhythms in Economic Development and Political Behaviour*. Baltimore and London: John Hopkins University Press.

Berry, T. (1999) *The Great Work: Our Way into the Future*. New York: Bell Tower.

Blackburn, R. (2011) 'Crisis 2.0', *New Left Review*, 72(November/December), pp. 33–62.

Boldeman, L. (2007) *The Cult of the Market: Economic Fundamentalism and its Discontents*. Canberra: Australian National University Press.

Bradfield-Moody, J. and Nogrady, B. T. (2010) *The Sixth Wave: How to Succeed in a Resource-Limited World*. Sydney: Vintage Books.

Broadberry, S. (2007) *Recent Developments in the Theory of Very Long Run Growth: A Historical Appraisal*. Warwick: Warwick University, Department of Economics Research Papers.

Brynjolfsson, E. and McAfee, A. (2014) *The Second Machine Age: Working, Progress, and Prosperity in a Time of Brilliant Technologies*. Paperback. New York: W.W. Norton & Co.

Buckley, T. (2019) *Over 100 Global Financial Institutions Are Exiting Coal, with More to Come*. Cleveland, OH: Institute for Energy Economics and Financial Analysis.

Burke, M. J. and Stephens, J. C. (2018) 'Political power and renewable energy futures: A critical review', *Energy Research and Social Science*, 35, pp. 78–93.

Castells, M. (1997) *The Information Age Volumes 1, 2 and 3*. Oxford: Blackwell.

Debor, S. (2018) *Multiplying Mighty Davids: The Influence of Energy Cooperatives on Germany's Energy Transition*. New York: Springer.

Drucker, P. (1993) *Post-Capitalist Society*. Oxford: Butterworth-Heinemann.

Eisenstein, C. (2013) *The More Beautiful World Our Hearts Know Is Possible*. Berkeley: North Atlantic Books.

Escobar, A. (2015) 'Degrowth, postdevelopment, and transitions: A preliminary conversation', *Sustainability Science*, 10(3), pp. 451–462. doi: 10.1007/s11625-015-0297-5.

Fagerberg, J. (2003) 'Schumpeter and the revival of evolutionary economics: An appraisal of the literature', *Journal of Evolutionary Economics*, 13(2), pp. 125–159.

Fischer-Kowalski, M. (2011) 'Analysing sustainability transitions as a shift between socio-metabolic regimes', *Environmental Transition and Societal Transitions*, 1, pp. 152–159.

Fischer-Kowalski, M. and Haberl, H. (2007) *Socioecological Transitions and Global Change: Trajectories of Social Metabolism and Land Use*. Cheltenham, UK: Edward Elgar.

Fischer-Kowalski, M. and Swilling, M. (2011) *Decoupling Natural Resource Use and Environmental Impacts from Economic Growth*. Report for the International Resource Panel. Paris: United Nations Environment Programme.

Foxon, T. J. (2011) 'A coevolutionary framework for analysing a transition to a sustainable low carbon economy', *Ecological Economics*, 70(12), pp. 2258–2267. doi: 10.1016/j.ecolecon.2011.07.014.

Frase, P. (2016) *Four Futures*. London: Verso.

Freeman, C. and Louca, F. (2001) *As Time Goes by: From Industrial Revolutions to the Information Revolution*. Oxford: Oxford University Press.

Geels, F. (2011) 'The multi-level perspective on sustainability transitions: Responses to seven criticisms', *Environmental Innovation and Societal Transitions*, 1(1), pp. 24–40.

Geels, F. (2013) 'The impact of the financial-economic crisis on sustainability transitions: Financial investment, governance and public discourse', *Environmental Innovation and Societal Transitions*, 6, pp. 67–95.

Geels, F. W. (2002) 'Technological transitions as evolutionary reconfiguration processes: A multi-level perspective and a case-study', *Research Policy*, 31(8/9), pp. 1257–1274.

Geels, F. W. (2005) *Technological Transitions: A Co-evolutionary and Socio-Technical Analysis*. Cheltenham, UK: Edward Elgar.

Geels, F. W. (2010) 'Ontologies, socio-technical transitions (to sustainability), and the multi-level perspective', *Research Policy*, 39(4), pp. 495–510.

Geels, F. W. and Schot, J. (2007) 'Typology of sociotechnical transition pathways', *Research Policy*, 36(3), pp. 399–417. doi: 10.1016/j.respol.2007.01.003.

German Advisory Council on Climate Change. (2011) *World in Transition: A Social Contract for Sustainability*. Berlin: German Advisory Council on Global Change.

Gore, C. (2010) 'Global recession of 2009 in a long-term development perspective', *Journal of International Development*, 22, pp. 714–738.

Gowan, P. (2009) 'Crisis in the heartland', *New Left Review*, 55, pp. 5–29.

Grin, J., Rotmans, J., Schot, J., Geels, F. and Loorbach, D. (2010) *Transitions to Sustainable Development: New Directions in the Study of Long Term Transformative Change*. New York: Routledge.

Haberl, H., Fischer-Kowalski, M., Krausmann, F., Martinez-Alier, J. and Winiwarter, V. (2011) 'A socio-metabolic transition towards sustainability? Challenges for another great transformation', *Sustainable Development*, 19, pp. 1–14.

Harvey, D. (2009) 'The crisis and the consolidation of class power: Is this really the end of neoliberalism?', *Red Pepper*. Available at: https://www.redpepper.org.uk/Their-crisis-our-challenge/ (Downloaded 26 August 2019).

Hausknost, D. and Haas, W. (2019) 'The politics of selection: Towards a transformative model of environmental innovation', *Sustainability*, 11(2), p. 506. doi: 10.3390/su11020506.

Jackson, T. (2009) *Prosperity Without Growth? The Transition to a Sustainable Economy*. London: Sustainable Development Commission.

Janicke, M. (2012) '"Green Growth": From a growing eco-industry to economic sustainability', *Energy Policy*, 48, pp. 13–21.

Jessop, B. (2016) *The State: Past Present Future*. Cambridge: Polity Press.

Köhler, J. (2012) 'A comparison of the neo-Schumpeterian theory of Kondratiev waves and the multi-level perspective on transitions', *Environmental Innovation and Societal Transitions*, 3, pp. 1–15.

Kondratief, N. D. (1935) 'The long waves in economic life', *Review of Economic Statistics*, 27(1), pp. 105–115.

Korten, D. (2007) *The Great Turning: From Empire to Earth Community*. Oakland, CA: Berrett-Koehler Publishers.

Krugman, P. (2012) *End This Depression Now*. New York and London: W.W. Northon & Company.

Lawhon, M. and Murphy, J. T. (2011) 'Socio-technical regimes and sustainability transitions: Insights from political ecology', *Progress in Human Geography*. doi:10.1177/03091 32511427960.

Lipsey, R. G., Carlaw, K. I. and Bekar, C. T. (2005) *Economic Transformations: General Purpose Technologies and Long Term Economic Growth*. Cambridge, MA: MIT Press.

Mason, P. (2015) *Postcapitalism: A Guide to Our Future*. London: Penguin.

Mazzucato, M. (2011) *The Entrepreneurial State*. London: Demos.

Mazzucato, M. (2015) 'The green entrepreneurial state', in Scoones, I., Leach, M. and Newell, P. (eds.) *The Politics of Green Transformations*. London and New York: Routledge Earthscan, pp. 133–152.

Mazzucato, M. (2016) *Rethinking Capitalism: Economics and Policy for Sustainable and Inclusive Growth*. West Sussex: John Wiley & Sons.

McKinsey Global Institute. (2011a) *Resource Revolution: Meeting the World's Energy, Materials, Food, and Water Needs*. London: McKinsey Global Institute. Available at: www.mckinsey.com/client_service/sustainability.aspx (Accessed: 10 May 2011).

McKinsey Global Institute. (2011b) *Urban World: Mapping the Economic Power of Cities*. London: McKinsey Global Institute.

Meadowcroft, J. (2011) 'Engaging with the politics of sustainability transitions', *Environmental Innovation and Societal Transitions*, 1, pp. 70–75.

Nelson, R. R. and Winter, S. G. (1982) *An Evolutionary Theory of Economic Change*. Cambridge, MA: Harvard University Press.

Oxford Economics. (2015) *Assessing the Global Transport Infrastructure Market: Outlook to 2025, PwC*. London: PricewaterhouseCoopers. Available at: www.pwc.com/outlook2025.

Pearson, P. J. G. and Foxon, T. J. (2012) 'A low carbon industrial revolution: Insights and challenges from past technological and economics transformations', *Energy Policy*, 50, pp. 117–127.

Peetz, D. and Murray, G. (2012) 'The financialisation of corporate ownership and implications for the potential for climate action', in *The Necessary Transition*. Brisbane: Griffith Business School.

Perez, C. (2002) *Technological Revolutions and Financial Capital: The Dynamics of Bubbles and Golden Ages*. Cheltenham, UK: Edward Elgar.

Perez, C. (2007) *Great Surges of Development and Alternative Forms of Globalization*. Working Papers in Technology Governance and Economic Dynamics. Kattel, R. (ed.). Norway and Estonia: The Other Canon Foundation (Norway) and Tallinin University of Technology (Tallinin).

Perez, C. (2009) 'The double bubble at the turn of the century: Technological roots and structural implications', *Cambridge Journal of Economics*, 33, pp. 779–805.

Perez, C. (2010) *The Financial Crisis and the Future of Innovation: A View of Technical Change with the Aid of History*. Working Papers in Technology Governance and Economic Dynamics. Norway and Tallinin: The Other Canon Foundation & Tallinin University of Technology.

Perez, C. (2013) 'Unleashing a golden age after the financial collapse: Drawing lessons from history', *Environmental Innovation and Societal Transitions*, 6, 9–23. doi: 10.1016/j. eist.2012.12.004.

Perez, C. (2016) *Capitalism, Technology and a Green Golden Age: The Role of History in Helping to Shape the Future*. WP 2016–1.

Pierson, P. (2000) 'Increasing returns, path dependence, and the study of politics', *American Political Science Review*, 94(2), pp. 251–267.

Polanyi, K. (1946) *Origins of Our Time: The Great Transformation*. London: Victor Gollancz Ltd.

Rifkin, J. (2011) *The Third Industrial Revolution: How Lateral Power Is Transforming Energy, the Economy and the World*. New York: Palgrave MacMillan.

Rip, A. and Kemp, R. (1998) 'Technological change', in Rayner, S. and Malone, E. L. (eds.) *Human Choice and Climate Change*. Oxford: Oxford University Press, pp. 327–399.

Rosenberg, N. and Frischtak, C. R. (1983) 'Long waves and economic growth: A critical appraisal', *American Economic Association Papers and Proceedings*, 73(2), pp. 146–151.

Saldinger, A. (2019) 'Development finance institutions grapple with their growing role', *Devex*, 19 March.

Schot, J. and Kanger, L. (2018) 'Deep transitions: Emergence, acceleration, stabilization and directionality', *Research Policy*. doi: 10.1016/j.respol.2018.03.009.

Schumpeter, J. A. (1939) *Business Cycles: A Theoretical, Historical and Statistical Analysis of the Capitalist Process*. New York: McGraw Hill.

Schwab, K. (2017) *The Fourth Industrial Revolution*. Great Britain: Portfolio Penguin.

Smith, A., Stirling, A. and Berkhout, F. (2005) 'The governance of sustainable socio-technical transitions', *Research Policy*, 34, pp. 1491–1510.

Smith, A., Voss, J. P. and Grin, J. (2010) 'Innovation studies and sustainability transitions: The allure of the multi-level perspective and its challenges', *Research Policy*, 39(4), pp. 435–448.

Smith, N. (2008) *Uneven Development: Nature, Capital, and the Production of Space*. Athens, GA: The University of Georgia Press.

Standard Chartered (2010) *The Super-Cycle Report*. London: Global Research Standard Chartered. Available at: http://Gerard.Lyons@sc.com.

Stiglitz, J. (2010a) *Freefall: Free Markets and the Sinking of the Global Economy*. London: Allen Lane.

Stiglitz, J. (2010b) *The Stiglitz Report: Reforming the International Monetary and Financial Systems in the Wake of the Global Crisis*. New York: New Press.

Streeck, W. (2014) 'How will capitalism end?', *New Left Review*, 87(May–June), pp. 35–64.

Swilling, M. and Annecke, E. (2012) *Just Transitions: Explorations of Sustainability in an Unfair World*. Tokyo: United Nations University Press.

The Global Commission on the Geopolitics of Energy Transformation. (2019) *A New World: The Geopolitics of the Energy Transformation*. Abu Dhabi: International Renewable Energy Agency.

Turner, A. (2016) *Between Debt and the Devil: Money, Credit and Fixing Global Finance*. Princeton, NJ: Princeton University Press.

United Nations Development Programme. (1998) *Human Development Report 1998*. New York: United Nations Development Programme, p. 5, November 2006.

United Nations Environment Programme. (2011a) *Towards a Green Economy: Pathways to Sustainable Development and Poverty Eradication*. Nairobi: United Nations Environment Programme.

United Nations Environment Programme. (2011b) *World Economic and Social Survey 2011: The Great Green Technological Transformation, World Economic and Social Survey*. New York: United Nations.

United Nations Environment Programme. (2018) *Global Trends in Renewable Energy Investment 2018*. Nairobi: United Nations Environment Programme. Available at: http://fs-unep-centre.org/publications/global-trends-renewable-energy-investment-report-2018.

Verspagen, B. (2005) 'Innovation and economic growth', in Fagerberg, J., Mowery, D. C. and Nelson, R. R. (eds.) *The Oxford Handbook of Innovation*. Oxford: Oxford University Press, Chapter 18.

Von Weizsacker, E., Hargroves, K. C., Smith, M. H., Desha, C. and Stasinopoulos, P. (2009) *Factor Five: Transforming the Global Economy Through 80% Improvements in Resource Productivity*. London: Earthscan.

World Bank. (2017) *Investments in IDA Countries: Private Participation in Infrastructure*. Washington, DC: World Bank.

5

TOWARDS RADICAL
INCREMENTALISM[1]

Introduction

It is time now to explore what the discussions in the previous chapters mean for
our understanding of change. As I argued in Chapters 1 and 2, what matters are
the 'passions for change' in the age of sustainability. Using distinctions drawn from
transdisciplinary research, systems knowledge is a necessary condition for change
but not a sufficient condition. Nor is target knowledge good enough, especially
if packaged in the form of grand normative statements that mask the contested
meanings and interpretations that lie beneath the surface. What matters is transformation knowledge about the contested passage(s) from the present to particular
desired futures. More specifically, this is deep knowledge about the evolutionary
potential of the present.

Inspired by the African *Ukamian* perspective, Chapter 2 proposed a metatheoretical framework – referred to as Metatheory 2.0 – drawn from the synthesis achieved by Bhaskar et.al. (2016) of the three most influential contemporary
interdisciplinary metatheories. With complex adaptive systems at its core (Preiser,
Biggs, de Vos and Folke, 2018), this framework provided the foundation for a
meso-conceptual synthesis of a relational theory of the state (drawn from Jessop,
2016) and a thermodynamic theory of the economy (drawn from Ayres, 2016)
that is appropriate for understanding the increasingly complex sustainability age.
In essence – and at its most succinct – the result is a conception of collibratory
governance for managing non-equilibrium economies. This is then institutionally expressed in the emerging commons-based peer-to-peer (CBPP) economies.
Practically speaking, states need to be brought back in to set the terms for a
wide range of complex collaborative arrangements that have emerged over the
past decades to negotiate agreements that address an extraordinarily wide range
of issues in the social, economic, political, environmental, physical-material and

cultural fields (referred to as 'frame 3' policymaking by Schot and Steinmuller, 2016).

Chapter 4 proposed that long-wave theory might be useful for comprehending the current polycrisis and the underlying dynamics of change. It was suggested that asynchronic waves of transition may brought us to the threshold of a second deep transition. We are living through the interregnum between the post–World War II long-term development cycle that ended in 2009 and whatever comes next. The future may be fiction, but long-wave theory helps to build a sense of history and the patterns that may to repeat themselves.

The challenge, of course, is whether the deep transition can also be a just transition. An unjust deep transition is a distinct possibility. All will depend on how the terms of the deep transition are contested within the polity. While Chapter 7 explores this in greater depth with reference to how we rethink the polities of developmental states, this chapter and the next explore the micro-level dynamics of these contestations within particular geographical spaces. Given that over half of the global population now live in urban settlements, it follows that so much about the evolutionary potential of the present is expressed and contested within urban spaces. This chapter addresses this discussion and provides the foundation for the next more empirical chapter on ecocultures.

What follows is a discussion that proposes a way of thinking about incremental change that connects futuring and experimentation. The end result is a conception of change commensurate with the implications of the preceding chapters. The final section suggests a way of thinking about the politics of incrementalism. It is a conception of change that infuses all subsequent chapters.

Our urban world

It is a well-known fact that over 50% of the world's population now lives in urban settlements that vary in size from small towns of a few thousand people to metropolises of over 20 million. We live, therefore, in a majority urban world. What is less recognized outside urban research circles is that the UN estimates that the size of the urban population could nearly double over the four decades 2010–2050 (Department of Economic and Social Affairs, 2012). In other words, if these estimates are correct, the urban populations that have evolved over centuries will nearly double in size across existing and new settlements in four decades. This megatrend brings into focus the two greatest contemporary challenges: inequality and environmental unsustainability. Inequality is reflected in the fact that one in three urbanites lives in slums (United Nations Centre for Human Settlements, 2003), and environmental unsustainability is reflected in the fact that 75% of GHG emissions stem from urban settlements (Dodman, 2009), 60% of global resources (excluding water) are consumed in these settlements (Swilling, Hajer, Baynes, Bergesen, Labbe, et al., 2018) and if the 100-year trend of urban de-densification continues, urbanization will spread from the current coverage of 1 million km^2 in 2005 to 2.5 million km^2 by 2050, with much of this additional land being our most productive farmland (Swilling, Hajer, Baynes, Bergesen, Labbe, et al., 2018).

While major crises in the past have historically stimulated the desire to reimagine futures, never before has it been necessary to anticipate futures where the majority live in urban settlements with specific spatial consequences for inequality and environmental sustainability. Indeed, many initiatives to reimagine the future take this majority urban world for granted, thus failing to recognize that this extraordinary spatial transformation has major implications for what we imagine futures to be and, more importantly, how these futures are imagined. More specifically, it will be impossible to appropriately anticipate futures in ways that inform actions in the present, without reimagining the urban spaces where most of us now live and where the large bulk of material production and consumption takes place. Indeed, imaginaries of the future of urban space should be at the centre of anticipatory thinking.

It will be argued that urban spaces offer unique opportunities for manifesting in practice future-oriented thinking that is used to shape actions in the present (this chapter reworks the material in Swilling, Pieterse and Hajer, 2018). Urban spaces – and larger or fast-growing cities in particular – tend to get shaped by constant reinventions of the evolutionary potential of the present as expressed in a wide variety of imaginaries: policy, strategy and planning documents as well as artistic, fictional, aesthetic and visual media that respond to the modalities of urban governance, market dynamics, cultural shifts and socio-demographic changes as individuals and households make locational choices. However, this uniqueness is inadequately captured by two dominant ways of conceptualizing this dynamic that tend never to meet, namely futuring and experimentation.

By futuring we mean the wide range of practices that have emerged over recent decades to explore the future, including forecasting, foresight and, more recently, anticipatory thinking (Poli, 2018). By experimentation we mean the wide range of activities that are described using various terms, such as innovation (Verspagen, 2005), niche experiments (Smith and Raven, 2012), social innovation (Murray, Caulier-Grice and Mulgan, 2010; Moulaert, MacCullum, Mehmood and Hamdouch, 2013), 'urban living labs' (Marvin, Bulkeley, Mai, McCormick and Palga, 2018) and urban experimentation (Evans, Karvonen and Raven, 2016).

This chapter will propose that in a majority urban world where so much about the future will be determined by what happens within urban spaces (especially major cities), it will be necessary to synthesize futuring and experimentation to gain a better understanding of the dynamics of incrementalism. Urban spaces have emerged as arenas for expressing selected futures in practical small- and large-scale urban experiments. These can potentially coalesce into implementable alternatives that get captured in ever-evolving sets of imaginaries. This is what lies at the centre of what we will call towards the end of this chapter the "politics of urban transformation".

The next section will flesh out what we mean by futuring and experimentation and why a synthesis is needed if we are interested in anticipatory thinking and incrementalist actions. This is followed by a discussion that focuses on ways of framing urban futures in the global North and the global South. This is done by discussing two key representative texts that in contrasting ways help focus the discussion of this vast research territory. To illustrate the argument, the next section provides an overview of the way urban futures are depicted in a selection of influential global reports. With a thorough discussion of futuring in place, the next section compares two contrasting ways of understanding urban experimentation – one that is appropriate to the global North, and another that is appropriate to the global South. The penultimate section then connects the arguments about particular conceptions of futuring and experimentation in order to elaborate a particular interpretation of anticipatory thinking that is appropriate for the diverse urban contexts spread out across the global North and South. This, in turn, is taken further to reflect on the political implications of this mode of anticipatory thinking. Drawing on the work of Roberto Unger, it is proposed that radical incrementalism inspired by anticipatory thinking can drive a transformative urban politics.

Futuring and experimentation

During the lead up to 2015 and subsequently, there has been a proliferation of initiatives to reimagine alternative futures. From elite forums such as the World Economic Forum (WEF) to more social movement-based forums like the World Social Forum (WSF), there are initiatives to articulate these futures in a wide range of conceptual languages. Science-based bodies like the Intergovernmental Panel on Climate Change (IPCC), Intergovernmental Science-Policy Platform on Biodiversity and Ecosystem Services (IPBES) and the International Resource Panel (IRP) have all in one way or another engaged in scenario-building exercises. The German Advisory Council on Climate Change called on the world to make a conscious commitment to another "great transformation" to create an ecologically sustainable and socially just world as significant as the agricultural transformation 13,000 years ago and the industrial transformation 250 years ago (German Advisory Council on Climate Change, 2011). In the academic community, new discussions have emerged about "transformative innovation policy" (Schot and Steinmuller, 2016) and "techniques of futuring" (Hajer and Pelzer, 2018). And most important of all, across the world there are literally thousands of examples of city-wide vision-driven initiatives to reimagine the future of urban spaces, built structures and infrastructure systems (Broto and Bulkeley, 2013; Bulkeley and Broto, 2013; Evans and Karvonen, 2014; Karvonen and van Heur, 2014; May and Perry, 2016; Eadson, 2016; Evans, 2016; Evans, Karvonen and Raven, 2016; Hodson and Marvin, 2016; Caprotti and Cowley, 2017; Hodson, Evans and Schliwa, 2018). The consolidation of global alliances of cities such as United Cities and Local Governments (UCLG), C40 League, ICLEI and others has created a set of accepted "best practices" which include the necessity for every city leader to be able to express a future vision for his/her city. By way of only one visually iconic example, Paris has its *Re-Imagine Paris* program which is symbolized by a graphic of the Eiffel Tower lying on its side.

Periods of great uncertainty and crisis have tended to intensify two related trends: a desire to reimagine the future to create pathways out of the crisis and a proliferation of experiments to figure out real-world alternatives that address the perceived causes of the crisis. Karl Polanyi's renowned conception of a "double movement" captured how a multiplicity of collaborative innovations counterbalanced the individualistic materialism promoted by laissez-faire capitalism during the century that culminated in the two World Wars (that were partly catalysed by reactions by those who never benefitted from the growth in unprecedented prosperity for the few) (Polanyi, 1946). His argument was confirmed when the emergence of this multiplicity of innovations culminated in the consolidation of the welfarist vision that inspired the post–World War II "golden era" prior to the rise of neoliberalism from the 1980s onward.

The struggle against Apartheid in South Africa was not just about the struggle for the vision of a free non-racial democratic South Africa as the sole dynamic of change that culminated in the first democratic election in 1994 (Marais, 1998); this vision was also rooted in the build-up of prefigurative experiments in inter-racial engagement and democratization during the decade prior to 1994 (Swilling, 1999). Our current global conjuncture is no exception, but now the complexities are greater and arguably the stakes could not be higher. Polanyi's double movement is reflected in the way solidarities are getting built around (often localized) alternatives to the environmentally destructive and socially unjust consequences of neoliberalism and its manifestations in corporate-led globalization (Hawken, 2007; Swilling and Annecke, 2012 – Chapter 11; Pretty, Boehme and Bharucha, 2015). Urban spaces are where Polanyi's two movements are most visible and also most directly in conflict with one another.

At the same time, it is well known that there is a flowering of urban experimental initiatives from around the world that express the desires and visions of local actor networks that coalesce around alternative spaces (ecovillages, eco-neighbourhoods, eco-cities) and alternative socio-technical systems (from cycling to organic agriculture, to renewable energy, to fair trade coffee, waste recycling, green spaces, energy-efficient buildings, mass transit systems and so on). These are starting to be documented in various ways (Broto and Bulkeley, 2013; Allen, Lampis and Swilling, 2015; Pretty, Boehme and Bharucha, 2015; Sengers, Berkhout, Wieczorek and Raven, 2016). A useful definition of experimentation has been articulated by Sengers et al.:

> An inclusive, practice-based and challenge-led initiative designed to promote system innovation through social learning under conditions of uncertainty and ambiguity.
>
> *(Sengers, Berkhout, Wieczorek and Raven, 2016:21)*

The literatures on futuring and the literatures on experimentation generally do not connect. This probably has much to do with the fact that the futures/anticipatory thinking community tends to be made up of systems thinkers and modellers interested in systemic narratives from which the rationale for action is derived. Those interested in experimentation are social scientists, technologists, utopians, institutionalists, social activists and social entrepreneurs interested in the micro-dynamics of real-world change. They tend to be impatient with grand visioning that often results only in more noise from the chattering classes.

This chapter proceeds from the assumption that our understanding of the pathways to sustainable futures will depend on a deep appreciation and synthesis of these two ways of thinking about the futures. Futuring needs to be informed by an appreciation of the real-world dynamics of actual experimentation to ensure that constraints on what is imaginable are not inbuilt from the start; and experimentation needs to be informed by the imaginable probabilities that could unfold over the long term.

Indeed, experimenters often embed implicit longer-term visions of the future into their ambitions for the potential of their experiments. They often justify their actions with reference to game-changing dynamics that their experiments are aimed at addressing (Jorgensen, Wittmayer, et al., 2016). And those in the futuring business may often be inspired by what they have witnessed in microcosmic experiments, but they might need to recognize that the present is significant in its own right rather than a mere burning platform between the past and future.

That said, it needs to be immediately acknowledged that while futuring and experimentation are in their different ways interrogating alternatives to the present, the social forces united by a desire to prevent change are gathering in strength. The rightward political shift in many developed countries and the rise of corrupt authoritarianism in many developing countries coupled to spreading fundamentalisms seems to be pushing the world into an era of greater instability, conflict and suffering. This is not an appropriate context for long-termism and experimentation to flourish. Fear, short-termism, survivalism, parochialism, anti-intellectualism, intolerance and misogyny (see Chapter 9) are more likely to thrive in this kind of context (Mishra, 2018).

A variety of futuring approaches and modes of experimentation are the emergent techniques for making sense of an opaque future from within an increasingly complex, uncertain and ambiguous present. Roberto Poli distinguishes between three modes of futuring (Poli, 2014): forecasting, foresight and anticipation. Forecasting is predictive and quantitative. Forecasting cannot envisage major discontinuities because projections are commonly continuations of curves compiled from historical trends. Modelling is the tool of choice.

Foresight is more qualitative and attempts to build narratives of the future, that is, stories of possible futures rather than predicting the future. The aim is to enable actors to select a possible future and then reverse engineer to the present (for a comprehensive application of forecasting and foresight in urban planning, see Hopkins and Zapata, 2007; Hall, 2014). For both forecasting and foresighting, the present is a burning platform between the past that needs to be transcended and a desired future rather than a context that needs to be deeply interrogated for clues to the future.

Anticipatory thinking has a deep appreciation of complexity, context, the need for discontinuity and impredicativity,[2] and therefore the need to appreciate the "thick present" as a key source of the future via actions taken in the present (Poli, 2018). While anticipatory thinking valorizes action in the present as the source of desired futures, those writing in this field have a limited appreciation of what actual real-world action is all about mainly because they engage inadequately with the literature on experimentation. By contrast, the literature on experimentation tends to be impatient with futuring, but this is mainly because futuring is associated with forecasting and foresight rather than anticipation.

Anticipatory thinking about the evolutionary potential of the experimental dynamics of the present is where futuring and experimentation can meet. It is this conception of futuring rather than the forecasting/foresight modes of futuring that will be used from this point onward – indeed, unless referred to otherwise, the word "futuring" is used in some places in the rest of this discussion as a substitute for this mode of anticipatory thinking.

Histories of anticipation

Cities, urban spaces, built forms, infrastructures and settlements are the emergent outcomes of conscious anticipatory intent by a vast array of actors operating within specific contexts. This section illustrates this by discussing two particularly significant texts: the work by Dunn, Cureton and Pollastri (2014) that relates mainly to the regulated formal urban systems located typically in the global North and expanding enclaves in the global South and the work by Simone and Pieterse (2017) that relates to the least regulated highly informalized cities located mainly in the global South.

By contrasting these two texts, the differences in the way urban imaginaries are shaped across different contexts are illustrated. In the more regulated cities of the global North (and expanding enclaves in the global South), what gets built is the product of the imaginaries of professional designers who produce designs for property developers. These, in turn, get approved by the relevant authorities in accordance with some sort of long-term strategic plan for the city. Where these elites are tightly coupled in powerful urban coalitions, a mutually reinforcing narrative ensures that long-term visions reflect the short-term financial interests of the developers and the design imaginations of the professionals. Where these design, developer and regulatory elites are loosely coupled, competitive, or even conflictual, dynamics emerge.

By contrast, in the least regulated highly informalized cities located mainly in the global South, built form and urban space are the emergent outcomes of multiple micro-level actions by households that "quietly encroach" their way into urban systems in a variety of ways (Bayat, 2000). Collective action is rare but does happen to resist relocation or control of land by dominant interests. Vast sprawling informal settlements emerge alongside wealthy enclaves of formally built structures which, in turn, could be either informally regulated or the outcome of formal governance and planning procedures. In other words, in contrast to formal regulated cities typically found in the global North and enclaves in the global South, urban futures in the global South tend to embody the incrementalist imaginaries of millions of poor (and in some cases less poor) households whose primary aim is to secure a foothold for survival in rapidly expanding cities and urban settlements (Simone, 2001; McFarlane, 2011; Simone and Pieterse, 2017).

In recent years, the growing literature on urban futures addresses both contexts, including histories of urban futures (Alison et al., 2007; Brook, 2013), futuring in urban planning in developed world contexts (Hopkins and Zapata, 2007; Hall, 2014) and how to think about the futures of cities in the global South (Pieterse, 2008; Simone and Pieterse, 2017).

Notwithstanding the fundamental distinction between formal regulated urban spaces and informal unregulated spaces that exist on the two ends of a continuous spectrum between these extremes, it is possible to argue that there are histories of urban imaginaries that have shaped what has occurred in practice across this formal–informal spectrum of contexts (Hall, 2014: Chapter 1). The most comprehensive assessment from a global North perspective was commissioned by the UK Government Office for Science as a contribution to the Foresight initiative. Entitled A Visual History of the Future (Dunn, Cureton and Pollastri, 2014), this report analysed 94 influential visual images of future urban imaginaries,[3] including Ebenezer Howard's famous classic image of the 'Garden City'(1902); Eugene Henard's 'The Cities of the Future'(1911); LeCorbusier's 'Radiant City' (1924); Raymond Hood's 'Century of Progress International Exposition' (1932–1933); Geddes' 'World of Tomorrow' at Futurama, New York World Fair (1939); Kevin Lynch's 'The Perceptual Form of the City', Boston (1954–1959); Albert Speer's Nazi vision of 'Planning for the World Capital Germania' (1939); Buckminster Fuller's 'Dome over Manhattan' (1960); an image from Kubrick's 'Clockwork Orange' (1971); an image from Ridley Scott's 'Blade Runner' (1982); Rem Koolhaas' 'Asian City of Tomorrow' (1995); Marcos Novak's 'Mutable Algorithmic Landscapes' (2000); Alfonso Cuaron's 'Children of Men' (2006); an image from Blomkamp's film 'District 9' (2009); Zaha Hadid Architects' 'One North Masterplan, Singapore, 2001–2021' and OMA's 'Eneropa EuroGrid – Extract from Roadmap 2050: A Practical Guide to a Prosperous, Low-Carbon Europe' (2010). Without claiming these images were literally translated into reality, the authors argue they shaped futuristic cultural perceptions of the city during key moments across the decades. They are, in fact, a history of successive imaginaries of urban futures.

To make sense of the diversity of imaginaries, six major paradigms were derived from an analytical clustering process, and each image was located within a timeline from 1900 to 2014. The result is a systematic overview of the iconic images that reflect how urban futures were imagined over the decades since 1900 (Figure 5.1).

The paradigms derived from the 94 visual urban imaginaries reflected in the first diagram (each number corresponds to a figure in the report) are described as follows:

- regulated cities: urban visions that integrate aspects of rural/country/green living;
- layered cities: portrayals that have explicit multiple but fixed levels typically associated with different transit mobilities;

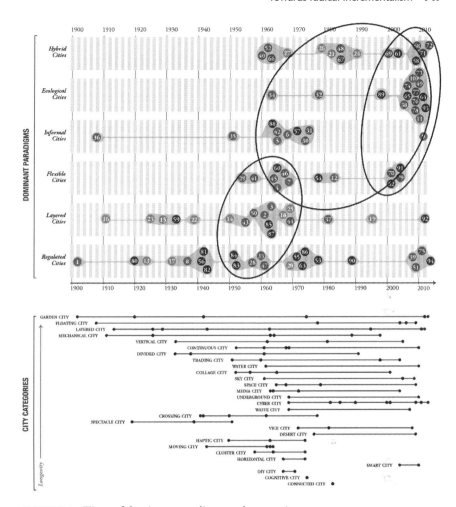

FIGURE 5.1 Time of dominant paradigms and categories.

Source: Dunn, Cureton and Pollastri, 2014:126 – circles added

- flexible cities: urban depictions that allow for plug-in and changes but still fixed in some manner to context;
- informal cities: presentations of visions that suggest much more itinerant and temporary situations and include walking, nomadic and non-permanent cities;
- ecological cities: illustrations of cities that demonstrate explicit ecological concerns, renewable energies and low or zero carbon ambitions;
- hybrid cities: urban visions that deliberately explore the blurring between physical place and digital space, including augmented reality and "smart" cities.

The 94 images were also clustered into 28 "city types" which are reflected in the second diagram. The remarkable endurance of the Garden City image is clearly reflected in the way it has survived the entire time period but now morphed into images of more ecological cities. Significantly, the morphing of the Garden Cities imaginary into its more ecological reincarnation is now complemented by the rise of the Cyber City and Smart City types in the second diagram corresponding with the emergence of ICTs and the Information Age. When read together, the circles that I imposed on the first diagram show a clear shift from the imaginaries of the Regulated and Layered Cities of the 1930s–1970s period of welfarist Fordism premised on cheap oil and the start of mass mobility, to the fusing of the Flexible, Informal, Ecological and Hybrid Cities from the 1970s onward corresponding with the shift to neoliberalism and post-Fordism. The strong concentration of Ecological Cities imaginaries from 2000 onward, reinforced by Hybrid Cities and Flexible Cities imaginaries, reveals how far the discourses of environmental crisis have influenced post-2000 urban imaginaries.

It is debatable whether the Ecological/Hybrid/Flexible City imaginary provides the basis for a break from neoliberalism and post-Fordism following the global impact of the economic crisis after 2007. This debate was the focus of the 2016 International Architecture Biennale Rotterdam (IABR) that linked urban imaginaries to questions about the Next Economy (Brugmans, van Dinteren and Hajer, 2016). As discussed later in this chapter, it is also the question addressed by the German Advisory Council on Global Change (GACGC) that links the need for a "great transformation" to the need to reimagine urban futures (German Advisory Council on Climate Change, 2011; German Advisory Council on Global Change, 2016).

Unsurprisingly, the UK's Foresight project on cities ignored the burgeoning literature on the urban dynamics of the global South (Swilling, Khan and Simone, 2003; Simone and Abouhani, 2005; Simone, 2006; Pieterse, 2008; Edensor and Jayne, 2012; Parnell and Oldfield, 2014; Allen, Lampis and Swilling, 2016; Simone and Pieterse, 2017). On the one hand, this is defensible on the grounds that it was the first urbanization wave (1750–1950) that generated the intellectual, aesthetic and axiological foundations of the urban imaginaries that have shaped not only the urban dynamics of the global North but also the urban dynamics of the global South where the contextual conditions are very different but where elites aspire to imitate the "Western" ideal.

On the other hand, it is now generally accepted that the global South has started to generate its own set of "untamed" urban imaginaries (Allen, Lampis and Swilling, 2016). As argued by Simone and Pieterse in their appropriately titled book *New Urban Worlds: Inhabiting Dissonant Times*, Asian and African contexts defy the simplistic categorizations, binaries, codified logics, hegemonic rationalities and neatly demarcated time-space modalities that have been accepted as norms in the formalized regulated spaces of the global North (Simone and Pieterse, 2017). It is worth quoting them at length:

> Cities across Africa and Asia move towards and away from each other in significant ways. No longer, if ever, coherent actors in themselves, cities as

social and administrative entities, nevertheless, attempt to posit themselves as dynamic engines of economic growth and social transformation. Urbanisation as a process once embodied by the city-form, now takes on varying shapes and sizes, expanding cities into megalopolises, shrinking them into shadows of former selves, or articulating a vast range of places and resources in tight relationships of interdependency. Cities become the venues for all kinds of countervailing tendencies: where narrowing and expansion, ambiguity and precision, dissipation and consolidation, embodiment and digitalization, movement and stasis are all intensified; and sometimes become indistinguishable from each other. Urbanisation is something that seems to increasingly make itself, something independent from its once familiar function as an arena where different things were made, articulated and prompted into new synergies. Associations with density, social diversity, churn, and the circulation of disparate experiences through each other no longer seem to hold as key criteria for designating something as 'urban'. Differentiations between local and global, public and private, exterior and interior, intensive and extensive appear to fold into, and sometime collapse upon each other. The very organisation of meaning, with its boundaries of here and there, self and other, citizen and stranger become both more pronounced and more subject to erasure. More and more the urban seems to be a confounding story.

(Simone and Pieterse, 2017:85)

What does it mean to think politically about African and Asian futures within an urban environment of such seeming paradox? This is the question that Simone and Pieterse attempt to address in their book, particularly by considering how urban politics and programmatic interventions to build imaginaries of urban futures in specific contexts might operate simultaneously through inventions at the level of municipal and metropolitan systems and acupunctural interventions at the level of neighbourhoods or districts. This double approach assumes that the conventional rules of the game – home and property ownership, formal taxation systems and standardized outlays of infrastructure – are inadequate to the realities in which urban life is actually lived. This is life not layered through orderly scales and sectors but rather assumes multiple spatial forms. As such, interventions, policies and mobilizations inspired by future imaginaries must be capable of resonating across disparate terrains and vectors of impact.

Prevailing conceptions of agency as either state-centric or market-centric or civil society-centric do not fully grasp how the contemporary times of deep crisis and uncertainty can be deployed to concretize an adaptive urbanism agenda. The sheer scale of technological innovation that is required to address the dramatic imperatives of resource decoupling, decarbonization and the restoration of biodiversity in the processes of production and consumption points to the role of public-private research and development coalitions to forge these innovations. Similarly, the volume of soft regulation and anticipation that will have to be borne by the state is undisputable. And it is equally obvious that unless popular culture and mediated aspirations find different resonances through a social enactment of new patterns and forms of consumption, there is hardly any impetus for technological innovation or politically charged regulation to achieve a set of imagined futures.

So, how can we think in more entangled and dynamic terms about these imperatives that pay homage to institutional divisions but also reach for a much more enmeshed unconventional conceptualization of urban change? Theoretically and practically, this will only emerge when we systematically articulate the dense registers and sensibilities of the street with the technocratic utopias of future times, which in turn points back to a series of political choices and possibilities in the precarious now. No matter how dire conditions may appear, forging a new political imagination requires a generous engagement with the molecular details of urban life. These details are not only those of the street but of institutions as well and the interfaces among them.

This generosity is reflected in a capacity to redescribe conditions in ways that extricate the details from serving as locked-in evidence of particular dispositions and instead treat them as secretions that may mix and congeal in ways that go beyond our available vocabularies. We must always act and intervene. But for too long, the urban has been primarily experienced predicatively as the context for intervention and redemption, without paying attention to all of the impredicative resonances among seemingly discordant things and times that have in some ways adapted themselves to each other all along.

Instead of translating this conception of doubleness into a particular fixed visual imaginary of Asian and African futures, Simone and Pieterse propose a set of heuristics that make anticipatory thinking-acting possible in highly complex, largely informalized and unregulated environments. These heuristics connect visioning and policymaking to the turbulent dynamics of the street in ways that recognize the futility of searching for consensus in a context where at best agreement will be provisional. By transforming this apparent instability into a catalyst for longer-term futuring, what Simone and Pieterse have done is ensure that futuring is rooted in experimentation, while making sure that the significance of experimentation is never limited to the claims about the possible that get established in the codified imaginaries of policy frameworks.

Significantly, Dunn, Cureton and Pollastri (2014) and Simone and Pieterse (Simone and Pieterse, 2017) end up concluding on a normative note that a contextually adapted combination of ecological, hybrid and 'smart' urbanism would be the most appropriate reference points for assembling imaginaries of urban futures across all contexts.

Contextual specificity and resonance

The *Weight of Cities* report referred to in Chapter 3 concludes that because cities are so diverse, there is no 'one-size-fits-all' solution. Instead, what is appropriate are modes of urban collibratory governance described in this report as 'entrepreneurial governance' of a multiplicity of urban experiments (Swilling, Hajer, Baynes,

Bergesen, Labbe, et al., 2018). This argument is remarkably similar to a perspective in a report on the urban transition by the GACGC. Using the German word *Eigenart* to capture a way of thinking that respects the socio-cultural-spatial specificity of each urban context, the GACGC insists on the

> resulting plurality of urban transformation pathways: every city must seek "its own way" to a sustainable future. This *Eigenart* (a German word meaning "character") is not only hugely important for creating urban quality of life and identity, it is also an indispensable resource in the sense of developing each city's specific potential for creativity and innovation. . . . Sustainability is a universal target system; the ways of getting there will be many and varied.
>
> *(German Advisory Council on Global Change, 2016:3–5)*

What the GACGC calls Eigenart is also remarkably similar to what Simone and Pieterse refer to as "resonance". Both words – *Eigengart* and *resonance* – can be understood as different ways of describing systems that are inherently 'impredicative'. As opposed to predicative systems, impredicative systems are relatively self-referential complex adaptive systems because their responses to the impact of external factors cannot be predicted in advance. Instead, responses will be conditioned by the context-specific configuration of the internal dynamics of the system that can result in a wide range of very different responses to the same external factors. By contrast, a predicative system is a system whose responses to an external determining factor can be more or less predicted in advance. This impredicative property is best explained by Simone and Pieterse's explanation of resonance (also applicable to *Eigengart*):

> Resonance is both the modality and by-product of people, materials, and places "feeling each other out", of attending to each other, of being drawn or repelled in the midst of so many things to which attention could be drawn. In other words, resonance is the affective process of people and things associating with each other, of having something to do with each other, of acting as components in the enactment of operations larger than themselves and their own particular functions and histories. When things resonate with each other there is a connection that proceeds, *not from the impositions of some overarching [external] map or logic, but from a process of things extending themselves to each other*. It is a matter of institutions, practices of knowledge production, and different tacit ways of doing things finding concrete opportunities to take each other into consideration. This process of resonance is critical to urban development work.
>
> *(Simone and Pieterse, 2017:95 – emphasis added)*

No matter the term used – *impredicativity, Eigengart* or *resonance* – the substance is similar: urban systems have their own internal logics that shape/condition the way urban systems respond in contextually specific ways to the impacts of external factors. Some cities will prepare for climate change, others will not; some cities will limit the impact of finance capital, others will not; some cities will prepare for water shortages, other cities faced with the same threat will not and so on.

Anticipating the Habitat III summit on cities that took place in Quito in October 2016, the GACGC report advocated a new "social contract for urban transformation" comprising a new "polycentric" system of urban governance that is similar to the IRP's notion of entrepreneurial governance, a set of "transformation action fields" that are similar to the IRP's "interventions" and a "normative compass" that connects generality and contextual specificity.

The GACGC's "polycentric responsibility architecture" would entail recognizing cities in national constitutions, granting cities rights to self-government, further decentralizing power to cities in accordance with the subsidiarity principle, securing adequate funding, more effective governance capacities, inclusion of cities in national decision-making, enabling urban communities to influence the transformation process and implementing charters for urban transformation at all levels of governance. This is identical to the vision articulated by the UCLG's flagship publication launched at the Quito summit (United Cities and Local Governments, 2016). Echoing the IRP report, the "transformative action fields" used by GACGC include decarbonization, mass transit, denser urban form, adaptations to climate change, poverty reduction, more effective land use, more resource-efficient material flows and improved urban health. The normative compass links a commitment to sustain natural life support systems, ensure social inclusion and promote Eigenart (German Advisory Council on Global Change, 2016:18).

Anticipatory thinking is ideally suited for addressing this challenge of connecting the general need for urban transitions with the contextual specificity of each pathway. By invoking the future to make sense of what is already emerging and by invoking the present to make sense of potential emerging futures, it becomes clear why space matters when it comes to deciphering the complex connections between futuring and experimenting. Because physical and symbolic structures are so fused in urban spaces, this is where anticipation – understood as mutually reinforcing futuring and experimentation processes – is most directly and clearly manifested.

Now, against the background of the earlier discussion about an anticipatory approach to the future, it is appropriate to discuss urban experimentation per se.

Urban experimentation

The extensive literature that has now documented the mushrooming of urban experiments of various kinds around the world is sufficient to substantiate the conclusion that there is a rapidly growing worldwide movement of urban experimentation (Beatley, 2000; Beatley and Newman, 2009; Beatley, Boyer and Newman, 2009; Suzuki, Dastur, Moffatt and Yabuki, 2009; Broto and Bulkeley, 2013; Swilling, Robinson, Marvin and Hodson, 2013; Allen, Lampis and Swilling, 2016; Evans, Karvonen and Raven, 2016; Affolderbach and Schulz, 2018). Major institutions like the C40 League, ICLEI, UCLG, Cities Alliance, major research institutions and others have managed to raise funding to employ large numbers of full-time staff to support these urban experiments. The best definition of an urban experiment from a recent edited collection on The Experimental City is repeated here:

> An inclusive, practice-based and challenge-led initiative designed to promote system innovation through social learning under conditions of uncertainty and ambiguity.
>
> *(Sengers, Berkhout, Wieczorek and Raven, 2016:21)*

In light of the discussion in this chapter, one could ask "social learning about what?" Quite often experimenters do have a particular imaginary of the future, but they choose to realize this by experimenting in practice to prove that this imaginary can work before advocating a new policy regime.

Before addressing the definition by Sengers et al., it needs to be noted that urban experimentation is not self-evidently a good thing. For some, like Maarten Hajer in the Foreword to The Experimental City, urban experimentalism is a positive development that is part of a larger societal turn toward "experimental governance" (Sengers, Berkhout, Wieczorek and Raven, 2016: iii) that has given rise to what he has called the "energetic society" that has emerged as old-style top-down regulatory governance has been forced to give way to empowered societal stakeholders with unprecedented access to information and participatory platforms (Hajer, 2010). The "linguistic turn of 'experimental cities' is", he argues, "a profound move towards a new way of thinking about social change in which cities are places of hope" (Evans, Karvonen and Raven, 2016:xix).

However, for others such as May and Perry, experimentalism represents an abrogation of distributional justice that has been co-opted by economic and scientific elites because it "enables the future to be placed in a process that seeks to reconfigure the present in the name of the imaginary. The result is that a politics of the present is suspended in the name of a possibility that benefits the few" (May and Perry, 2016:39).

For those who share May and Perry's perspective, collaborative futuring and experimentalism are delusionary ways of including subordinate groups into the imaginaries of the powerful. This is achieved by insisting on the virtues of consensus and the possibility that power differentials can be discursively overcome. From this perspective, inclusion through participation becomes just another way to ensure exclusion. Hajer's sense of hope, however, can only be justified if collaborative processes are facilitated in ways that explicitly create spaces for hearing voices of the traditionally unheard and less powerful (Freeth and Annecke, 2016).

To start the discussion about experimentation, we must return to the definition of urban experimentation offered by Sengers et al. who boldly claim that "in principle, all experiments can be assessed against this definition" (Sengers, Berkhout, Wieczorek and Raven, 2016:24). It is this claim that will be contested via a contrast between what urban experimentation means in formalized regulated urban environments versus what it means in informalized unregulated urban environments. Sengers, Berkhout, Wieczorek and Raven (2016) have in mind the former, while Simone and Pieterse (Simone and Pieterse, 2017) have in mind the latter.

The conception of urban experimentation proposed by Sengers, Berkhout, Wieczorek and Raven (2016) suggests that experiments are the outcome of purposive intent by a defined set of societal actors who aim to implement in practice an alternative in space that explicitly anticipates that its replication will be beneficial to society in general. It is ideally suited to formal regulated urban environments in the cities of the global North and enclaves in the global South. This conception of the replicability of an experiment is what they mean by "system innovation" (2016:21). By inclusion, they mean ensuring that a variety of societal actors are included so that innovations align with "diverse interests and values" (2016:21). "Practice-based" interventions are not ones that are tested in the lab and controlled by private interests who want to ensure that the benefits of their investments in innovation accrue back to themselves and not others in a market competitive environment. They are, instead, tested in public settings where benefits accrue to many interests – a process uniquely suited to urban spaces that benefit from agglomerations of various kinds.

The publicness and therefore the absence of private investment in these experiments is what makes it necessary to ensure state support for these experiments. However, what is at stake is not just technological innovation but also social learning whereby new modes of social organization and collaboration emerge that are both enabled by and enable the technological innovations (2016:22). Urban experiments tend to address "societal challenges" that have normally been articulated by "the diverse network of social actors involved" rather than by a particular business or state agency investigating the potential of new market opportunities (2016:22). And finally, experimentation tends to emerge in contexts characterized by deep uncertainty and ambiguity.

But this does not dampen Sengers et al.'s optimistic rationalism that allows them to insist that it is possible to ensure that "[t]he design of experiments therefore needs to be adaptive to ambiguity and uncertainty" (2016:26). In short, consistent with their definition, even ambiguity and uncertainty can be anticipated and an appropriate "design" put in place to adapt when required. Such confidence in the capacity of rationalist modalities to "tame" the city disintegrate when confronted by the vast churning rapidly accelerating dynamics of the informal unregulated urban dynamics of the global South (Allen, Lampis and Swilling, 2015).

When it comes to the realities of the unregulated informalized cities of the globalized South with their limited infrastructures that cannot support their fast-growing populations, a very different conception of experimentation starts to emerge. Everyone who lives in cities and urban settlements needs to somehow access basic urban services, especially energy, waste disposal, water, sanitation and mobility. For historical reasons, the generally accepted technologies and institutions that have made this possible evolved first in the industrializing cities of Western Europe and North America. The result was centrally managed public monopolies with professionally run highly regulated bureaucracies mandated to deliver uniform services in a given area to almost everyone, including cross-subsidization where required.

Although these conditions do not apply in many cities and urban settlements in the global South, the conventional service delivery system has nevertheless been regarded as the norm by both international aid agencies and local policy elites. Failure is thus defined as anything that deviates from this norm despite contextual differences. Although neoliberalism from the 1980s onward resulted in the privatization of many urban services in the global North, this in fact reinforced the highly regulated nature of the resultant service delivery systems (albeit of a far more complex institutional configuration of interacting public and private agencies). This is the context that explicit intended purposive urban experimentalism takes as a point of departure (which is, in turn, implicit in the discussion by Sengers et al. already cited).

This formalized regulated urban regime is not applicable to many cities and urban settlements in the global South. There is now a substantial body of literature that has demonstrated how complex, heterogonous, hybridized and hodge-podged many urban systems in the global South have become (Swilling, Khan and Simone, 2003; Simone, 2004; Edensor and Jayne, 2012; Parnell and Oldfield, 2014; Allen, Lampis and Swilling, 2016; Simone and Pieterse, 2017). In essence, unlike formalized regulated urban systems, space and time have not been transformed into predictable regulated routines of daily urban life in the "untamed urbanisms" of the global South. This sociocultural-economic heterogeneity has, in turn, resulted in a diversity of hybridized formal and informal service delivery systems that are appropriate for fast-changing rapidly expanding and inherently unstable urbanization processes. Sylvia Jaglin provides the following apt description:

> Service provision in southern cities is a combination made up of a networked infrastructure, deficient in varying degrees and offering a rational service, and of private sector commercial initiatives, whether individual or collective, formal or informal, which are usually illegal in respect of the exclusive contracts of operators officially responsible for the service. These services fill the gaps in the conventional service and, depending on the type of urban area, target either the well-off clientele or poor clientele excluded from the main networks because of lack of resources, geographical remoteness or illegal status. These delivery configurations have one thing in common: the conventional network does not always reach the end user.
>
> *(Jaglin, 2014:438)*

Significantly, she concludes: "[I]n heterogeneous cities, the diversity of service needs has been a vector for innovation" (Jaglin, 2014:439). In other words, urban experimentation in these contexts is not a marginal niche activity but a defining feature of the way entire hybridized urban service delivery systems work in practice! Here experimentation is implicit and emergent, not explicitly intentional and purposive as is the case in a fixed formal regulated environment.

It would be a mistake, however, to comprehend emergent experimentation as a mere divergence from the conventional universal service delivery model; nor is it a temporary step/phase along a developmental pathway towards the final realization of this ideal. Instead, a diversity of interconnected hybridized service delivery configurations is a totally different urban service delivery approach, and it is here to stay in fast-growing complex heterogeneous cities and urban settlements concentrated in the global South.

Jaglin provides an example of what is commonly found in the urban energy sectors of the global South: an ever-changing interdependent set of conventional, community-based, illegal and stand-alone non-grid systems – in short, the co-dependence of formal and informal systems within a wider evolving partially self-organizing experimental urbanism that would be almost impossible to regulate even if capacitated governance institutions were in place (Figure 5.2)

Institutional hybridity, as the logical response to contextual heterogeneity, is effectively the emergent outcome of an endless multiplicity of experiments in daily life that constantly change and recompose. This results in a particular socio-cultural pattern of urban living and acting.

When conditions do not seem to be taking you anywhere, where you constantly battle to keep your head above water and where most of the efforts you make, both individually and as part of a larger collective, at best only manage to repair break-downs of all kinds, then the particular format or mode of living the urban is often characterized by indifference: inhabitants wait for seemingly inevitable displacements or eagerly jump for opportunities to acquire new assets, new property, new lifestyles, if the price is right.

What is important, though, is the capacity to keep going. What was productive about many instances of self-constructed urbanization was the experimental way in which the things that were built could be translated into each other in many

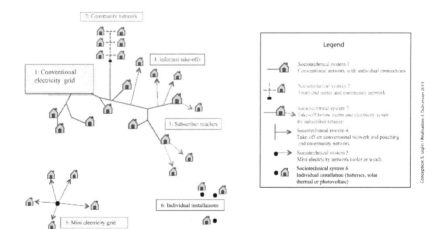

FIGURE 5.2 Household access to energy: example of a delivery configuration.

Source: Jaglin, 2014

different ways – what some have called urban assemblages (McFarlane, 2011). In this way, inhabitants had recourse to concrete exemplars of different versions of themselves, as well as different versions of what was possible in an endless process of improvisation, experimentation, failure, collapse, recomposition and provisional stabilizations. The character of the self-constructed was a space for many characters, a space where the many could become one, and the one many in a back and forth movement that ensured that there were a sufficient number of different ideas and ways of doing things in circulation. But at the same time, these differences did not rule out people paying attention to each other and, as a result, making them an integral part of the stories they would weave out of their own lives.

What then does anticipatory thinking and practice mean in these kinds of informal unregulated spaces where hybridity is the emergent outcome of an extraordinary capacity for permanent experimentation as the primary driver of city-making? Simone and Pieterse answer this question by concluding as follows:

> [I]t is essential to keep both "the systemic" and "the acupunctural" in view when urban interventions are designed and operationalised. Molecular actions may seem acupunctural but they are unlikely to find sustenance unless they feed into and off broader systemic actions that can generate durable transformations over time. . . [W]e explored our understanding of the assemblage of inter-dependent systems calling out for a strategic politics of transition. Similarly, large-scale ambitions need to be tempered by the micro impacts and ramifications that they will carry in tow. However, it is much harder to accentuate the cultural and popular significance of the systemic, since everyday life gains shape through the intimacies of acupunctural actions. *The masculanist claims that only large-scale systemic interventions that shift the political economy of access and citizenship count as real politics amounts to hubris if it is unable to recognise the power of micro-transformations in the domain of everyday living and psychological dispositions.*
> (Simone and Pieterse, 2017 – emphasis added)

In short, just as futuring is expressed along a continuum from forecasting to foresighting, to anticipation, so does experimentation materialize across a continuum of contexts. These range from the reflexively explicit intentional purposive experiments in mature formal regulated urban environments concentrated mainly in the global North (but existent in enclaves across the global South) to the implicit emergent provisional experimentation as a "way of life" in the informalized unregulated 'untamed' urban environments of the global South (Allen, Lampis and Swilling, 2016). If anticipation is about invoking futures to inform action in the present, then action in the present as experiments along this continuum is what anticipation means in the cities and urban settlements of the world.

From anticipation to transformative politics

The argument thus far is that anticipatory behaviours and practices are being reshaped by the advent of a majority urban world dominated now by new agglomerations of people, knowledge, cultures, institutional arrangements, economic resources, natural materials and built forms. Cities are gigantic, highly complex emergent outcomes of imaginaries applied in space by many different actors to produce the physical and cultural spaces for living, working, learning, praying, creating, playing, healing and dying. Visualizations of urban futures have shifted from the regulated/layered city to the ecological/cyber city over the past century. As the crisis of these spaces deepens, futuring and experimentation have become the primary activities of anticipatory thinking and practice. They emerge as intentional purposive actions in more formal regulated environments, and they are embedded in everyday urbanism in less formal unregulated environments.

Anticipatory practice in the city is where futuring (as constructions of future urban imaginaries – "the systemic") and experimentation (as "the acupunctural") meet and fuse in specific spaces. If futuring is about invoking future imaginaries to inform action in the 'precarious now', that is anticipatory practice. If experimentation is about implementing in miniature either implicit or explicit imaginaries of possible futures, that is anticipatory practice. It is this kind of anticipatory practice that can also be understood as a particular form of 'radical incrementalism'.

So the question this raises is whether this conception of anticipatory thinking and practice as the conjoining of futuring and experimentation can become the core of a new transformative politics to drive radical incremental change (for a contribution that addresses the politics of experimentation, see Voss and Simons, 2018). Is the process of change envisaged by anticipatory thinking and practice fundamentally different to the prevailing conceptions of change, that is, reform at the margins or revolution? Is experimentation more than merely tinkering to improve systemic efficiency? And can experimentation be as radical as revolutionary change? Can anticipation become the basis for transformative politics? Is it, in other words, at the very core of a theory of change appropriate for the age of sustainability?

To answer these questions, we need to turn to the theory of change developed by the Brazilian-born Harvard Professor of Law Roberto Unger, who has already asked these questions. Writing in the late 1990s, Unger's largely underappreciated body of work stems from his observation that "a vigorous underground experimentalism has begun to change production and learning" (Unger, 1998:4). He sets out to develop a theory of "democratic experimentalism" that aims to liberate the future from what he calls "structure fetishism" in order to establish a transformative politics of radical reform. Unlike the views represented by May and Perry cited earlier, Unger sees great potential in experimentalism, but only if we do away with a particularly dominant tradition in Western political and economic theory that holds us hostage to mental maps that exclude the radical potential of future-oriented experimentation – an exclusion that renders hope naïve. To this extent, Unger's critique of Western political and economic theory has much in common with those writers discussed in Chapter 1 who contributed to the development of Metatheory 2.0 and the related African notion of *Ukama*.

For Unger, most conceptions of change are based on a pattern of thinking inherited from Western political and economic theory. In this paradigm, 'structures' – primarily institutional configurations – are derived from grand narratives (liberal democracy, free market economics, social democracy, socialism), and then we derive practices and action from our institutional configurations. They are, in Unger's words, "frozen politics". If we want a future that is more democratic, emancipatory and more equal, then we assume that we must replace the existing structures – or institutional arrangements and practices – that stand in the way of such a future. However, as per Metatheory 1.0, this can only be done if we first develop an alternative grand narrative (that gets expressed in future imaginaries), derive a new set of institutional arrangements from this narrative (i.e. a new structure) and then derive the new practices from these arrangements – a reductionist pattern of thinking that Metatheory 2.0 has attempted to transcend. Unger's critique proceeds from his consideration of the implications of this way of thinking for our theory of change.

For Unger, the result of this 'structuralist' binary pattern of thinking is that a viable future is invariably a *new* indivisible set of institutional configurations that are counterposed to the *old* equally indivisible set of institutional configurations. To justify this view, history is interpreted as a series of changes from one 'structure' to another – from feudalism to capitalism and from capitalism to socialism. For instance, if markets are the cause of the problem in the existing paradigm, then the alternative must be to do away with market mechanisms in the new paradigm and so on. Unsurprisingly, experimentalism in this context is thus regarded at best as a marginal (and hence 'naïve') activity – marginal and naïve because experimentalism is regarded as an ineffective counter-force to the 'structural conditions of existence' of existing modes of production and consumption.

The choices that follow are, therefore, either marginal tinkering to slightly reform existing practices to at best improved institutional performance/efficiency or revolutionary activity to totally replace the prevailing grand narrative, institutions and practices. I lived through a revolution and for the decade leading up to 1994 debated these options endlessly. Unfortunately, we did not realize at the time that they are not the only options. For Unger, neither of these is capable of delivering "radical reform": the former because it reinforces the status quo and the latter because revolutionaries end up using violence to force institutional change which they incorrectly assume can effectively transform institutions, and usually all in one dramatic move. Change in South Africa happened in ways we did not fully comprehend. Unsurprisingly, this left us unprepared for the tasks that followed on the morrow of the democratic transition on 27 April 1994.

For Unger, the solution lies in abandoning "structure fetishism" and the notion that institutional arrangements are indivisible. When this happens, you enter the world of the "democratic experimentalist" who taps into the 'impredicative' dynamics – or 'resonances' – of the system and redirects them into untapped potential capacities for radical change.

Structure fetishism, Unger argues,

> denies our power to change the quality as well as the content of our practices and institutions: the way in which we relate to our structure-defying and

structure-changing freedom. Structure fetishism finds expression and defence in an idea, hallowed in the history of social thought, that opposes interludes of *effervescence, charisma, mobilization, and energy* to the ordinary reign of institutionalised routine, when, half asleep, we continue to act out the script written in the creative intervals. An extreme version of structure fetishism is the political via negative that celebrates rebellion against routinized institutional life as the indispensable opening to authentic freedom while expecting that institutions will always fall again, Midas-like, upon the insurgent spirit. . . . [S]tructure fetishism represents an unwarranted denial of our power to change society, and, therefore, ourselves.

(Unger, 1998:26 – emphasis added)

This profound insight into the consequences of structuralist and reductionist thinking expresses the strongest reason why we need an appropriate theory of change premised on the wider reconceptualization of reality presented in Chapter 2. The yearning across the globe for the power to change the quality as well as the context of our practices must surely be what motivates the appeal of alternative metatheoretical paradigms that open up new ways of seeing the 'big picture' and ways of acting to bring about change.

However, even without a fully developed metatheory (as proposed in Chapter 2), the alternative proposed by Unger is to recognize that institutional arrangements are in fact divisible and therefore experimentalism can achieve radical reform "part-by-part and step-by-step". This understanding of change as an emergent outcome must be grasped by those political coalitions that gain ascendance in the polity, either as governing parties or as social partners allied with governing parties. Echoing Hajer's notion of the "energetic society" (Hajer, 2010), Unger argues,

The point of acknowledging radical reform to be the dominant mode of transformative politics is to associate the idea of discontinuous, structural change with the practical attitudes of the person who forever asks: What is the next step? There are basic institutional arrangements and enacted beliefs in a society. . . . Although these formative arrangements are connected, and although some arrangements cannot be stably combined with others, the institutional order of society changes part by part and step by step. . . . It is the combination of parts and the succession of steps, reaching far beyond the starting point, and changing along the way our understanding of our interests, ideals, and identities, that makes a reform project relatively more radical. It is the direction in which the steps take us that make it more or less democratic.

(Unger, 1998:19)

Making a similar point about action, but from a different perspective, Braidotti argues,

If power is complex, scattered and productive, so must be our resistance to it.

(Braidotti, 2013:27)

In short, by snapping the connection between grand narratives and seemingly indivisible institutional configurations, Unger is able to see institutional configurations as amalgams of a range of divisible component parts that have their own respective logics that do not all derive from one central hegemonic logic derived, in turn, from a grand narrative. This, in turn, makes it possible for what he calls "democratic experimenters" to reassemble alternative institutional configurations by recombining those parts that are useful (e.g. the market, with social entrepreneurship and ecological restoration, or new modes of participatory learning, social entrepreneurship and school systems for children) and dispensing others (e.g. debt-driven development or teacher-driven top-down "teaching" underpinned by authoritarian control). This constant context-specific reassembling is not arbitrary: guided by an emergent ever-evolving urban imaginary, it inspires the experimenter to act without the structuralist burden of having to justify the means by the end.

A "step-by-step" trajectory of change with plausible milestones for achieving radical reforms is then envisaged. Those who are inspired by a future imaginary that is not captured by "structure fetishism" but also always risks asking "What is the next step?" are the true futuring experimenters – they invoke the future for action in the present, and they invoke the evolutionary potential of the present to map out plausible implementable radical reforms over time. Inspired by imaginaries of the future that are themselves ever-changing, experimentation understood as a "part-by-part and step-by-step" process of change is the essence of the transformative politics that drives the incrementalist dynamics of deep transition in general and a just transition in particular. It is most evident at the level of urban politics because this is where the symbolic and physical dynamics of change are most evident in their connectedness. But it is equally valid for polities at all levels of governance.

Three notes of caution are required. Firstly, incrementalism must not be interpreted as a linear process that unfolds unproblematically over time. There are constant geographically uneven reversals, stagnant periods of relatively little change and periods of great 'creative destruction' that are akin to social revolutions.

Secondly, just as all experimentation is not about radical incremental change, nor is all futuring liberated from "structure fetishism". Many experiments can be just about tinkering to improve system efficiencies or just about survival rather than change in less regulated informal cities. To contribute to a transformative politics, these experiments need to involve collectives of large groups and networks, and they need to express an alternative future imaginary either implicitly or explicitly in their actions in the present (for a similar but more technocratic argument, see Schot and Steinmuller, 2016).

Thirdly, it needs to be acknowledged that many futuring activities are trapped by "structure fetishism": they are simply thinly disguised, facilitated or model-based processes to justify a preconceived grand narrative that is required for constructing a new institutional configuration. The best indicator is when these futuring exercises leave participants with nothing to do in the present other than wait for the new grand narrative to be adopted while they repeatedly reconfirm for themselves and others that things are getting worse or else they must just go out to convince others to adopt the grand narrative on the assumption that if enough people reach some shared cognitive agreement about a desired future, things will change. The future imaginaries that get built this way do not, in short, get expressed in experiments that recombine "part-by-part and step-by-step", and as a result they have a limited capacity to inspire sustained and organized action over time.

To be truly transformative, futuring and experimentation need to become mutually self-reinforcing practices that can, over time, consolidate a shared sense of directionality that accepts as a minimum the need for a deep transition. Ideally, this should include a normative commitment to a just transition. In practice, most transformative processes are inspired by future-oriented experimenters or experimentally oriented futurists depending on where the emphasis lies in their respective discourses and methodologies. The next chapter discusses a range of experiments that are inspired by – and also inspire – a commitment to a deep and just transition. These are by no means static models. They are highly contested and driven largely by people who cannot help continuously asking "What is the next step?" Surely the most radical person in the room is the person who asks this question, not the person who claims to have mastered the fundamental contradictions of global capitalism?

This reality of emergent transformative processes discussed in this chapter and elaborated empirically in the following chapter raises questions about the kinds of research practices that can build what was referred to in Chapter 1 as transformative knowledge appropriate for the conception of radical incrementalism developed in this chapter.

Conclusion

The extent and complexity of the multifaceted polycrisis that we all face seems fertile ground for the emergence of a new grand narrative capable of mobilizing the multitude behind a viable alternative. Ironically, in a majority of urban world, it is from the multiplicity of highly diverse urban spaces across the developed and developing countries that new coalitions are emerging that use the context-specific dynamics of the urban spaces they know well to catalyse a wide range of neighbourhood and city-wide experiments that are to a larger or lesser extent inspired by a shared imaginary of the future. Simultaneously, because the built environment is a designed environment, there is a growing need to reimagine the futures that designers need to invoke to justify their design proposals, hence the shift in visual imagery towards the intersection between the ecological and cyber city.

This double imperative to both experiment in the now and reimagine implementable futures that seems so implicit in the dynamics of the city (as both artefact and symbol) is what this chapter has reflected on. In a majority of urban world, urban spaces are the most useful context for grasping this double anticipatory imperative. It holds the key to an understanding of what a potentially mainstream transformative politics could be in practice at any spatial scale. Following Unger, this potential cannot be realized if "structure fetishism" retains its grip on our cognitive maps of the future. Remaining attached to a theory of change that regards institutions as indivisible wholes derived from grand narratives obliterates the space for a "part-by-part and step-by-step" conception of radical reform that seems to be the strategy of choice of countless future-oriented experimenters and experimentally minded futurists from all over the world. It is time that the significance of their practices is understood, appreciated and supported.

Notes

1 Some arguments, paragraphs and ideas for this chapter are drawn from Swilling, M., Pieterse, E. and Hajer, M. (2018)'Futuring, experimentation and transformative urban politics', in Poli, R. (ed.) *Handbook for Anticipation*. New York: Springer. doi: 10.1007/ 978-3-319-31737-3_24-1.
2 The internal elements of a predicative system tend to respond in predictively similar ways to the same external determinants. By contrast, impredicative systems are relatively self-referential complex adaptive systems because their responses to the impact of external factors cannot be predicted. Instead, responses will be conditioned by the context-specific configuration of the internal dynamics of the system that can result in a wide range of very different responses to the same external factors.
3 Dates after each of the names mentioned refer to when they were generated and not to a bibliographical reference.

References

Affolderbach, J. and Schulz, C. (2018) *Green Building Transitions: Regional Trajectories of Innovation in Europe, Canada and Australia*. New York: Springer.

Alison, J., et al. (2007) *Future City: Experiment and Utopia in Architecture*. London: Thames &Hudson.

Allen, A., Lampis, A. and Swilling, M. (2015) *Untamed Urbanisms: Enacting Productive Disruptions, Untamed Urbanisms*. London and New York: Routledge.

Allen, A., Lampis, A. and Swilling, M. (2016) *Untamed Urbanisms*. New York: Routledge.

Ayres, B. (2016) *Energy, Complexity and Wealth Maximization*. Berlin: Springer.

Bayat, A. (2000) 'From "dangerous classes" to "quiet rebels": Politics of the urban subaltern in the global south', *International Sociology*, 15(3), pp. 533–557.

Beatley, T. (2000) *Green Urbanism*. Washington, DC: Island Press.

Beatley, T., Boyer, H. and Newman, P. (2009) *Resilient Cities: Responding to Peak Oil and Climate Change*. Washington, DC: Island Press.

Beatley, T. and Newman, P. (2009) *Green Urbanism Down Under: Learning from Sustainable Communities in Australia*. Washington, DC: Island Press.

Bhaskar, R., Esbjorn-Hargens, S., Hedlund, N. and Hartwig, M. (eds.) (2016) *Metatheory for the Twenty-First Century: Critical Realism and Integral Theory in Dialogue*. New York: Routledge.

Braidotti, R. (2013) *The Posthuman*. Cambridge: Polity Press.

Brook, D. (2013) *A History of Future Cities*. New York: W.W. Norton & Company Inc.

Broto, V. C. and Bulkeley, H. (2013) 'A survey of urban climate change experiments in 100 cities', *Global Environmental Change*, 23(1), pp. 92–102.

Brugmans, G., van Dinteren, J. and Hajer, M. (2016) *The Next Economy IABR 2016*. Rotterdam: International Architecture Biennale Rotterdam.

Bulkeley, H. and Broto, C. (2013) 'Government by experiment? Global cities and the governing of climate change', *Transactions of the Institute of British Geographers*, 38, pp. 361–385.

Caprotti, F. and Cowley, R. (2017) 'Interrogating urban experiments', *Urban Geography*, 38(9), pp. 1441–1450.

Department of Economic and Social Affairs, United Nations. (2012) *World Urbanization Prospects: The 2011 Revision*. New York: United Nations.

Dodman, D. (2009) 'Blaming cities for climate change? An analysis of urban green house gas emissions inventories', *Environment and Urbanization*, 21, pp. 185–201.

Dunn, N., Cureton, P. and Pollastri, S. (2014) *A Visual History of the Future*. London: Foresight, Government Office for Science.

Eadson, W. (2016) 'State enrolment and energy-carbon transitions: Syndromic experimentation and atomisation', *Environment and Planning C: Government and Policy*, 34(8), pp. 1612–1631.

Edensor, T. and Jayne, M. (2012) *Beyond the West: A World of Cities*. London and New York: Routledge.

Evans, J. (2016) 'Trials and tribulations: Problematizing the city through/as urban experimentation', *Geography Compass*, 10(10), pp. 429–443.

Evans, J. and Karvonen, A. (2014) 'Urban laboratories: Experiments in reworking cities', *International Journal of Urban and Regional Research*, 38(2), pp. 2437–2453.

Evans, J., Karvonen, A. and Raven, R. (2016) *The Experimental City*. London: Routledge Earthscan.

Freeth, R. and Annecke, E. (2016) 'Facilitating social change', in Swilling, M., Musango, J. and Wakeford, J. (eds.) *Greening the South African Economy*. Cape Town: UCT Press, Chapter 21.

German Advisory Council on Climate Change. (2011) *World in Transition: A Social Contract for Sustainability*. Berlin: German Advisory Council on Global Change.

German Advisory Council on Global Change. (2016) *Humanity on the Move: Unlocking the Transformative Power of Cities*. Berlin: WBGU.

Hajer, M. (2010) *The Energetic Society: In Search of a Governance Philosophy for a Clean Economy*. The Hague: PBL. Available at: www.pbl.nl/sites/default/files/cms/publicaties/Energetic_society_WEB.pdf.

Hajer, M. and Pelzer, P. (2018) '2050 – An energetic odyssey: Understanding "techniques of futuring" in the transition towards renewable energy', *Energy Research and Social Science*, 44, pp. 222–231.

Hall, P. (2014) *Cities of Tomorrow*. 4th ed. Oxford: Blackwell.

Hawken, P. (2007) *Blessed Unrest: How the Largest Social Movement in History Is Restoring Grace, Justice and Beauty in the World*. London: Penguin.

Hodson, M., Evans, J. and Schliwa, G. (2018) 'Conditioning experimentation: The struggle for place-based discretion in shaping urban infrastructures', *Politics and Space C*, pp. 1–19.

Hodson, M. and Marvin, S. (2016) *Retrofitting Cities: Priorities, Governance and Experimentation*. London: Earthscan.

Hopkins, L. D. and Zapata, M. A. (2007) *Engaging the Future: Forecasts, Scenarios, Plans and Projects*. Boston, MA: Lincoln Institute of Land Policy.

Jaglin, S. (2014) 'Regulating service delivery in southern cities', in Parnell, S. and Oldfield, S. (eds.) *The Routledge Handbook on Cities of the Global South*. London and New York: Routledge.

Jessop, B. (2016) *The State: Past Present Future*. Cambridge: Polity Press.

Jorgensen, M., Wittmayer, J., et al. (2016) *Transformative Social Innovation Theory: Synthesis Across Social Innovation Case Studies*. Aalborg University Copenhagen, Dutch Research Institute for Transition.

Karvonen, A. and van Heur, B. (2014) 'Urban laboratories: Experiments in reworking cities', *International Journal of Urban and Regional Research*, 38(2), pp. 379–392.

Marais, H. (1998) *South Africa: Limits to Change*. London: Zed Books.

Marvin, S., Bulkeley, H., Mai, L., McCormick, K. and Palga, Y. (eds.) (2018) *Urban Living Labs*. New York: Routledge.

May, T. and Perry, B. (2016) 'Cities, experiments and the logics of the knowledge economy', in Evans, J., Karvonen, A. and Ravan, R. (eds.) *The Experimental City*. New York and London: Routledge, pp. 32–47.

McFarlane, C. (2011) *Learning the City: Knowledge and Translocal Assemblage*. Oxford: Wiley Blackwell.

Mishra, P. (2018) *Age of Anger: A History of the Present*. London: Penguin Random House.

Moulaert, F. D., MacCullum, A., Mehmood, A. and Hamdouch, A. (2013) *The International Handbook on Social Innovation: Collective Action, Social Learning and Transdisciplinary Research*. Camberley: Edward Elger.

Murray, R. J., Caulier-Grice, J. and Mulgan, G. (2010) *The Open Book of Social Innovation*. London: NESTA.

Parnell, S. and Oldfield, S. (2014) *The Routledge Handbook on Cities of the Global South*. London and New York: Routledge.

Pieterse, E. (2008) *City Futures*. Cape Town: Juta.

Polanyi, K. (1946) *Origins of Our Time: The Great Transformation*. London: Victor Gollancz Ltd.

Poli, R. (2014) 'Anticipation: A new thread for the human and social sciences?', *CADMUS*, 2(3), pp. 23–36.

Poli, R. (ed.) (2018) *Handbook for Anticipation*. New York: Springer.

Preiser, R., Biggs, O., de Vos, A. and Folke, C. (2018) 'A complexity-based paradigm for studying social-ecological systems', *Ecology and Society*, 23(4), p. 46. doi: 10.5751/ES-10558-230446.

Pretty, J., Boehme, S. and Bharucha, Z. (2015) *Ecocultures: Blueprints for Sustainable Communities*. New York: Routledge.

Schot, J. and Steinmuller, W. (2016) *Framing Innovation Policy for Transformative Change: Innovation Policy 3.0*. Sussex, UK: Science Policy Research Unit (SPRU).

Sengers, F., Berkhout, F., Wieczorek, A. J. and Raven, R. (2016) 'Experiments in the city: Unpacking notions of experimentation for sustainability', in Evans, J., Karvonen, A. and Raven, R. (eds.) *The Experimental City*. London: Routledge Earthscan, pp. 15–31.

Simone, A. (2001) 'Between ghetto and globe: Remaking urban life in Africa', in Tostensen, A., Tvedten, I. and Vaa, M. (eds.) *Associational Life in African Cities: Popular Responses to the Urban Crisis*. Stockholm: Elanders Gotab.

Simone, A. (2004) *For the City Yet to Come: Changing African Life in Four Cities*. Durham, NC and London: Duke University Press.

Simone, A. (2006) 'Pirate towns: Reworking social and symbolic infrastructures in Johannesburg and Douala', *Urban Studies*, 43(2), pp. 357–370.

Simone, A. and Abouhani, A. (2005) *Urban Processes and Change in Africa*. London: Zed Press.

Simone, A. and Pieterse, E. (2017) *New Urban Worlds: Inhabiting Dissonant Times*. London: Polity Press.

Smith, A. and Raven, R. (2012) 'What is protective space? Reconsidering niches in transitions to sustainability', *Research Policy*, 41, pp. 1025–1036.

Suzuki, H., Dastur, A., Moffatt, S. and Yabuki, N. (2009) *Eco2 Cities: Ecological Cities as Economic Cities*. Washington, DC: The World Bank.

Swilling, M. (1999) 'Rival futures', in Judin, H. and Vladislavic, I. (eds.) *Blank: Architecture, Apartheid and After*. Rotterdam: NAI Publishers.

Swilling, M. and Annecke, E. (2012) *Just Transitions: Explorations of Sustainability in an Unfair World*. Tokyo: United Nations University Press.

Swilling, M., Hajer, M., Baynes, T., Bergesen, J., Labbe, F., et al. (2018) *The Weight of Cities: Resource Requirements of Future Urbanization*. Paris: International Resource Panel. Available at: www.internationalresourcepanel.org.

Swilling, M., Khan, F. and Simone, A. (2003) '"My soul I can see": The limits of governing African cities in a context of globalisation and complexity', in McCarney, P. and Stren, R. (eds.) *Governance on the Ground: Innovations and Discontinuities in Cities of the Developing World*. Baltimore and Washington, DC: Woodrow Wilson Centre Press and Johns Hopkins University Press, pp. 220–250.

Swilling, M., Pieterse, E. and Hajer, M. (2018) 'Futuring, experimentation and transformative urban politics', in Poli, R. (ed.) *Handbook for Anticipation*. New York: Springer. doi: 10.1007/978-3-319-31737-3_24-1.

Swilling, M., Robinson, B., Marvin, S. and Hodson, M. (2013) *City-Level Decoupling: Urban Resource Flows and the Governance of Infrastructure Transitions*. Available at: www.unep.org/resourcepanel/Publications/AreasofAssessment/Cities/tabid/106447/Default.aspx.

Unger, R. M. (1998) *Democracy Realized*. London: Verso.

United Cities and Local Governments. (2016) *Co-creating the Urban Future: The Agenda of Metropolises, Cities and Territories*. Barcelona: United Cities and Local Governments.

United Nations Centre for Human Settlements. (2003) *The Challenge of Slums: Global Report on Human Settlements*. London: Earthscan.

Verspagen, B. (2005) 'Innovation and Economic Growth', in Fagerberg, J., Mowery, D. C. and Nelson, R. R. (eds.) *The Oxford Handbook of Innovation*. Oxford: Oxford University Press, Chapter 18.

Voss, J. and Simons, A. (2018) 'A novel understanding of experimentation in governance: Co-producing innovations between "lab" and "field"', *Policy Sciences*, pp. 1–17. doi: 10.1007/s11077-018-9313-9.

6

EVOLUTIONARY POTENTIAL OF THE PRESENT: WHY ECOCULTURES MATTER[1]

Preface

It is early on a mild November morning in 2004 and the Toi Market in Nairobi is starting up. There are around 2000 stalls, almost all engaged in the sale of foodstuffs to people from surrounding poor communities. Stalls are made of wooden poles and a mix of plastic and metal materials provides cover. There is no grid structure but rather a dense network connected by pathways wide enough for one person. The floors are raw, muddy and risky. No tourists to be seen in this market and Nairobi's middle class does not shop here. The market has been going since 1991 and has enjoyed the usual highs and lows that poor people face when it comes to winning space to conduct ordinary life in highly contested and congested urban systems. Sixty per cent of the stallholders own their own stalls, the rest rent their stalls from owners of the stalls. Sellers buy their goods from the central market, or from local farmers, and some are farmers themselves selling directly to the public. This market was set up by poor people who need to make a living by selling the goods that poor communities need for daily survival. What holds this market together is a self-organizing savings and loan system that is controlled by the stallholders themselves. The key leader is your classic male community organizer: charismatic, shrewd, high energy, confident, supersensitive about his perceived integrity, mindful of the need for inclusiveness of his inner leadership group and as alert as a mamba to every inflection and signal from those surrounding him as he moves down the narrow aisles between the stalls or chairs the large member's meetings. The savings group has 800 members, of which 596 had loans in November 2004. After 2 p.m. every day, the collectors (members who volunteer for the task on a rotating basis) visit those stallholders who are members to collect the daily savings. On a good day, they collect from 500 stallholders – most days the number of savers varies between 300 and 500. Each saver has his/her own book and the collector has a book – the transactions are carefully recorded in full view of everyone in both books. It's a routine transaction, conducted with a certain conscious absent-mindedness interspersed with chatter about matters or problems of the day. The connection made, the relationship renewed, the fact that the system has survived another day has been registered – precious certainty when all else can change in an instant. On Thursdays all members are supposed to meet in a hall on the outskirts of the market that the savings group has built from their own funds. Usually, 250 pitch up. The leader sits in the middle of a table upfront, various people with roles on both sides of him. The rest are seated in church-style rows on wooden benches, some with their backs against the side walls which are adorned with names of members on flipchart paper, messages and posters. The purpose is to pool the savings collected and to agree to the loans required. This happens three Thursdays out of four – on the fourth Thursday, the savings collected are banked and not loaned out. The reserve is built up steadily – more investment in certainty. The savings collected are loaned out for short periods – one to two months at a fixed rate of 5%. Normally 1,500 Kenyan Shillings for one month, 3,000 for one and a half months or 4,500 for two months. The process of requesting, approving and dispensing the loan happens simultaneously in full view of everyone at the meeting. The trans-action, once again, carefully recorded in the books of borrower and lender. The group has also borrowed money from the AMT Fund managed by the Pamoja Trust, an NGO that is part of the family of institutions that comprises the Shack/Slum Dwellers International (SDI) movement. This money is loaned at 13% for periods of between six and eight months, and the amounts involved are between 15,000 and 30,000 Kenyan Shillings. A borrower must be a saver and must gradually prove his/her creditworthiness by starting with a small loan over a short period. As s/he proves capable of repaying, s/he is authorized to borrow larger amounts over longer periods of time. This means when the collector comes around, the member contributes his/her daily saving plus repayment of a small short-term loan at 5% plus repay-ment of a larger longer-term loan at 13% – there is a column in the saver's book and the collector's book for recording all three transactions. The Thursday meetings will also listen to reports from Committees – there is a Loan Committee, Collectors Committee, Audit Com-mittee and, significantly, a Welfare Committee (which attends to social problems and issues). And no overhead! – it's all run by the members, hence no need to waste money paying others to do the work or for fancy bureaucratic systems that need professionals to operate them. They estimate they have dispersed 5 million Kenyan Shillings. There are some bad debts, but most defaulters are seen as people who have genuine problems and need help and not as miscreants deserving punishment (which is the approach associated with the much loathed micro-lenders). Some have died, others left the market for good. Collective profits are ploughed into projects, including the building of a toilet block, a composting system and extensions to the hall. Now whereas many such markets can be found across all African cities, this one has a sense of pro-gress beyond bare survival for one simple reason – cash surpluses are retained, reinvested and recirculated, thus limiting leakage via banking and other financial institutions who normally take their cut via administrative charges, low interest on deposits, high interest on loans (in the unlikely event of a poor person getting a loan) and lending out savings to richer people in the formal economy who live elsewhere. Participatory self-organization is therefore the key to controlling the flow of cash through the market in a way that builds the total amount of cash in local circulation stimulating a sense of growth and cohesion. It literally sucks in and holds huge cash flows on a daily basis. The alternative is all too familiar in Africa: people make an effort, cash leaks out to the benefit of others, things start to collapse as internal tensions erupt along socio-cultural faultlines and blame replaces trust as that suffocating sense of scarcity over-whelms even the most decent people. The Toi Market represents in microcosm an alternative logic, a living example of self-organizing pro-poor development that does not need parasitic financial institutions or to waste precious human resources accounting to external agencies or employing professionals.

Introduction

Anybody who has encountered in a direct and personal way a wide multiplic-ity of local development contexts in the global South will recognize this story of the Toi Market in Nairobi. The remarkable combination of effective engaged and embedded leadership, social ingenuity, entrepreneurship, organizational viabil-ity, experimentation and vision-building about a potential better future is what incrementalism is all about in practice. Inspired by this story, this chapter con-nects the previous chapter on incrementalism with the wider discussion about deep and just transitions in Chapter 4. Informed by the metatheoretical framework in Chapter 2 and following the logic of Polanyi's 'double movement' discussed in Chapter 3 (Polanyi, 1946), the connection between incrementalism and transition is demonstrated in the cases that follow. The 'double movement' is reflected in the way ecocultures have emerged as the potential foundation for the future while in parallel rampant ecologically destructive debt-financed consumerism became the dominant culture of the late twentieth century and into the new millennium.

The aim of this chapter is to describe and illustrate a wide range of contexts drawn from the global South, very few of which are generally known about or referred to in the literature on alternatives (partly because this literature tends to be dominated by writers from the global North). Short descriptions are used that can never do justice to the complexities of the actual cases. Nevertheless, these short evocative qualitative descriptions are suggestive of wider trends across similar contexts. They reinforce the argument that there are a set of common characteris-tics that seem to replicate themselves across a wide range of contexts that are not directly linked to one another.

This chapter provides a brief set of analytical descriptions of future-oriented experiments from across the global South. They all reveal the potential of incremen-tal change and how this could tip the balance of forces in favour of a longer-term just transition. The challenge, however, is not about 'upscaling' these experiments so that they can result in 'structural change', but rather the challenge is about replicat-ing them across space and time so that they accumulate into emergent society-wide ecocultures that demonstrate in practice what a just transition could look like. For this to happen, however, it will be necessary to discern the shared and common characteristics of these various initiatives and how they can potentially affect the overall *directionality* of the deep transition. What follows is a proposed framework that refers to these initiatives as *ecocultural assemblages* that demonstrate to a greater or lesser degree a proposed set of five primary characteristics.

Ecocultural assemblages

The notion that there are an emergent set of *ecocultures* has begun to attract the attention of academic researchers (see Pretty, Boehme and Bharucha, 2015). This is a much broader set of local formations to those identified in Latin American writing on post-development which focuses on emerging 'communal', 'relational' and 'pluriversal' alternatives to neoliberal capitalism, extractivism and developmentalism (Escobar, 2015). However, instead of seeing ecocultures as relatively stable and resolved socio-cultural phenomena, for the purposes of this discussion ecocultures will be understood as *emergent ecocultural assemblages* – a formulation that blends the notion of assemblage (as used by McFarlane, 2011) with the notion of emergence (as used in complex adaptive systems theory). Emergent ecocultural assemblages can, therefore, be understood as dynamic learning processes expressed in spatially specific ever-changing provisional (re-)configurations of people, cultures, infrastructures, buildings, materials, ecosystems and natural resource flows. As the cases reveal (discussed later), these can take many forms – from the light green techno-fix for the rich of Songdo and Bangalore's green gated communities to the deep green socially just ecovillages of Ecobarrio and the Lynedoch EcoVillage (as discussed in the Introduction and Swilling and Annecke (2012 – Chapter 10)).

Cases were selected that can be regarded as representative of trends rather than completely unique and therefore unreplicable. Although there were serious limits to the evidence available, each case selected needed to reflect some of the following 'trace elements' of an ecocultural sensibility:

1   evidence of commons-type learning processes expressed in the way a particular natural resource is managed and learning institutions established, with special reference to how socio-technological innovations are developed to more sustainably use resources, recycle wastes and restore ecosystems;
2   existence of adaptive leadership capabilities characterized by a sensitive engagement with the local context, a profoundly relational approach and a commitment to innovation and a sense of trust in the emergent outcomes;
3   institutional arrangements that encourage, incentivize and stimulate sustainability-oriented innovations, including the requisite degree of trust between actors located across different sectors and a commitment to dialogical and relational approaches to conflict resolution and community building;
4   a tendency to value family and community life, limit conspicuous consumption as a mode of identity and avoid self-destructive or abusive personal behaviours;
5   an awareness of the need to focus on improvements in the quality of life of the poorest participants through education, participation and a degree of tolerance of diversities of ethnicity, belief and class.

None of the cases explicitly illustrate the institutional form referred to in Chapter 2 as a 'commons-based peer-to-peer' (CBPP) mode of socio-economic organization. However, they do illustrate what is a core property of CBPP, namely the application of a relational sensibility in contextually specific ways to address a particular set of social and/or ecological challenges. In almost every case, the collaborative voluntary contribution to a shared knowledge commons is clearly evident. Without this, none of these initiatives would have been able to achieve what they achieved. To this extent, this chapter provides an overview of the types of proto-ecocultural assemblage that could easily underpin the kinds of ICT-enabled CBPP formations referred to in Chapter 2. It is only a matter of time before there are a sufficient number of progressive-minded coders and standardized downloadable CBPP-type templates to enable grassroots movements to transform themselves into CBPP-type economies with high potential for replicability. This chapter, therefore, shifts the lens away from CBPP formations per se to the underlying social dynamics that emerge across many different contexts without in any way being related to one another.

Three categories of initiatives are discussed.[2] The first, *Resisting Disconnections*, are largely rural initiatives that have actively and consciously established themselves to resist the disconnections from nature instigated by modernity and the destruction of natural systems to serve mainly urban-based economies.

The second category groups together *Green Urbanism* initiatives. Motivated by a desire to 'minimize damage', these initiatives are top-down initiatives by governments and/or property developers who are responding to either the rising cost of conventional urban infrastructures and/or the rising demand amongst elites for green low-carbon or even 'carbon-free' environments.

In contrast to green urbanism initiatives, the third category refers to a mixed bag of initiatives as examples of *Liveable Urbanism*. These are profoundly urban initiatives aimed at creating distinctive urban ecocultural assemblages, frequently with the poorest urban dwellers in mind. The underlying assumption is quite often the view that formal conventional urban infrastructures and formal dwellings cannot provide affordable and viable living environments for the urban poor. Alternatives were found that were drawn from the rapidly expanding repertoire of socio-technical solutions generated by a wide range of ecocultural movements, ecological design initiatives and research institutions.

Resisting disconnections

This section reviews a range of initiatives located within rural/peri-urban contexts because so-called rural spaces quite often offer spaces for innovation that may be hard to find in urban environments where land prices, planning regulations or restrictions on access to land often prohibit innovation. The cases cited here created spaces for niche innovations that helped demonstrate new approaches to buildings, infrastructures, ecosystems and food production that have influenced and inspired many urban-based ecocultural movements. These more rural initiatives are worth categorizing into 'new initiatives' created by groups with an explicit agenda to build new ecocultural assemblages; 'traditional initiatives' where existing traditional communities adopt practices that preserve their cultures, skills sets and connections to nature and 'project initiatives' which are essentially initiatives by various types of actors aimed at reorganizing existing rural communities around a range of ecocultural practices.

New initiatives

Four initiatives are worth referring to here: Sekem in Egypt (founded in 1977), Auroville in India (founded in 1968), Gaviotas in Colombia (founded in 1971) and Picaranga in Brazil (founded in 2000). All were motivated by highly idealistic leadership groups who wanted to directly and explicitly intervene in degraded/threatened natural systems in order to restore them to sustain human livelihoods and, in some cases, facilitate exports into local and international markets.

Auroville is located near Pondicherry in the state of Tamil Nadu, south India (this account is derived from Dawson, 2006:24–26, plus formal interviews and less formal discussions with founders and participants). Initiated in 1968 by a core group that adhered to the philosophies of guru Sri Aurobindo, it currently has about 1,700 residents from 35 countries. The aim is to grow to 50 000. It began in what was then almost desert-like conditions resulting from decades of deforestation and destructive farming. To restore the ecosystems, dams and swales were built along the contours to replenish the aquifers and contain erosion during the monsoon, while over 2 million trees were planted. Most of Auroville is now a lush forest and 135 hectares are farmed to produce most of Auroville's food requirements. The restoration of the ecosystems as a bio-physical knowledge commons established the foundation for the 'development' and well-being of the Auroville community. With a strong spiritual centre, the community has evolved a collectivist ethic that has restrained inequalities. Leadership is non-hierarchical and services are provided by local enterprises. Most of the 125 enterprises operating within Auroville are collectively owned, benefitting from the voluntary contributions to learning and systems. Extraordinary technological advances were achieved that were way ahead of their time, including the use of solar power, biogas-based cooking, stabilized earth block construction and various effective ecosystem management techniques. None of this know-how was individually owned or used for private accumulation. Although located in a rural setting, it is effectively evolving into an economically viable and ecologically sustainable 'town' of interconnected clusters connected closely to the land. Major research centres that have collated what has been learnt in the knowledge commons disseminate the tried and tested technologies across international networks.

Founded in 1971 in the inhospitable Llanos Badlands in the Colombian region of Vichada by Colombian social entrepreneur Paolo Lugari, Gaviotas followed a similar trajectory to Auroville (this account is based on Weisman, 2008). Initially surviving off donor funds, the Gaviotas community merged local Indian families with middle-class professionals committed to building a new ecoculture. Although the focus in the first decade and a half was on building up alternative rural development technologies that donors funded to be replicated elsewhere (e.g. the Gaviotas water pump), by the 1990s the founders discovered that a certain variety of pine could thrive in the desiccated soils. Eventually, 1.5 million pine trees were planted which, in turn, created the conditions for the re-emergence of the indigenous forest that used to grow in the region. This became the most valuable bio-physical knowledge commons of this community. Not owned by any individual, the collective care of this commons provides the economic basis for the survival of this community. Harvesting the resin from the collectively managed pine forest generated the income needed for the community to thrive.

Sekem started in 1977 as the brainchild of Ibrahim Abouleish.[3] Motivated by the values of anthroposophy (originating in the works of Rudolf Steiner), his Muslim faith and a belief in biodynamic farming methods, Abouleish decided to establish a commercially viable agricultural enterprise on 70 hectares of desert located 60 km northeast of Cairo. The experiment worked, resulting in a bio-physical knowledge commons managed by a social enterprise not geared towards the enrichment of shareholders. This resulted in the establishment of what is now a large conglomerate of agro-ecological businesses involved in land reclamation, organic farming, food production for sale into local and international markets, phyto-pharmaceuticals and textile production from organic cotton. All this is rooted within a thriving community spread across different sites that includes Waldorf Schools, vocational training centres, an association of biodynamic farmers spread across Egypt and a private university called the Heliopolis University which was established in Cairo in 2012. Employing nearly 2000 staff, Sekem defines itself as a social enterprise and its vision is "sustainable human development". In 2011 it made a gross profit of 96 million Egyptian Pounds. Profits are deployed to fund social, cultural and educational activities.

Picaranga EcoVillage is located on the picturesque coast of Bahia state in Brazil, 6 km from the tourist centre of Itacare. Started in the early 2000s by a family, it had 160 inhabitants by 2010, including local Brazilians (who are in the majority) and people from 20 different countries. The Center for Art and Human Development where "people can learn and experience the miracle of life" and the "holistic free school" for students of 3–18 years of age are clearly the institutions at the heart of this "ecovillage for nature lovers". Buildings are constructed from "local materials", electricity is generated from solar and wind power and solar hot water heaters, septic tanks for treating sewage and a permaculture garden complete the picture of an ecovillage that has made considerable progress in ten years. A core focus is the protection of the 120 hectares of virgin tropical forest adjoining the ecovillage.[4] Once again, like Auroville and Sekem, the pattern is clear: the collective management of a forest and learning is the basis for the survival of the community.

All four cases demonstrate how rural or peri-urban environments provided the space for ecocultural networks to coalesce around new technologies for restorative farming, more sustainable modes of human settlement and holistic patterns of cultural development. All four were explicitly aware of their social justice commitments and the importance of building up a knowledge commons as repositories for what was learnt. This knowledge commons then becomes the basis for outward-oriented teaching and learning. Auroville, Sekem and Gaviotas have all become international icons of actually existing micro-systems that explicitly connected ecological sustainability and social justice commitments.

Traditional, transformed

The four cases described here are traditional rural villages that faced a crisis of some sort. This triggered the emergence of responses that reinforced the resilience of these villages. In all four cases, greater resilience entailed voluntary contributions to a knowledge commons and investments in new socio-technical innovations that both created and were shaped by various context-specific ecocultural assemblages. The four cases are Hivre Bazar in Maharashtra state, India[5]; Mbam and Faoune villages in Senegal (Dawson, 2006–2010:26–28) (Dawson, 2006:26–28); Da Ping Village in China[6]; and Ghandi Nu Gam in India.[7]

Hivre Bazar is located in the barren expanses of central Maharashtra. A small village of 257 families, Hivre Bazar has managed to break the patterns of drought, poverty, alcoholism and rural-urban migration that plague villages in the rest of this region. However, 20 years ago, it was a village in crisis and on the verge of collapse caused by migration to the cities, poverty and social disintegration. About this time Popatroa Pawar completed his postgraduate studies in Pune. He was 26. Instead of staying in the city, he decided to go back to Hivre Bazar, his home village, to find solutions to the crisis. Mobilizing shramdaan (voluntary labour), his first project was to build watersheds along the contours and replant the trees (like in Auroville). The replenished water table enabled irrigation which, in turn, made it possible for farmers to diversify their crops and increase yields. Average incomes increased by 400% between 1991 and 2009, 60 families that migrated to the cities returned, social improvement and cultural renewal programmes were introduced, formal sanitation improved health and eliminated mosquitos and education improved (especially for girls). The Gram Sansad (Village Parliament) has played a central role as both a knowledge commons and a leadership forum, and there is a pervasive sense of prosperity without significant inequalities. Hivre Bazar is held up by many now as a successful rural development model.

Earthquakes in Da Ping Village in China and Ghandi Nu Gam in India triggered ecological design responses that reinforced the long-term resilience of these two rural villages. The earthquake that hit Sichuan Province in 2008 destroyed nearly all the buildings in Da Ping Village. Instead of rebuilding using the generally accepted modern brick and concrete methods that have become pervasive in China, the local authorities decided to accept advice from the Green Building Research Centre that this poor village of rice farmers should be rebuilt using traditional building methods in order to "bring greater harmony with the local environment". A thorough participatory process (which included residents paying 50% of the costs of their homes) was initiated to build 200 houses and 11 public buildings using local wood and bamboo, salvaged materials and renewable energy (including biogas). The end result was homes that had more space and were more comfortable than what would have been provided using conventional industrial methods. A new Community Centre was also built that became the centre of a new village management structure that continued to promote an ecoculture of sustainable living, organic farming, ecological literacy and environmental restoration. A new tourism industry emerged because of the publicity generated by the unique approach which has helped stimulate the local economy. The approach started to be replicated in other villages in the Province. Although the initial idea came from an external agency, the success of the community has resulted in a knowledge commons that allows learning to be replicated across other communities.

After an earthquake devastated Gujarat in 2001 killing 20,000 and leaving millions homeless, it was discovered that the traditional adobe 'bhungas' (circular dwellings) survived while the modern brick and concrete rectangular buildings favoured by younger people collapsed. The NGO called Vastu-Shilpa Foundation (VSF) for Studies and Research in Environmental Design decided to assist Ghandi Nu, one of the worst hit villages. The project constructed 455 traditional bhungas together with three schools, community buildings, production centres, religious shrines, an electricity network using renewables, a water harvesting system and a sewage treatment system that redirected nutrients into biomass/food production. VSF set up an office to ensure that villagers participated fully in design and construction. This office became the centre of this knowledge commons in this village. This included extensive training. All materials were locally sourced, including local wood for doors and windows.

The Senegalese villages of Mbam and Faoune located to the South of the capital Dakar are formally registered as EcoVillages by the Global EcoVillage Network (GEN). They both applied to become EcoVillages after they were hit by a major drought in the mid-1980s that nearly destroyed the mangroves of the Sine-Saloum Delta. Reinforced by in-migration of people who harvested the mangroves for firewood, the breakdown of this natural desalinator threatened the rice paddies that had hitherto flourished in the salt-free lands near the lagoons and rivers. Like Popatroa Pawar in India and inspired by Ghandi, Demba Mansare returned from his studies to his home village, Faoune, during the 1980s drought to provide assistance. He established COLUFIFA, which is an acronym that translates from the French as Committee to Put an End to Hunger, that now works in hundreds of villages across West Africa. COLUFIFA coordinated an extensive programme of learning, skills development and institution building, all aimed at building new knowledge capabilities in communities forced to adapt to changing natural and social conditions. Permaculture design, reforestation, biogas production, agro-ecological farming and environmental conservation all formed part of a new learning repertoire driven by the brutal realities of survival in a rapidly changing region. The outcome was a knowledge commons that became the social capital that generated the funding for replication in the region.

Like the intentional initiatives, the participants in these transformed traditional villages found they had more to gain from working with nature in a collective way on alternatives than by adopting so-called modern building or farming technologies that simply replicate a known solution. The resulting ecocultural assemblages not only attracted publicity, but their learning was packaged into a knowledge commons and transmitted to others either via NGO partners and/or their own institutions.

Project initiatives

Although not as profoundly transformative as the initiatives referred to thus far, four rural project initiatives are worth referring to. In all cases, an external NGO played a role using traditional NGO strategies: awareness raising, skills training, technical advice, project management and financial support. The well-known limits to the transformative impact of an externally funded agency apply in these cases, in particular with respect to social empowerment and the limited nature of the knowledge commons that gets left behind. The primary beneficiaries of the learning processes in these arrangements tend to be the external agency that initiated the project in the first place.

- Situated in the middle of the Calakmul Biosphere Reserve, the Calakmul Rural Housing Programme in Mexico was initiated in 2004 by Échale a tu Casa. Over 1000 self-built homes were constructed for poor rural households using adobe bricks stabilized using a patented additive. Renewable energy helped reduce deforestation.[8]
- In the cold desert areas of the Western Himalayas, energy for heating is a major challenge. The French NGO Groupe Energies Renouvelables, Environnementet Solidarités (GERES) intervened to retrofit 550 houses by 2011 that demonstrated the advantages of solar gain, thermal mass and insulation. Whereas in winter indoor temperatures dropped to −10 degrees or below forcing families to live together in one room with an indoor stove causing respiratory diseases, average temperatures in the retrofitted homes do not drop below 5 degrees.[9]
- The Better Life Association for Comprehensive Development operates in the Minia governate, Egypt. Using a rights-based approach that champions the interests of the poor in villages located along the Nile, it initiated in 1997 the Local Housing Movement programme which works with local communities to improve and develop their housing, basic services, security of tenure, construction skills and training opportunities. Working with quarry workers, fishermen, low-income farmers and female-headed households in Minia, 400 new houses had been built and nearly 600 houses improved by 2010. The programme also made it possible for 5,900 households to gain access to potable drinking water and latrines in their homes.[10]
- The traditional methods of home building in the Sahel region of Africa (where 150 million people live) are no longer viable because the use of locally available timber is either illegal or too expensive or just not viable because the trees are gone. To avoid the alternative of using commercial timber and corrugated iron, the *Association la Voute Nubienne* (AVN) began a programme of training

masons to build vaulted earth brick houses (that do not require timber) using an ancient Nubian technique not previously used in the Sahel. Starting in Burkina Faso, the programme has spread to Mali, Senegal, Togo, Ivory Coast and Guinea.[11]

All 12 initiatives described in the previous sections demonstrate how rural and peri-urban communities managed to learn from processes initiated by a desire to work with nature in more restorative ways. Ecocultural assemblages of various degrees of sophistication emerged that were able to demonstrate the viability of settlement, farming and ecosystem management technologies that are usually overlooked as viable development strategies by development agencies and communities.

Green urbanism: minimizing damage

As argued elsewhere (Swilling, 2011), although green urbanism has its roots in a diverse range of movements (from the 'hippie' ecovillages of the 1960s to the planned solar towns of Western Europe, to pioneer cities like Curitiba, to UN Programmes to promote 'sustainable cities'), today it is the official ideology of a section of the property development industry that accepts the mainstreaming of sustainability and low-carbon consumption. The globally entrenched institutionally well-resourced Green Building movement is the most visible and influential expression of this movement. In contrast to the restorative mission of the cases considered in the previous section, the overall mission of the green urbanism movement is to 'minimize environmental damage'. Many countries now have regulatory frameworks that in some way enforce aspects of the 'green building' codes, including some in the global South.

To 'minimize damage', green urbanism is usually expressed in large-scale top-down technocratic interventions driven by either states or developers, or in some cases public-private partnerships. The three cases described here illustrate these variations: Songdo (South Korea) and the Lagos Bus Rapid Transit (LBRT) are public-private partnerships, whereas Bangalore's green gated communities are developer driven. The key question, of course, is whether these can be defined as ecocultural assemblages. They do fundamentally redefine the relationship between resource use and consumption, making it possible for richer people to believe that it is possible to continue to consume but without destroying the planet. What is distinctive is that this is achieved by technocratically reconfiguring the infrastructures that conduct resource flows through the urban system with consumers defined in market terms as individuals willing to 'buy into' the system. To this extent, they are market-driven urban ecocultural assemblages delivered through technocratic interventions that break from traditional modes of urban consumption such as dependence on the private car, fossil fuels, unrecycled wastes and inefficient toxic buildings. What the Asian cases lack, of course, is a commitment to social justice. But this cannot be blamed purely on the fact that they are market-based private sector initiatives. Indeed, this rendition of them as purely private sector is also problematic: Songdo, for example, received public sector support of various kinds. The LBRT clearly has had a major impact from a social justice perspective – millions of Lagosians have had the benefit of access to affordable mass transition.

Songdo is a greenfield development in South Korea that aimed to create a user-friendly International Business District (IBD) that co-locates global businesses with large Asian markets in a free trade aerotropolis that is both 'green designed' and coordinated by the most advanced smart grid technologies available (Kuecker, 2013; see Songdo case study in Appendix to Swilling, Robinson, Marvin and Hodson, 2013; Baek, 2015).[12] It is located on 1,500 acres of land reclaimed from the Yellow Sea, near Incheon International Airport. Developed by a public-private partnership between Gale International and Korea's POSCO E&C and designed by Kohn Pedersen Fox, the 9.3 million m^2 master plan includes commercial office space, residences, retail shops, hotels as well as civic and cultural facilities. When fully developed by the original target date of 2015, this new city was planned for 80,000 apartments, 4.6 million m^2 of office space and 0.93 million m^2 of retail space. The initial estimates of the total investment stood at US$35 billion. By 2012, over 100 buildings were complete. According to Statistics Korea, Songdo's population in August 2014 was 79,395.

Although constantly referred to in mainstream fora as a green development model, the extensive investments in energy efficiency were only intended to reduce consumption by 14% compared to conventional developments of a similar size. Surprisingly, solar energy is not widely deployed. Water consumption was estimated to be 20% less and sourced from a desalination plant.

The real innovations in Songdo are in mobility: all major facilities were planned to be no more than 12.5 minutes-walk apart, cycle paths connect everything, a pool car system was formally institutionalized (with cars parked in underground garages as far as possible) and hydrogen-powered buses. In reality, private car use remains. A high percentage of waste is recycled, and guidelines for non-toxic building materials were prescribed (although information on adherence is not available). Of the total area, 40% is reserved for parks and waterways. Added together, these infrastructure innovations mean that although daily urban living in Songdo will not be very different to most conventionally planned developed country cities, resource flows were ideally supposed to be very different. Given the vast number of new cities or city extensions sprouting up all over Asia (especially China and India) and elsewhere in the developing world, the Songdo approach is a step in the right direction.

In practice, however, the actual intended resource efficiencies of the Songdo model did not really materialize because the target market was high-end residential and business markets. The result is that higher than average incomes negated the intended resource efficiencies, in particular, with respect to water and energy. Compared to average consumptions levels per capita in Seoul, Songdo residents use more energy and water per capita, and their output of sewage and waste is only slightly less (see Table 6.1).

To this extent, although it may not be fully realized in practice, the design imagination behind Songdo certainly has in mind a new urban ecocultural assemblage

TABLE 6.1 Summary of material consumption in South Korean cities

City	Year	Water	Sewage	Electricity	Waste	LNG	CO_2
		kilolitre/year	kilolitre/year	kWh/year	ton/year	m³/year	ton/year
Seoul	2011	110.49	176	4,455	0.327	0.396	Not available
	2012	110.42	181	4,523	Not available	0.407	Not available
Incheon	2011	123.26	Not available	7,769	0.279	0.356	Not available
	2012	Not available	117	7,813	0.261	0.336	Not available
Songdo	2013	146.55	131	14,179	0.110	0.396	Not available

Source: Baek, 2015:89

targeted mainly at middle and higher income earners. To put it crudely, Songdo is a fusion of the artificiality of Dubai and the green ambitions of the design glitterati which must compete to deliver the 'largest and greenest' to global elites who have developed a new desire for living green low-carbon lives. The fact that design can do little to affect income levels in a new development suggests that innovative designs must be complemented by measures that ensure that developments do not cater exclusively for the rich and that inequalities are reduced. The dominance of private sector actors meant that a knowledge commons was not built.

By contrast, the Lagos BRT is a technocratic intervention by a public-private partnership in one of the most congested and poorly planned cities in the world (derived from Case Study Annex of Swilling, Robinson, Marvin and Hodson, 2013). Study tours by key officials and representatives of the bus driver union to cities in Latin America with well-developed BRT systems (Bogota, Santiago and Curitiba) helped to build a shared locally rooted understanding and vision for the project. Unlike Songdo, it does not only cater for middle and upper income groups. Starting February 2007, 22 km of BRT lanes were constructed in 15 months at a cost of $1.7 m/km, just less than a third of the average cost for such systems elsewhere in the world. Funding came from a $100 million loan from the World Bank and a $35 million grant from government. Key to success was an extensive stakeholder engagement and public awareness programme that sold the concept as an indigenous Lagosian initiative and not another external venture that would be hijacked by local bureaucrats. These consultations included intensive negotiations with the bus driver union to allay fears that their members would be excluded from the new job opportunities because they were not skilled enough. By the end of its first year of operation, 195,000 passengers were using the buses on a daily basis. Journey times for those who switched to the BRT were halved. Despite problems with maintenance and operations that has reduced the number of buses in service, the Lagos BRT is widely regarded as an African success story that clearly demonstrates how an infrastructure investment can change urban consumer behaviour in ways that benefit some of the poorer members of society. The lack of attention to maintenance would suggest that internal learning has been limited. However, there is insufficient information available on how internal learning worked within this project.

Bangalore's industrial manufacturing base has traditionally been the primary driver of this rapidly expanding city's economic growth. However, over the last two decades IT and outsourcing have become equally, if not more, significant. This, in turn, has attracted a new cohort of skilled wealthy professionals and entrepreneurs in search of high-end accommodation. Property developers responded by building numerous gated communities using traditional resource and energy-intensive designs. This, however, posed a problem for institutionally weak government authorities who cannot deliver the energy, waste, water and sanitation infrastructures required by these wealthy communities. Property developers could not secure building approvals.

Spotting the opportunity, a company called Biodiversity Conservation India Limited (BCIL) positioned itself as a company that can deliver 'green living' to those who can afford it. Using all the green technologies pioneered by generations of true believers in a green techno-fix solution, BCIL started building 91 self-sufficient villas on the edge of Bangalore in 2003 and named it *Towards Zero Carbon Development (or the T-Zed development)*. Completed in 2007, BCIL has continued to initiate similar T-Zed developments in various parts of Bangalore. The T-Zed developments are intended to be self-sufficient in water by using rainwater harvesting systems and boreholes. The use of biogas digesters, solar energy, green roofs and passive cooling systems reduced the requirements for externally sourced electricity. Waste recycling, composting and the growing of organic food are also included, as are requirements to use building materials with a low embodied energy.

BCIL is now a growing commercially successful property development business that serves as a model for other businesses. This growth is a response to rising demand for green housing in low-density gated communities on the edges of the city.

Ironically, what BCIL realized is that green design enables the rich to secede from the city into self-sufficient 'ecocultures for the rich', thus exacerbating sprawl and fragmentation – the main factors that caused the problem in the first place. A positive spin-off, though, is the growth in the number of design professionals who understand ecological design and in the number of contractors who can build in this way. There is no reason why these capabilities cannot now be deployed within a CBPP-type platform for the purposes of building higher-density socially mixed and mixed-use inner-city brownfields developments. By contributing their learning to a knowledge commons where improvements can be made, each of them benefits which, in turn, improves their individually owned businesses.

Songdo, the Lagos BRT and Bangalore's green gated communities are representative examples of the kinds of green urbanism interventions that break with some of the conventional urban design assumptions underlying the patterns of resource and energy-intensive urban development that many regard as the desired norm in industrializing countries in the global South. If the cities of the global South are to become more sustainable in a predominantly urban world, learning capabilities will need to be developed that could be deployed to design niche innovations that over time coalesce into more meaningful urban ecocultural assemblages that can also be more socially inclusive. A CBPP-type platform would greatly enhance these capabilities.

All three interventions rapidly built new sustainability-oriented learning capabilities, but only interventions like the Lagos BRT have the potential to foster socially inclusive ecocultural consumption behaviours. There will, however, undoubtedly be more interventions in future across the cities of the global South like Songdo and Bangalore's green gated communities which will generate 'ecocultures for the rich'. The danger is that this generates a political backlash against green design as an 'elite solution' rather than just the way it is applied by opportunistic property developers.

Towards liveable urbanism

Liveable urbanism refers to profoundly urban ecocultural assemblages that combine equity and ecological restoration. Most of these initiatives tend to emphasize the needs of the urban poor while recognizing that working with rather than against nature provides the most effective way to deliver affordable liveable urban environments. To illustrate this argument, two categories of initiatives will be reviewed. The first can be referred to as urban struggles by organized formations to either resist interventions that threaten the environmental quality of their lives, or they are struggles to secure access to key natural resources that urban living depends on. The second refers to project-based initiatives by urban communities, often supported by an NGO. Here the focus is more on the technologies, learning capabilities and modes of social organization best suited for building ecocultural assemblages in complex and challenging urban environments.

Urban struggles

Urban struggles are usually about public policies or plans. Either these policies/ plans are resisted by a particular set of actors (before and/or after inception) or they are advocated as alternatives to the existing (lack of) policies/plans.

One of the most dramatic and sustained struggles over urban environmental resources in a developing country city took place around the Billings Dam in Sao Paulo, Brazil (case derived from Keck, 2002). Initially built as a water reservoir in the 1920s (and named after the engineer who designed it), it was eventually surrounded by elite housing over the next four decades. However, by the 1970s, it started to take delivery of the sewage of a rapidly expanding city that lacked a sanitation plan. This prompted opposition by the elite neighbourhoods. Despite well-organized opposition and court action, the Billings Defense Committee that took on the authorities had failed by the early 1980s to halt plans to use the dam for managing sewage. It was replaced after democratization in the 1980s by the Movement in Defense of Life (MDV) led by a former trade unionist and Workers Party member who represented working class communities that had, by then, moved into the area as property prices went into decline as the middle class moved away from a sewage-filled dam.

However, the heavy emphasis on privatization of public utilities during the first years of the post-militaristic democratic era that began in the mid-1980s meant there was no interest in committing scarce funds to resolve what was defined as an environmental dispute articulated by left-wingers. By the 1990s, land invasions and illegal property developments exacerbated the pressures on the water basin and the Billings Dam got more polluted than ever. After the electoral victory of the Workers Party in the 1990s, new participatory structures were established (Basin Committees) to co-opt civil society formations into water planning processes. A new generation of environmental movements emerged to take advantage of these opportunities, but now they were pitted against housing and land movements championing the interests of the urban poor who had by this time occupied areas that the environmental restoration of the Billings Dam would depend on.

The struggles over the future of the Billings Dam over many decades have not resulted in an adequate solution. The complex dynamics continue. However, what is significant about this story is how a natural resource became the focus of collective action by very different groups, initially middle class, later on the urban poor, and eventually a mixed alliance of environmentalists. The institutionalization of the Basin Committee to facilitate joint planning was significant, but did not result in an implementable solution. Nevertheless, a negotiated plan did emerge that has become the de facto knowledge commons of the joint participatory planning process. Government policies get measured against this plan.

Slum Dwellers International (SDI)[13] is an international confederation of largely women-led urban slum dweller movements in over 40 countries in the global South. Originating in Mumbai and Cape Town, this global urban social movement has managed to craft an approach that transforms urban struggles into positive programmes of state-supported community-driven upgrading that demonstrates that slum dwellers can be agents of their own development. A particularly well-developed example is the Odisha Alliance that is active in the Indian state of Odisha. The partnership involves a support NGO called Urban Development Resource Centre (UDRC), the grassroots women's organization Mahila Milan, the Odisha/National Slum Dwellers' Federation (O/NSDF) and another support NGO called the Society for the Promotion of Area Resources Centre (Sparc). This alliance is active in 225 settlements in five cities in Odisha and in three cities in the state of West Bengal. By combining community savings, mutual learning between organized groups and construction of model houses that are affordable and adapted to local needs, a unique power base was created for negotiating with state actors who often lack the bureaucratic capacity to implement anything in informal settlements. Sixty model houses were built and two state-supported programmes were underway by 2012, with 400 houses under construction. Because the legitimation of the designs stemmed from existing informal settlements and not the state, designs for incremental upgrading were accepted that did not need to adhere to conventional architectural or engineering standards. To save costs by sourcing local building materials and managing sewage at neighbourhood level, the result was far more sustainable than what a conventional urban development process would have delivered.[14]

SDI has a particularly well-developed and sophisticated strategy for building the knowledge commons for what is effectively a global social movement of women living in informal settlements. Firstly, they believe that poor people learn best from other poor people experiencing the same conditions. Hence, they invest in peer-to-peer learning via organized visits of groups from one settlement to another, often in another country. Secondly, they establish savings groups in poor communities, with a preference for collecting daily savings as a way of renewing relations on a daily basis. Thirdly, they believe in building up boundary objects within communities to catalyse expectations and hope. These can be proto-types of houses, sanitation blocks or whatever. Finally, each country has an NGO that employs professional staff for managing training, knowledge collation and dissemination and fund raising. At the global level, there is a global leadership group and fund holding instruments. It is easy to see how this organizational configuration could be converted into an ICT-enabled CBPP-type structure. Almost all the pieces, especially the documentation of innovations, are in place.

A good example of a similar development was the construction of adobe houses and eco-san toilets in Lilongwe, Malawi, by a particularly well-organized SDI affiliate.[15] After waging a campaign to secure land for housing, a well-organized movement of backyard shack dwellers eager to escape rising rentals secured land from the Municipality on condition no additional funding or services would be requested by the occupiers of the land. Traditionally, Malawians have built houses made from clay bricks baked using charcoal. However, with deforestation caused by rapid urbanization, charcoal had become too expensive as the sources of supply became increasingly remote which increased the cost of transporting the material into the urban centres. After visiting the Lynedoch EcoVillage in 2010 to learn about the alternative of using unfired adobe bricks, the women's groups – supported by a local NGO – built 800 adobe brick houses. This was later on supplemented by urine-diversion toilets managed by local entrepreneurs who generated an income from users and from the sale of composted wastes.

Since 1980, the Orangi Pilot Project (OPP) has been operating in Karachi's informal settlements (known as Katchi Abadis), starting in Orangi Town. Although involved in a wide range of developmental initiatives, the OPP is most well known for its successful campaigns to secure government support for its Low Cost Sanitation Program. This program combines social organizing, savings, capacity building, sophisticated technical design and campaigns to influence government to create low-cost sanitation systems that work well in Katchi Abadis. In essence, this involves convincing government to finance the installation of primary bulk sewer lines, with communities funding, building and maintaining the secondary connections at neighbourhood level. This has proved so successful that it led to the cancellation in 1999 of a US$100 million Asian Development Bank loan for conventional sanitation systems. The OPP's Low Cost Sanitation Program has now extended to service all of Orangi town, resulting in significant health and environmental improvements in the neighbourhoods it has affected. Over 105,670 poor households invested Rs. 118.7 million in secondary, lane sewers and sanitary latrines, and government has invested Rs. 807.2 million on main disposal lines. The model has been replicated by 421 other settlements in Karachi, along with 32 cities/towns and 93 villages covering a population of more than 2 million (Hasan, 2006; Annexure to Swilling, Robinson, Marvin and Hodson 2013).

Urban projects

Although most urban projects involve some sort of organized engagement (and even prolonged struggles) with state actors, they do not necessarily require a change in policies/plans in order to be implemented. They are usually opportunity-driven initiatives that emerge within unique relatively controlled, bought or captured spaces either as a result of the efforts of an organized community, intervention by an outside agency or actions of a leadership group of some sort.

Ecoovila is an ecovillage in the Brazilian city of Porto Alegre (case derived from Dawson, 2006–2010:32–36). Started in 2001 by professionals influenced by the permaculture approach, the aim of Ecoovila is to provide an affordable, socially inclusive and eco-friendly alternative to mainstream Brazilian architecture and urban design. The aim was to build homes for 28 families on a 2.6 hectare plot in the centre of the city. After prolonged negotiations with municipal officials, by 2006, 20 homes had been built. All the eco-design features were present: passive solar gain, central fireplace, solar hot water heaters, passive cooling systems (including grass roofs), use of local building materials (cob, bamboo and adobe bricks), organic vegetable gardening and an on-site biological sewage treatment plant that produces an effluent that can be used for irrigation. The overriding aim is to build a vibrant community. Unsurprisingly, like similar places all over the world, Ecoovila has received a lot of publicity and it receives many curious visitors, some of whom are inspired to initiate their own projects.

Located in the crime-ridden Colombian city of Cali, Ecobarrio defines itself as the "first ecovillage in Latin America". It comprises a total of 270 mainly self-built homes inhabited mainly by poor urban families plus recreational facilities and community gardens. Started by the Federación Nacional de Vivienda Popular (FENAVIP) in the early 2000s and funded with government subsidies for the poor, Ecobarrio now includes individual and collective vegetable gardens and community service facilities such as a community centre, drugstore, restaurant and shops. Waste materials were used to make the cement-based building materials. Two large areas, each measuring approximately 1,200 m², have been set aside for the creation of an 'active' recreational park with sports facilities and a 'passive' park for leisure activities. Ecobarro also includes a 'Native Germoplasm Bank' for the cultivation of 12 endangered species of native fruit trees, an integrated system for the management of solid waste, organic agricultural production cooperatives as part of a regional food security project and cultural programmes for young people. There is a strong emphasis on income generation through 'work cooperatives', community self-management, conflict resolution and participation, including a savings scheme introduced by FENAVIP.[16]

The Dajopen Waste Management group is a cooperative that was started in 2008 by 30 people (20 women, 10 men) living in an informal settlement in Kitale, Kenya. They each contributed start-up capital from their meagre savings to set up and register the cooperative. Their aim was to generate incomes for themselves by cleaning up the environment through recycling, including composting of organic wastes for use as inputs into organic food production. DWM members collect waste locally or buy it from street families. The products they make for sale from these recycled materials include roofing and floor tiles made out of recycled plastic; baskets, mats and ropes made from plastic bags; jewellery and briquettes made from paper waste; water filters made from saw dust mixed with clay; fencing posts; and organic fertilizers, liquid fertilizers and biocides made from bio-degradable waste materials. They also conduct training courses in waste management and organic farming for Government agencies, municipalities and NGOs for which they get paid a fee. In order to cover the running costs of the project, each member contributes 30% of their sales to the group's collective reserve fund and retains the remaining 70%. At the end of the year, approximately 20% of this collective fund is distributed amongst members and 80% is reinvested[17]; in other words, a classic CBPP prototype.

Quite often the most significant way to improve the lives of the urban poor is to ensure that they remain in close proximity to the services and employment opportunities that are usually concentrated in the inner city. In many cities, the urban poor are concentrated on the peripheries which means only fundamental devaluations of the inner city property markets will result in the inclusion of the urban poor. Without state intervention or a massive disaster (financial, military, natural), this outcome is unlikely. As oil prices rise, the increasing cost of transportation to the city further disconnects the urban poor from the local economy. However, there are some cities where the urban poor are concentrated in the dilapidated historic inner cores (originally built for the wealthy in the nineteenth and early twentieth centuries who later fled to the suburbs when this became trendy and profitable) but increasingly threatened with evictions caused by gentrification processes as the wealthy return to the inner cities.

Santos, Brazil, is a good example of this phenomenon. A social movement led by the Associação Cortiços do Centro, Condomínio Vanguarda (Association of Tenements in Central Areas – ACC) represents the residents of the so-called cortiços or tenement-style slums. Under a 'right to the city'/'right to housing' banner, the ACC has successfully championed the rights of the urban poor to remain in the inner city. In 2007, after spending time learning from the successful experiences of other grassroots organizations and examining legislation and potential funding streams for housing with the assistance of a group of volunteer architects, the ACC was able to obtain a 6,000 m^2 plot of land in the Santos city centre from the state and have it classified as a 'Special Zone for Social Interest' (ZEIS), enabling access to funding. The following year, funding was approved for the construction of the first 113 housing units and additional funding was secured in 2010 for the second phase of the project, involving the construction of an additional 68 units through a system of mutual aid and self-management of resources. Technical assistance was obtained from NGOs and volunteers. A number of environmental features were incorporated into the project, including rainwater harvesting, waste recycling, use of recycled building materials and the use of solar energy. The ACC approach defines the project as a good example of a collective participatory approach.[18]

At the outset of this section, it was argued that liveable urbanisms were particular ecocultural assemblages that valued both equity and ecology. The first four cases illustrated how this was expressed through a wide range of struggles over public policies and plans that affected everyday urban living. In Sao Paulo, many decades of struggle to protect the city's key water resources reflected competing visions for how to embed urban infrastructures within a city-region's ecosystems – an ecological issue with profound implications for equity. The remarkable stories from Odisha, Lilongwe and Karachi all suggest that when organized, the urban poor can become the agents of their own development in ways that can win state support, especially in environments where the state lacks the capacity for effective delivery on its own. This kind of bottom-up collibration makes possible emergent outcomes that are often socio-technical innovations that are not only more inclusive of the urban poor but if replicated can also potentially shape the evolutionary future of informal settlements in ways that break quite fundamentally from conventionally designed resource and energy-intensive urban development trajectories. This kind of replication, however, will depend on an institutionalized ICT-enabled CBPP formation that is not simply a depository of learnings but an active platform for constantly adding and upgrading the innovations.

Although all four *urban projects* shared a commitment to equity and ecological restoration, Ecoovila remains an example of a middle-class urban ecoculture, and the ACC initiative is primarily about social inclusion by empowering the inner city urban poor. While Ecobarra was almost entirely dependent on state funds and DWM was not, both shared a profound commitment to building ecocultures that benefitted the urban poor – one by building a state-subsidized urban ecovillage, the other by promoting micro-enterprises using recycling waste materials.

Some lessons for builders of future ecocultures

The three clusters of initiatives reviewed have continuously evolved through intentional and unintended learning conditioned in part by the ideals of the participants and in part by their engagements with the complexities of their respective contexts. As the inevitable frictions between ideals and contexts instigate many small and some large compromises, each follows a unique context-specific trajectory with significant lessons for those who in future will face the challenge of purposive design of socially just and more equitable human settlements. None of these cases are likely to arrive at a point where they can be depicted as exemplifying a particular set of preconceived characteristics that will make it possible to define them as 'ecocultures'. Nevertheless, using the five 'trace elements' of an ecocultural sensibility discussed at the outset of this chapter, it is possible to conclude this chapter by drawing out some generic lessons that may usefully inform the coming debates about the future prospects for building up ecocultures from below.

Socio-technical innovations

Evidence of socio-technical innovations is undoubtedly the most visible and immediate outcome of all these initiatives. Motivated by an awareness of the need to mitigate emissions and/or restore biodiversity and/or minimize resource repletion, key actors in all cases initiated innovations that addressed these ecological challenges. From the more rural cases where soils were restored to the use of appropriate technologies to benefit poorer communities, to the hi-tech solutions offered in Songdo and Bangalore, what is clear is that experimentation with a wide range of technologies is taking place across many different contexts.

As more is learnt from these experiments, so improvements will be made and applications on scale will become more likely. Diffusion from centres of (often unrelated) niche innovations is how new technologies in the past have spread until they become the new dominant approach. As the social and economic costs of unsustainable resource use mount, so more niche innovations are likely to spread. Many more niche innovations like the ones discussed have already emerged, but they must still coalesce into alternative socio-technical regimes.

Finally, and most significantly of all, in most cases there is evidence of a knowledge commons created from the voluntary contributions of the participants. In some cases, this is the lasting enduring legacy that, in turn, creates the basis for ongoing impact beyond the local project. The extent to which this knowledge commons in each case has been digitally enabled in the form of a CBPP-type configuration is unknown. However, it is pretty clear how this could happen.

Adaptive leadership

As far as the existence of adaptive leadership capabilities is concerned, from the evidence available it seems like leadership through partnerships is a common theme across all these cases. This, of course, is unsurprising. Innovation often means that the required knowledge and capacity required to bring about an innovation rarely exists entirely within a single institution, especially if a particular disciplinary set (e.g. engineering) is dominant. Facilitating inter-institutional and interdisciplinary cooperation is notoriously challenging which inevitably creates an opportunity space for those with the capabilities to do this kind of work. The traditional alpha male-type leader is not appropriate for this kind of leadership. The key leadership capability relates to being able to distinguish between technical and adaptive challenges: the former can be dealt with by replicating tried and tested usually technocratic approaches; the latter requires the empowerment of the relevant affected and/or involved actors to figure out solutions via discursive often women-led processes (Heifetz, 1995).The difficulties are greater in environments that involve poor people with limited education who may be suspicious of professionals or educated leaders.

Trust-building, no matter the context, is often key because trust allows for a diversity of inputs and greater defence of the end result. Anecdotal and some documented evidence from some of the cases suggest that trust ebbs and flows over time as the project expands, leaders move on, new institutions get built, new actors enter the fray and the wider policy and market conditions change. What matters is the capacity to manage this dynamic.

Institutional arrangements

The diverse institutional arrangements evident across all the cases makes it impossible to claim that one or other form of private, public or civil society-based mode of organization is most suitable for promoting ecocultural innovations, especially within the urban context. Nor is it as simple as saying that private sector organizations only drive more up-market developments (like Songdo and Bangalore's green gated communities) – they also played key roles in the Lagos BRT, Sekem and even the Dajopeng initiative. In all these cases, however, the modes of collibratory governance involving state agencies, private sector organizations and NGOs (as discussed in Chapter 2) were necessary. The success of the Lagos BRT, for example, was because so much effort was made securing support from trade unions, different state institutions and communities. The LBRT is probably one of Africa's most successful examples of collibratory urban governance.

The key role NGOs play as facilitators of niche innovations and collibration is reflected in the majority of cases reviewed. This is often because they can access grant funding or secure policy protection – both of which create spaces that are relatively protected from market pressures and political interference for a period of time – a key condition for allowing innovations to mature (Seyfang and Smith, 2007; Smith and Raven, 2012). However, in quite a few cases the NGOs involved were either international NGOS or local NGOs who had accessed international donor aid funding (e.g. Western Himalayas, Odisha and Lilongwe) or else they were community-based nonprofits mixing together local resources and global funds (e.g. Auroville, Hivre Bazar).

It is noticeable that state agencies rarely initiate and lead ecocultural innovations. However, they can play a key enabling and supportive role. In the Da Ping, Lagos BRT and Orangi Pilot Project cases, the state has played a critical role in the enabling and funding of the projects. The same was true of the Lynedoch EcoVillage (see Chapter 1). The changing nature of the role of the state as political dynamics change is clearly reflected in the Billings Dam case. But the capacity for initiating innovations in all cases lay outside the state.

Behaviour and values

Although very little information is available from the cases reviewed about behaviour and values, it is significant that ecological restoration of some kind is common to initiatives that focus on the needs of richer communities (e.g. Songdo, Banga-lore's green gated communities and Ecoovila) and the needs of poorer communi-ties (e.g. Lilongwe, Odisha, Ecobarrio). Some, of course, are quite mixed, such as Auroville, Gaviotas, Picaranga, Da Ping and the Lynedoch EcoVillage. It therefore follows that it is not possible to assume that the building of ecocultures is only of interest to – and affordable by – the middle class; or alternatively, that an ecocul-tural commitment is only valid if it focuses on the needs of the poor. What is very clear, however, is that there is effectively no room for traditional urban consumer cultures that combine a detachment from nature with conspicuous individualized consumption and high levels of waste. Even in Songdo, elite green consumption is coupled to some form of reconnection to nature via green landscaping and resource efficiency.

However, admittedly, Songo cannot in any way be regarded as a model for the rest of the world for the simple reason that the level of resource consumption per capita in Songdo (and possibly also Bangalore's green gated communities) remains unsustainable from a global perspective. By contrast, the struggle by Brazilians for the Billings Dam was clearly at certain times a multi-class struggle not only to preserve a key natural resource but also to help shape a set of ecocultural identities. While ecocultural behaviours and values could possibly emerge from co-location within designed 'green urbanism' developments, it should not be underestimated how collective struggles play a key role in shaping such identities (especially if they are multi-class identities).

Intolerance of Poverty

Finally, a wide range and diversity of ecocultural assemblages explicitly focus on the needs of the poor. This suggests that social movements and anti-poverty pro-grammes may have much to learn from cases like the ones reviewed here. Signifi-cantly, these cases reveal that ecologically designed solutions may not only produce a more affordable better quality outcome, they also seem to offer communities an opportunity to become better organized in order to realize the full benefits of, for example, a locally provided waste recycling or a shared renewable energy system.

Conclusion

The primary aim of this chapter was to demonstrate that there are a wide range of emergent ecocultural assemblages within and outside the city-regions of the global South. Some of the more mature initiatives are located outside formal city bounda-ries where they *resist disconnections* from nature driven by the agro-industrialization of the countryside and the biomass demands of the expanding city-regions. *Green urbanism* initiatives are top-down technocratic interventions that reconfigure urban infrastructures and the design of the built environment to achieve more sustainable outcomes. The initiatives described as examples of *liveable urbanism* are profoundly urban responses, most of them directly concerned with the livelihoods and well-being of the urban poor.

All the cases reviewed can be framed as ecocultural prototypes that anticipate a CBPP mode of production and consumption. To this extent, they – and thousands like them – could contribute to the realization of the next deep transition. All the cases respond to the resource limits of the industrial socio-metabolic regime; they can be understood as niches within the framework of socio-technical transition; they prefigure a new mode of consumption that could drive a sixth wave of technological change.

These cases confirm the validity of incrementalism as a force for change, especially if local experimentation is coupled to the process of building imaginaries of potential futures. In every case there was a core leadership grouping that held the vision and created a safe environment for experimentation.

Notes

1 Arguments and ideas for this chapter have been drawn from Swilling, M. (2015) 'Ecocultural assemblages in the urbanizing global south', in Pretty, J., Boehme, S. and Bharucha, Z. *Ecocultures: Blueprints for Sustainable Communities*. New York: Routledge, pp. 218–238.

2 Sources of information about the cases include personal observation of some cases, discussions with others who have experienced the cases, interviews with key leadership individuals, websites, project reports on www.worldhabitatawards.org and commissioned case study research prepared for a report on City-Level Decoupling by the Cities Working Group of the International Resource Panel (see Swilling, Robinson, Marvin and Hodson 2013).

3 This account is based on two visits in 2013 and 2014 plus informal discussions with the founders.

4 Sources: Personal communications with a visitor plus website http://directory.ic.org/22 097/Piracanga_EcoCommunity

5 Personal interview with a visitor plus (Kulkarni, 2009).

6 Professor Liu Jiaping, Green Building Research Centre, liujiaping@xauat.edu.cn

7 www.vastushilpa.org

8 www.echale.com.mx

9 www.geres.eu

10 www.blacd.org

11 www.lavoutenubienne.org

12 I visited Songdo in 2015.

13 www.sdinet.org

14 Personal interviews with various participants plus www.udrcalliances.org

15 Personal visits and interviews with various participants in 2010 in Lilongwe, followed by the coordination of an exchange programme for local leaders to the Lynedoch EcoVillage, and an architectural support team who spent time in Lilongwe advising family builders.

16 Case derived from www.fenavipvalle.com and general discussions with visitors to the project.

17 Case derived from www.worldhabitatawards.org plus correspondence with researchers familiar with the project at Kenyatta University, Nairobi.

18 http://forumcorticos.blogspot.co.uk/

References

Baek, I. (2015) *A Study on the Sustainable Infrastructure of the Songdo City Project: From the Viewpoint of the Metabolic Flow Perspective.*

Dawson, J. (2006) *Ecovillages: New Frontiers for Sustainability*. Totnes, UK: Green Books.

Escobar, A. (2015) 'Degrowth, postdevelopment, and transitions: A preliminary conversation', *Sustainability Science*, 10(3), pp. 451–462. doi: 10.1007/s11625-015-0297-5.

Hasan, A. (2006) 'Orangi Pilot Project: The expansion of work beyond Orangi and the mapping of informal settlements and infrastructure', *Environment and Urbanization*, 18(2), pp. 451–480.

Heifetz, R. (1995) *Leadership Without Easy Answers*. London: Belknap Press.

Keck, M. E. (2002) '"Water, water, everywhere, nor any drop to drink": Land use and water policy in Sao Paulo, Brazil', in Evans, P. (ed.) *Livable Cities? Urban Struggles for Livelihood and Sustainability*. Berkeley, London and Los Angeles: University of California Press, pp. 162–195.

Kuecker, G. D. (2013) 'Building the bridge to the future: New Songdo city from a critical urbanism perspective', in *SOAS University of London Centre of Korean Studies Workshop on New Songdo City and South Korea's Green Economy: An Uncertain Future*, London.

Kulkarni, S. (2009) 'An ideal village, an inspiring leader', Sun, 12 July. Available at: http://archive.indianexpress.com/news/an-ideal-village-an-inspiring-leader/488234/.

McFarlane, C. (2011) *Learning the City: Knowledge and Translocal Assemblage*. Oxford: Wiley Blackwell.

Polanyi, K. (1946) *Origins of Our Time: The Great Transformation*. London: Victor Gollancz Ltd.

Pretty, J., Boehme, S. and Bharucha, Z. (2015) *Ecocultures: Blueprints for Sustainable Communities*. New York: Routledge.

Seyfang, G. and Smith, A. (2007) 'Grassroots innovations for sustainable development: Towards a new research and policy agenda', *Environmental Politics*, 16(4), pp. 583–584.

Smith, A. and Raven, R. (2012) 'What is protective space? Reconsidering niches in transitions to sustainability', *Research Policy*, 41, pp. 1025–1036.

Swilling, M. (2011) 'Reconceptualising urbanism, ecology and networked infrastructures', *Social Dynamics*, 37(1), pp. 78–95.

Swilling, M. and Annecke, E. (2012) *Just Transitions: Explorations of Sustainability in an Unfair World*. Tokyo: United Nations University Press.

Swilling, M., Robinson, B., Marvin, S. and Hodson, M. (2013) *City-Level Decoupling: Urban Resource Flows and the Governance of Infrastructure Transitions*. Available at: www.unep.org/resourcepanel/Publications/AreasofAssessment/Cities/tabid/106447/Default.aspx.

Weisman, A. (2008) *Gaviotas: A Village to Reinvent the World*. Burlington, VT: Chelsea Green.

PART III

Making and resisting sustainability transitions

7

DEVELOPMENTAL STATES AND SUSTAINABILITY TRANSITIONS

Introduction

Since the onset of the global economic crisis in 2007/2008, two key trends converged in ways that require a new discussion about the connection between 'development' and 'sustainability transitions'(STs): the rise of the so-called BRICS-plus economies as most of the traditional OECD economies plunged into a prolonged depressive malaise (Bogdan, Hurduzeu, Josan and Vlasceanu, 2011; Van Agtmael, 2012; Pant, 2013), and the emergence of a global narrative that started with the 'green economy' discourse in 2009 (Geels, 2013; Death, 2014; Swilling, Musango and Wakeford, 2016) followed soon after by the adoption of the Sustainable Development Goals (SDGs) in 2015. The primary implication of the convergence of these trends over the period of nearly a decade is the need to rethink the relationship between development and ST. Although Scoones et al. make a significant contribution in this regard (2015), the focus in this chapter is not on the development-sustainability nexus in general. Instead, the focus is more on sustainability and 'developmental states' (DSs) – the latter usually regarded as having a historic mission to accelerate the development and modernization processes in the spirit of 'catch up' (Evans, 2010).

Building on earlier work with similar aims (see Swilling and Annecke, 2012: Chapter 5), this chapter argues that we need to draw on the well-established literatures on the DS and STs in order to better understand the challenge of combining development strategies and commitments to ecological sustainability that many states in the global South now face since the adoption of the SDGs.

Chapter 2 argued that there is a difference between a deep transition and a just transition. A deep transition is shaped by the asynchronous interaction between socio-metabolic, socio-technical, techno-economic and long-term development cycles. Whether this is also a *just transition*, I suggested, will depend on the struggles over the *directionality* of the deep transition. These struggles for political representation and voice will condense in – and refract through – the polity and thus shape policy choices. The stronger the organizational capacities and voices of the poor/marginalized (whoever that might be in each context) expressed both directly (via trade unions, social movements, cooperatives, associations, alternative communities, ecocultures, etc.) and indirectly (via NGOS, intellectuals, parties, media, etc.), the more their interests will be represented within the polity. And this, in turn, will affect the policies that get adopted and implemented by state apparatuses and non-state actors. As argued later, the adoption of positive policies without a reconfiguration of power relations within the polity will not result in a just transition.

As a step towards defining the goal of a just transition, a contemporary definition of development by Castells and Himanen may be useful:

> Development . . . is the self-defined social process by which humans enhance their well-being and assert their dignity while creating the structural conditions for the sustainability of the process of development itself.
>
> *(Castells and Himanen, 2014:29)*

This definition is useful because well-being rather than GDP per capita is at the centre. Following Sen (1999), this definition of development is not derived from an abstract categorical imperative but from the everyday processes of 'self-definition' via dialogical engagement, which are of course context-specific. However, these acts of becoming are inseparable from the wider process of structural change to ensure the longer-term sustainability of development processes. But the former is not conditioned by the latter – instead, the latter is the emergent outcome of the continuous struggles over the terms of the development process itself. Of course, this is not how 'Development' (with a big 'D') is usually officially defined in mainstream narratives (Nederveen-Pieterse, 2000). Nevertheless, this is, most certainly, 'development' (with a small 'd') as a process of mutual flourishing within communities of human and non-human beings. For some, this takes us into what is referred to in the Latin American literature as "postdevelopment" (Escobar, 2015), while for others it implies 'degrowth' (D'Alisa, Demaria and Kallis, 2015) or 'alternative development'. Either way, well-being and a relational perspective seems to be what is common across nearly all perspectives that break with 'Development' (big 'D').

Following this perspective,[1] a just transition can be defined as a set of complex highly contested socio-political processes that result in (a) significant improvements in well-being for all (including the eradication of poverty and reduced inequalities, in particular asset inequality), and (b) the simultaneous restoration of degraded ecosystems, decarbonization and radical improvements in resource efficiency. Achieving both via a just transition would require – and result in – far-reaching structural transformations that are, arguably, implied by the commitments embedded within the SDGs. An exclusive focus on the former will leave planetary systems to collapse, resulting in rising prices of increasingly scarce resources, starting with the most sensitive which is food, but also water, energy and other extracted materials.

Conversely, strategies that only focus on the ecological sustainability of the planetary systems will tend to neglect what is needed to build the capabilities of the poor to define their own solutions and the capabilities of the state to intervene where required. The rich will be the main beneficiaries of improved sustainability as they retreat into their decarbonized enclaves where they can drive hybrid cars, shop at organic stores and generate renewable energy on their rooftops. Where the fusion of these goals becomes most explicit is in the idea of a just transition expressed most concretely in the call for energy democracy (ED) as a way of reorienting the directionality of the global renewable energy revolution (see Chapter 8).

The problem, however, is that the literatures on DSs and the literatures on STs have evolved in parallel without much cross-over (for a key exception and seminal contribution, see Johnstone and Newell, 2017). This is an opportune moment to synthesize these literatures in order to conceptualize in a more detailed way the dimensions of a sustainability-oriented polity that holds in balance the developmental and sustainability agendas (in a way that avoids the usual 'trade-off' narrative introduced by the 'triple bottom line' approach).

We use a synthesis of the DS and ST literatures. While the DS literature has been widely used to address the development challenges of industrializing economies in the global South (Thompson, 2009; Leftwich, 1995, 2000; Bagchi, 2000; Jayasuriya, 2001; Mkandawire, 2001; Chibber, 2002; Rodrik, Subramanian and Trebbi, 2004; Kohli, 2006; Chang, 2007; Swilling, 2008; Edigheji, 2010; Evans, 2010; Noman, Botshwey, Stein and Stiglitz, 2011; Swilling, Musango and Wakeford, 2015), this literature has generally neglected to deal with environmental challenges in general and STs in particular. The ST literature generally has ignored development (with exceptions such as Swilling and Annecke, 2012; Scoones, Leach and Newell, 2015), but there is an emerging literature that has started to be used to address this lacuna, with a significant body of work already done on East Asian economies (Angel and Rock, 2009; Berkhout, Angel and Wieczorek, 2009; Rock, Murphy, Rasiah, van Seters and Managi, 2009) and now also starting to be applied in the South African context (Lawhon and Murphy, 2011; Swilling and Annecke, 2012; Baker, Newell and Philips, 2014; Baker, 2015).

Scoones, Leach and Newell (2015) have achieved a significant synthesis of the development economics and ST/transformation literatures, with specific reference to the 'politics of green transformations'. In a subsequent publication, Johnstone and Newell build on this foundation and offer a perspective on STs that draws on rich traditions in radical political economy (2017). This chapter builds on these works, the ST literature on East Asia (cited earlier) plus my previous work on synthesizing development, institutional and ecological economics to theorize the 'greening' of the DS (Swilling and Annecke, 2012: Chapter 4; Swilling, Musango and Wakeford, 2016).

To illustrate the argument, it will be argued that developmentally South Africa has not built a 'relatively autonomous' strong DS apparatus to pursue an employment creation-through-industrialization strategy (a la East Asia). Instead, a far less creative non-developmental welfarism to address inequalities using fiscal policy has emerged. Similarly, it will be shown, the mineral-energy complex (MEC) was accommodated rather than dismantled in the post-1994 polity and the related financialization of the economy (i.e. reliance on the growth of the finance sector) was encouraged. Secondary industrialization was the casualty, despite the vital role played by industrial trade unions in the struggle against apartheid. And yet, at the same time, since 2006, a vast array of environmental and sustainability-oriented policies have emerged, fusing into what has started to be referred to as the 'just transition' approach as one way of catalysing low-carbon developmentally oriented growth. But for this to materialize into a fully fledged ST, the self-same state capabilities that are needed to drive the structural transformations to promote investment-led 'developmental welfarism' (Khan, 2013:582) are needed to drive a ST. These, however, are lacking in the South African case.

The argument will be constructed as follows: firstly, a conceptual framework will be provided that draws from the DS and ST literatures. It will be argued that the DS and ST literatures share the view that deep-level 'structural transformation' is needed to achieve development and sustainability-oriented goals, respectively. To this extent, they align in different ways to the reference in the Preamble to the SDGs to a 'transformed world'. Secondly, a conceptual framework for understanding just transitions will be developed that is influenced by the application of the DS and ST literatures to the South African context. Finally, we conclude by suggesting that the adoption of the SDGs by the UN and the launch of the Future Earth programme that emphasizes 'transformation' provide an opportunity for widening the discussion of the challenge of achieving a just transition that can learn some valuable lessons from the South African experience.

Conceptual framework

As already noted, the DS and ST literatures agree on the need for deep structural transformation, but with two different ends in mind: for the DS literature, the end is accelerated economic 'Development' (big 'D') that substantially raises the average GDP per capita with a focus on industrialization and urbanization; while for the ST literature the end is a socio-technical transition that results in a low-carbon resource-efficient economy. Johnstone and Newell (referred to in subsequent pages) go a long way towards achieving such a synthesis, but like most work in the political economy field they neglect the institutional context. I will argue that the synthesis of these two literatures needs, rather, to open up a space for a more detailed discussion about governance for a just transition, with special reference to how we deepen our understanding of the dynamics of collibration within the polity.

As already indicated at the start of this chapter, the directionality of the deep transition will depend on power dynamics within the polity. To briefly repeat what was discussed in greater depth in Chapter 2, according to Jessop the polity is

> the institutional matrix that establishes a distinctive terrain, realm, domain, field, or region of specifically political actions. . . . Further, while the polity

offers a rather static, spatial referent, politics is inherently dynamic, open-ended, and heterogenous.

(Jessop, 2016:17)

Jessop's "strategic-relational approach" is useful for understanding the contemporary state and politics (2016). He argues (repeating parts of Chapter 2) that from the 1970s onwards, there has been a gradual 'de-centering' of the polity as the state 'retreated' (to use the word Strange (1996) used in the 1990s) as the primary driver of the policies that shape the future. The result is the emergence of a polity that is far more complex than what existed in the old state-centric polities. Polities have evolved into complex semi-institutionalized partner-based assemblages and dynamic sets of transactions. In the academic and policy literature, the notion of 'governance' emerged to capture this shift away from state-centric conceptions of the polity (Kooiman, 1993; Jessop, 2003; Hajer, 2009; Offe, 2009). Research agendas also moved away from a preoccupation with structures of government to the relational dynamics of gover*nance*. Out of these emerged what Schot and colleagues refer to as 'frame 3' policymaking: a multi-stakeholder, multi-goal process that requires a far more complex set of institutions and leadership skills (Schot and Steinmuller, 2016). Governance, however, is less about the long-term directionality of the polity and more about relational and dialogical management (Offe, 2009) of complexity over the short term. It is, in short, the outcome of political weakening and has, unsurprisingly, generated a reaction.

Jessop argues there is a counter-trend initiated by forward-looking political leaders with long-term visions who are interested in the "governance of governance", which is what he refers to as "collibration". This consists of a new set of intermediary public institutions that have the capacity and mandate to set the terms of governance so that the directionality of the relational polity is determined politically rather than via a constellation of negotiated deals.

Drawing on Jessop's conception of the polity and related notion of collibratory governance and Wilson's 'policy regime' theory, a more nuanced conception of the polity starts to emerge (Wilson, 2000). This is useful for making sense of the kinds of 'sustainability-oriented polities' that would be required to drive a just transition.

Following Jessop read together with Wilson, the polity can be understood as the space or arena that demarcates a specific constellation of political and quasi-political actors engaged in a defined set of contestations to influence policy outcomes and the actual roles of particular state apparatuses. This constellation of actors usually shares a sufficient consensus about a set of ground rules for conducting the business of everyday politics within and outside the formal institutions of the state apparatus. These actors (interests) subscribe to certain underlying beliefs about the legitimacy of the system, how institutions are controlled in various ways and the way cooperation and opposition works. They get organized into competing factions or alliances – and related policy networks – to secure advantages in the policymaking space and the wider polity.

In other words, a polity goes beyond the governing elite and the state apparatuses that the DS literature has tended to focus on. The nature and character of the polity demarcates the space within which the political game is conducted across various arenas (parliament, executive, media, civil society, judiciary, local/regional space economies, organized business, policy networks, personal networks and business sectors) in order to manage the overall stability of the political system and contest the direction of policy in ways that do not subvert the overall coherence of that political system. Some actors, however, operate within the polity on these terms but also either directly or indirectly act via proxies outside and against the polity and the way it is configured. These political hybrids have demands that could be revolutionary in nature (i.e. replacement of the polity by a completely different configuration of political and institutional arrangements) or reformist (i.e. for the reconfiguration of one or more fundamental dimensions of the polity such as reduced influence of big business or constitutional reform to reduce the powers of the Head of State).

Policy regime theory suggests that the polity has four dimensions (Wilson, 2000). The first of these is where *power relations* are played out and reproduced within the polity. This refers to how political power is constituted, distributed and maintained by the 'power elite' (Wright Mills, 1956), especially – but by no means exclusively – the governing party and its allies within and outside government. Second, there is usually a shared underlying *policy paradigm*, in particular at the sectoral level (e.g. coal-based energy generation) but also at the macro-economic level (e.g. a belief in neoclassical approaches to fiscal discipline and monetary policy). A policy paradigm incorporates a specific set of beliefs/assumptions which, in turn, determines how policy problems are defined (e.g. a faith in markets or a commitment to the SDGs). A shared policy paradigm is understood and narrated by the different policy actors who engage in the everyday business of politics, which is why a shared language emerges to enter into dialogue and negotiation. Third, there is the way *government and state institutions are organized and legitimized*. Although this reflects the underlying power relations and paradigm commitments, these power relations do not always determine how government is organized. Organizing principles get institutionalized and can have a relative autonomy and constitutional fixity that can be at odds at times with the underlying power dynamics. This happened in South Africa during Jacob Zuma's Presidency (2011–2017) when the Constitution came to be regarded as an obstacle in the way of 'radical economic transformation', and is arguably what is happening as the Trump Presidency acts against a wide range of constitutional norms. Fourth, there is the *policy content* of the policies themselves that are debated and adopted by policy actors within a given polity – this being the traditional focus of policy analysis. This framework is useful for revealing how fundamental policy change cannot take place without changes in the three dimensions of the policy that the fourth depends on: if power relations and the policy paradigm do not change, how can we expect the real substance of policy content to change?

The rise of governance since the 1970s (discussed in Chapter 2) is essentially about the organization and outward expansion of the polity to incorporate policy actors (in particular, the corporate sector) directly into policy processes rather than maintaining the illusion of autonomy. Much of this was to compensate for the weakening of the state as complexities mounted. However, democratic collibration is the counter-trend, referring to the way key actors within the polity seek – with varying degrees of success – to reduce the influence of corporates/special interests and harness/mobilize relational dynamics and, indeed, complexity more generally. It is this kind of 'governance of governance' that has the potential to guide and shape long-term structural transformation.

The key contribution made by policy regime theory is that it helps explain policy 'lock in' by referring to the complex interaction between all four dimensions of the polity, without assuming a priori that any one dimension determines any of the others. Determination, after all, is context-specific. This goes way beyond the traditional purview of the policy analysis community which is essentially locked into dimension 4 (policy content) and to some extent dimension 3 (organization of government). However, the experience of policymakers and a considerable body of research suggests that policies do in fact reflect underlying power dynamics (dimension 1) and paradigm commitments (dimension 2), and therefore unless these are changed, changes in dimensions 3 and 4 will be unlikely. Each context, however, will be different. In some instances, policy change (dimension 4) can drive changes in the other dimensions (especially policy shifts driven by global dynamics or local crises), while in other contexts nothing changes until the underlying power dynamics (dimension 1) change or the policy paradigm shifts in response to cultural or knowledge-related trends (e.g. the impact of environmental thinking). For example, the South African Government only dropped its commitment to nuclear power when power relations within the governing party shifted resulting in the replacement of the President (see Chapter 9).

In reality, polities tend to change in response to 'stressors' and 'enablers' (Wilson, 2000), often represented by some kind of external shock to the system (e.g. an 'upset' election, corruption scandal, economic crash, violent conflict, assassination, realignment of political forces in the governing party, mass uprisings, constitutional crisis, warfare, etc.). Stressors can emerge when new power players emerge and/or external dynamics force policy changes. Examples of enablers would be a paradigm shift, such as the gradual realization that climate change needs to be addressed in some way. As a consequence of a shock (or series of shocks) and the nature of the enablers and stressors, the dynamics and character of the polity will change via some contextually specific combination of power shifts (dimension 1), paradigm shifts (dimension 2), a legitimacy crisis (dimension 3) and organizational and policy change (dimension 4). A realignment of forces and dynamics within the polity can emerge from any one of these dimensions, although dimension 1 is where the most significant shifts will most often originate.

As will be shown in subsequent pages, developmentally oriented polities at the heart of Asian developmental states reflected the underlying constellation of forces (dimension 1) that were united behind a particular developmental paradigm (dimension 2). Government was reorganized from time to time (including nationalization and denationalization of industries), and long-term policies were crafted and implemented by relatively strong bureaucratic elites. STs would require a similar alignment to sustain a long-term commitment to structural transformation via a just transition, but to date there is no equivalent stable pattern for STs as in the case of the DSs. As will be shown, East Asian DSs were forced to adopt environmental policies (dimension 4) because of external pressures (environmental regulations adopted by trading partners) that caused a paradigm shift of sorts (dimension 2). The introduction of feed-in tariffs in Germany that triggered the renewable energy transition in Germany was a policy change (dimension 4) (Jacobsson and Lauber, 2006) that reflected a deeper shift in power as the greens and the environmental movement got stronger (dimension 1) and an environmental paradigm shift took place that transcended party-political divisions (dimension 2) (Aklin and Urpelainen, 2018).

As will be shown with respect to South Africa's renewable energy sector, the shift in this case was initially driven by policy change instigated by external pressures (World Bank, hosting of climate negotiations), which generated so much investment that it indirectly catalysed a change in power dynamics (dimension 1) within the polity and a paradigm shift away from nuclear (dimension 2). In short, this four-dimensional conception of the polity helps us decipher the dynamics that could influence the longer-term directionality of the deep transition. What kind of structural transformation (sustainability-oriented and/or developmental) eventually emerges will depend on whether a pro-just transition coalition emerges within the polity.

Developmental states

The defining feature of DSs is that they are primarily concerned with the structural transformation of modernizing economies (Evans, 1995; Kohli, 2006; Noman, Botshwey, Stein and Stiglitz, 2011). The legitimation of DSs is derived primarily from their ability to promote sustained growth and development via aggressive industrialization (Chibber, 2002). In practice, the underlying economic rationale for DSs has been an acceptance that markets left to their own devices will tend towards disequilibrium in unequal developing economies and therefore state intervention is a necessity (for a detailed discussion, see Chapter 2). However, at the ideological level, DSs were excellent at extolling the virtues of capitalism and even, when it suited them, the logics of neoclassical economics. As summarized by Khan (2008), their policy paradigm and orientation (dimensions 2 and 4) promoted sustained growth and development by deploying several unique abilities. These included the ability to extract and deploy capital productively, generate and implement national and sectoral plans and effect dynamic egalitarian and productivity-enhancing development programmes in land, education and training, small enterprise, infrastructure and housing sectors. In addition, DSs have been able to manipulate private access to scarce resources through, among others, financial sector re-engineering, subsidies, taxes, concessions and high levels of lending.

An authoritarian form of collibration was often pursued by states that were determined to tightly manage and cultivate a state-dependent national business class. The cultivation of close and productive relationships with business within a polity tightly managed by a dominant political elite was the norm. Interest groups were managed using corporatist arrangements, often in authoritarian top-down ways to impose the state's agenda versus the more consensual type of social corporatism that was pursued in South Africa after 1994. Thus, the East Asian DS was characterized by a capacity to coordinate the efforts of individual businesses by encouraging the emergence and growth of private economic institutions, target specific industrial projects and sectors, resisting political pressure from popular forces and, at times, also brutally suppressing them. These DSs often mediated and/or insulated domestic economies from (extensive) foreign capital penetration during the early stages and, most importantly, sustained and implemented a project of productivity improvement, technological upgrading and increased market share that broke them out of a path-dependent low-growth economic trajectory (Chang, 2007).

The institutionalization of the polity of the developmental state has received much attention since the 1990s. In a seminal contribution, Leftwich (1995) summarized how dimensions 1 (power relations) and 3 (organization of government) of typical DSs were configured:

- a 'determined developmental elite' committed to the modernization project;
- 'relative autonomy' from major capitalist economic interests who are always keen to capture the state;
- 'a powerful, competent and insulated economic bureaucracy' that enjoys the highest possible political support but operates without too much political interference;
- a 'weak and subordinated civil society' which means there are no rival centres of alternative policy formation;
- the 'effective management of non-state economic interests' via formal structured compacts, incentives and penalties, and
- accessible and usable institutions of 'repression, legitimacy and performance'.

Once DSs had consolidated an industrial base via technological capacity building, institutional functionality and human developmental capabilities, their focus shifted from the late 1990s onwards from massive investments in the material conditions of modernization to establishing the conditions required by the emergent knowledge economy created by the information revolution (Evans, 2006). New tasks emerged with major implications for the structures and logics of the polities that drove the initial phases of development. Together, these tasks clearly defined the slightly more consensual collibratory governance that emerged during the transition from accelerated heavy industrialization/urbanization to an emphasis on quality and greening – a process often associated with 'denationalization' of the 'commanding heights' of the economy (except for the China case).

Sustainability transitions

The conceptual structure of the multi-level perspective (MLP) – the most influential approach in the ST literature – was described in Chapter 4 and therefore will not be repeated here. In summary, the MLP makes a distinction between landscape pressures, socio-technical regimes and innovation niches.

Transition researchers characterize socio-technical regime change – or structural transformation – as being predicated on the ways in which shifting landscape pressures impinge on a regime and the extent to which responses to these pressures are coordinated, both from inside and outside the regime to accommodate or resist these pressures (Smith, Stirling and Berkhout, 2005). In this way the ST literature opens up the issue of governance interventions to facilitate regime transformation, but for some this has not hitherto been taken far enough (Meadowcroft, 2011; Geels, 2014; Hess, 2014; Johnstone and Newell, 2017). It is not only the objective reality of these pressures that matters but more importantly the adaptive capacity or the relationships, resources and their levels of coordination within the polity that shape responses to these pressures. This can be the outcome of historical processes (e.g. a gradual shift in consumer choices or evolution of new technologies) or purposively informed by a strategically aligned polity with a shared vision and capacity to implement a coherent set of policies. The ST literature is critical of the neoliberal assumptions about the virtues of the market, hence the constant insistence on a role for the state. As Johnstone and Newell put it,

> In short, within sustainability transitions literatures 'the state' has been an assumed but underconceptualised, secondary aspect in explorations of socio-technical transitions and niche development.
>
> *(Johnstone and Newell, 2017:74)*

Johnstone and Newell identify five implicit assumptions about the nature and role of the state in the ST literatures (Johnstone and Newell, 2017:74–76). Firstly, there is rising awareness of the key role of state institutions in accelerating transitions. The supportive role of state institutions in the rapid rise of renewable energy across all world regions is clearly a case in point (Mazzucato, 2015). Secondly, there is growing recognition of the political role played by coalitions of incumbents that resist STs. They can use state institutions, and equally state institutions can be used against them. Thirdly, it is becoming increasingly clear that state institutions will be required to actively destabilize and discontinue unsustainable socio-technical regimes (e.g. oil-based motor vehicle engines) – they are unlikely to wither away in the face of landscape pressures and niche innovations. Fourthly, the rising number of case studies of STs reveals how significant each context really is. Regimes, niches and transition pathways are profoundly embedded within – and shaped by – the dynamics of each specific context. How these dynamics pan out in South Africa will be very different in South Korea, Europe, or Brazil. Fifthly, despite the foregoing four trends, when the state is discussed in the ST literature, there is little recognition of its relational nature. Instead, "[s]trict dividing lines often persist in relation to 'state', 'market' and 'civil society'", and as a result the "processual and dynamic nature of the state in configuring geometries of power between different actors . . . remains largely unexplored" (Johnstone and Newell, 2017:75).

Drawing from cutting-edge thinking in contemporary political economy, Johnstone and Newell proceed to suggest five particular "dimensions of state power" (Johnstone and Newell, 2017:75) that align closely with Jessop's strategic-relational approach discussed in Chapter 2. They will be applied to the South African case discussed later in the chapter and in the subsequent two chapters. Firstly, assemblages of state institutions evolved *historically* in ways that are specific to each regional and national context. This is why an understanding of the dynamics of STs cannot be simplistically derived from global dynamics. Contextual specificity matters. The role of the 'mineral-energy complex' in the South African case is a case in point – how it manifests in South Africa will be different to how it manifests in Brazil. Secondly, the political economy of energy is such that there are *global geo-strategic interdependencies* that can enhance or constrain energy transitions at the national and local levels. For example, states use commercial and military means to secure and protect access to fossil fuel supplies in global markets in ways that can constrain the expansion of renewables at the national level. The former South African president's determination to procure a Russian nuclear power plant is a good example. But the drastic drop in prices of renewables in South Africa due to subsidies in other jurisdictions has the opposite effect.

Thirdly, there are always *multiple centres of power* in any state system. How competing conceptions of energy futures within the South African polity culminated in a serious political crisis during the course of 2017 is discussed in Chapter 9. Fourthly, a *relational approach to governance* means accepting there are no neat dividing lines between polities and societies. State institutions reflect and refract particular sets of interests, sometimes in contradictory ways. This often renders references to 'the state' as distinct from 'the market' and 'civil society' somewhat meaningless. The dependence of economic growth on affordable energy supplies, for example, often gives large energy providers a privileged place in the polity which they use to protect their positions and systems. Again, this is clearly reflected in the way the South Africa polity has worked.

Fifthly, there is "insufficient attention to the *material implications of certain technologies* in shaping institutional routines and practices that may influence the directionality of sustainability transitions" (Johnstone and Newell, 2017:78 – emphasis added). In short, as will be shown in the South African case (in this and Chapter 9), the dominance of the 'mineral-energy complex' not only retards the diversification of the economy and reinforces social exclusion but is also premised on the 'normalization' of a particular set of technological practices that get inscribed into policies, laws and regulatory regimes that serve to (wittingly and unwittingly) exclude potentially more productive and inclusionary alternatives.

As revealed in the South African case, all five dimensions of power identified here by Johnstone and Newell are germane to the wider discussion of the state's role in the development process. However, they do not address the crux of the matter, namely the inner dynamics of the polity itself.

Structural transformation, just transitions and the shaping of the polity

The purpose of the actually existing DSs was to drive the long-term structural transformation process of economic development in order to achieve a high level of human well-being with respect to income, education and health. Building on the ST literature, a ST can only be envisaged if a specific combination of state apparatuses facilitates a long-term structural transformation process that results in socio-technical transitions to more sustainable modes of production and consumption, with special reference to decarbonization, resource efficiency and ecosystem restoration. When these two conceptual frameworks and associated goals are combined, the result is a way of imagining a just transition.

A DS, however, is not merely defined by the goals it is committed to. The DS literature has paid considerable attention to the capacity of the state to instigate transformative developmental processes, paying special attention to the emergence of a (sufficiently uncorrupted) developmental bureaucracy and a well-entrenched policy management system. In short, this is about the way political leadership went about constructing and organizing a *developmentally oriented polity* and, in particular, quite an authoritarian mode of collibration. In some countries, including South Africa, a developmentally oriented form of collibration morphed into neo-patrimonialism (discussed further in Chapter 9; see also Khan, 2004; Pitcher, Moran and Johnston, 2009).

In contrast, the emphasis in the ST literature (and the Transition Management literature in particular) on collaborative stakeholder engagement reflects quite a sanguine view of governance. There is little appreciation of collibration as a specific responsibility of purpose-built state institutions mandated to set the goals and rules of the game for achieving long-term structural transformation via a just transition. Broadening out governance to improve stakeholder participation for its own sake ultimately makes little real difference other than creating an illusion of legitimation.

The ST literature on East Asia, however, is somewhat different. Angel and Rock show how the considerable governance capacity of the East Asian DSs to drive development in ways that contradicted the neoliberal script (because it was so interventionist) has become very useful for driving STs in response to environmental landscape pressures, in particular those globalization dynamics that require East Asian economies to be 'greened' (Angel and Rock, 2009; Rock, Murphy, Rasiah, van Seters and Managi, 2009).

However, Rock et al. correctly point out that landscape pressures in general are too diffuse and contradictory to be useful for isolating 'landscape variables in directing transition processes' (2009:242). As a solution, they proffer the notion of a 'socio-political landscape' to refer to the 'institutions, values and regulations broadly guiding an economy' (2009:242). This seems similar to the notion of the 'polity' as deployed in this chapter. However, given that the defining feature of landscape pressures is that they are long-term and slow moving, calling this a 'landscape' seems like a misnomer – the strategic coalitioning and political actions needed to guide structural transformation via a just transition are by no means slow moving and are not nearly as long-term as socio-technical landscape pressures like climate change, demographic change, resource depletion and values change. It therefore makes sense to retain the notion of the polity that recognizes the highly contingent nature of power dynamics and political action.

The conception of the polity advocated in this chapter addresses the challenge faced by the ST literature to conceptualize the role of politics and state, and it goes beyond the narrow institutionalist perspective that tends to pervade most accounts of the DS in the DS literature.

This analysis makes it possible to suggest that a just transition becomes a realistic outcome if a *developmental sustainability-oriented polity* emerges. This would entail a paradigmatic agreement within the polity that the overall goal of development is human well-being (income, education and health) within a sustainable world (decarbonization, resource efficiency and ecosystem restoration) (dimension 2). For this integrated goal to shape the direction of development, broader socio-technical landscape pressures would have to be seen by key actors within the polity as nudging historical processes in a way that reinforces the normative claims of these goal statements. Game changing dynamics emerging from niche innovations and experiments should also be coalescing around viable future alternatives and into alternative socio-technical regimes (Avelino, Wittmayer, O'Riordan, Haxeltine, Weaver, et al., 2014). However, the structural transformations needed for a just transition (at all four levels of transition) will only be achieved when a strategic coalition emerges that shares this paradigm (dimension 1). This would need to be supported by a programme for using state institutions to drive a just transition (dimension 3) based on an appropriate policy and legislative programme that is aligned with the overall goal (dimension 4).

The South African case will show that there are political and system shocks forcing a paradigm shift and some policy reforms, but overall the underlying balance of power remains largely unchanged but may be shifting as from 2018 onwards. Unlike the East Asian states, the South African state since 1994 has not been configured to drive either developmental modernization or a ST. It will be shown that an underlying unifying paradigm for reconstituting the South African polity around the triple goals of investment-led growth, developmental welfarism and sustainability has not yet emerged. However, since the ascendance to the Presidency of Cyril Ramaphosa in late 2018, there are potentially significant shifts underway that point in that direction, with the National Planning Commission leading a significant initiative to consolidate a consensus around a "just transition" pathway for South Africa. For this to be fully realized, however, new modes of collibratory governance will be required.

The South African case is useful for illustrating in practice how developmental and sustainability challenges connect in ways that are common across many developing countries, especially the emerging industrializing economies of Africa.

South Africa's transition

South Africa has a population of 53 million and has been a democracy since 1994. It has a strong democratic constitution, a relatively mature institutional regulatory structure, a solid core of economic infrastructures and a market economy. Between the 1960s and 1980s, the South African economy diversified by expanding the manufacturing sector, but following global trends from the early-1990s, it was the growth of the financial sector that became a key driver of growth, thus disincentivizing diversification (Mohamed, 2010; Black and Gerwel, 2014). According to the Gini coefficient, South Africa is one of the most unequal societies in the world (National Planning Commission, 2011). This, however, is based on a measurement of income inequality. Recent research on asset inequality based on data from two databases has reached an even more shocking conclusion about the class structure of South African society:

> [B]oth sources agree that wealth inequality is extreme: ten percent of the population own more than 90 percent of all wealth while 80 percent have no wealth to speak of; a propertied middle class does not exist.
>
> *(Orthofer, 2016)*

According to the 2017 Land Audit, white people own 26.6 million of the 37 million hectares of land classified as farmland and agriculture holdings (72%). Africans own 4% and make up 80% of the population (a mere 1.4 million hectares). That said, of the total land area of 121.9 million hectares, only 114 million hectares are registered as 'owned' by the Deeds Office (94%). The state owns 14% (17 million hectares) of this total. The difference between the land area (121.9 million ha) and the registered land (114 million ha) is 7.9 million hectares (five times what Africans formally own). This is communally occupied land mainly in the former bantustans – what the Land Audit euphemistically calls "unaccounted for" land. This excludes the million or so hectares owned by the Ingonyama Trust in KZN (which was classified as state-owned land, that is, part of the 17 million hectares of state-owned land). Without ownership, these families can do little to realize the potential value of the land; nor does the state assist with an alternative to private ownership that unlocks this value. No one, therefore, has a real interest in improving the quality of the soils in these areas which partly explains the high levels of soil degradation (5 million hectares of the 14 million hectares of high-value agriculture land).

In short, 90% of the wealth is owned by 10% of the population, and 72% of the productive privately owned farmland is owned by whites. Twenty years of democracy did not change this very much, with major political consequences including threats to the constitutional democracy itself.

Despite moderate growth between 1994 and 2007 and substantial real increases in fiscal expenditure (Swilling, Khan, Van Zyl and Van Breda, 2008), unemployment and poverty have persisted. The official rate of unemployment, based on the 'narrow' definition, has been around 25% for several years. Using a R524/month poverty line, 53% of the population lived in poverty in 1995, declining marginally to 48% by 2005. This decline was attributed largely to the impact of social grants, which now benefit more people than the number of people in formal employment (National Planning Commission, 2011). Many critical writers blame the failure of the state to initiate employment-creating industrial growth via structural transformation for the persistence of poverty and inequality since 1994 (Wolpe, 1995; Habib and Paday-achee, 2000; Bond, 2002; Gelb, 2006; Swilling, 2008; Hart, 2008; Mohamed, 2008, 2010; Freund and Witt, 2010; Marais, 2011; Netshitenzhe, 2011; Khan, 2013).

Chapter 5 of the National Development Plan calls for a "just transition to a sustainable development pathway". The case for a ST was put forward by the Minister in the Presidency and Chairman of the National Planning Commission (NPC), Trevor Manuel, in an address to the National Assembly in June 2011:

> Our economic path, our settlement patterns and our infrastructure all combine to place our country on an unsustainable growth path from a resource utilisation perspective. We are the 27th largest economy in the world but we produce more carbon dioxide emissions than all but eleven countries in the world. We are a water scarce country but we use our water inefficiently. We have to change these patterns of consumption and we have to learn to use our natural resources more efficiently. We must do this with appropriate consideration for jobs, energy and food prices.

What follows is first a brief summary of the explanation for why South Africa did not initiate the kind of developmental structural transformation that occurred in East Asia, despite rhetorical commitments to being (or wanting to be) a DS. Therefore, unlike in East Asia, the developmental phase has not left South Africa with the kind of polity that can select and drive a ST. However, that said, since 2006 South Africa has adopted a slew of environmental policies, and niche innovations have driven a fast-growing RE sector with an investment portfolio equal to nearly 5% of GDP over six years commencing in 2011. This presents the interesting prospect of a ST that is not state-driven in the same way that it is driven in East Asia (Baker, Newell and Philips, 2014; Msimanga and Sebitosi, 2014). This then raises an interesting challenge: can these developmental and environmental trajectories be fused together to comprehend South Africa's prospective just transition (which is, incidentally, exactly the term used in Chapter 5 of the National Development Plan and the theme of a series of national consultations initiated by the NPC in 2018).

South Africa's developmental trajectory

The South African Government formally defines itself as a 'developmental state' committed to the structural transformation of the economy to deal with the legacy of apartheid (Republic of South Africa. National Planning Commission, 2012). The government aims to stimulate economic growth primarily by increasing public investment in national infrastructure to stimulate private sector co-investments. This is complemented by an inflation target of 3–6%, a floating exchange rate, government deficits of 5–10%, extensive fiscal expenditure on education, health and welfare and incentives to promote raw material exports and expand the manufacturing sector. Despite all this, South Africa's economic growth rate is one of the lowest in Africa. For mainstream neoclassical economists, this is due to labour market rigidities, inflated government expenditures and energy shortages. While these are certainly a factor, the underlying resource drivers are as important, especially the rising cost of minerals, energy, water, waste and mobility (Swilling, Musango and Wakeford, 2016).

South Africa's key economic indicators for the post-1994 were cause for optimism. For instance, the average annual rate of economic growth between 1980 and 1993 was only 1.4%. By contrast, the average rate was 3.4% between 1994 and 2013, with an average annual high of about 5% between 2004 and 2007. National GDP by 2012 was 77% larger in real terms relative to 1994. Similarly, export growth per year increased from 2% to 5% between 1980 and 2000, cooling slightly to 4% between 2001 and 2007. South Africa's economy has struggled to regain its early momentum during the post-2007/2009 period. In 2015 economic growth slumped to little more than 1%, and in 2016 growth limped in at 0.5%. In 2016 debt reached almost 50% of GDP (from about 26% in 2008), the Rand continued its trend of depreciation – the nominal effective exchange rate of the Rand lost around 50% of its value between 2010 and 2017 – all culminating in the country's credit rating being reduced to one notch above junk in 2017. Growth oscillated between 0.6% and 1.1% in 2017 and in April 2017 rating agency *Standard & Poor* downgraded South Africa's credit rating to BB+ or junk status.

Following others writing in the ST tradition (Swilling and Annecke, 2012; Baker, Newell and Philips, 2014), the core structural problem of the South African economy is the dominant influence within the polity of the 'mineral-energy complex' (Fine and Rustomjee, 1996; Mohamed, 2010). The MEC refers to a coalition of interests that have a firm grip on energy production and extractive industries and their up- and downstream partners in the manufacturing sector. Up until 2009/2010, this set of interests had well-developed policy networks within the polity to protect their interests. This explained the politics of socio-technical lock-in that ensured that South Africa continued to be committed to energy- and carbon-intensive pathways and under-committed to supporting manufacturing that is not MEC dependent (Black and Gerwel, 2014). A purposive transition in these non-MEC sectors would require a substantial shift in the power relations within the polity to significantly reduce the policy leverage of the powerful mining and energy companies and in so doing respond to global landscape pressures relating to resource prices and competitiveness, as well as domestic labour strife. This would mean establishing a radical new paradigm that Latin Americans refer to as 'post-extractivism' (Economic Commission for Latin America and the Carribean, 2013).

At the core of South Africa's developmental failure is the fact that the negotiated settlement that took place between 1990 and 1994 after Mandela was released from prison left intact the basic structure of economic power. Since 1994, black elites were incorporated into the polity in return for maintaining the basic structure of economic power by limited state intervention in economic ownership and redistribution (Habib and Padayachee, 2000; Hart, 2008; Glaser, 2011). To rapidly stimulate growth and address inequalities, policy choices were made within the polity that reflected this power deal: neoliberal ideologies were adopted (Hart, 2008), financialization was promoted as a growth strategy (i.e. essentially debt-funded consumerism to expand the black middle class, ensuring that the financial sector grew faster than any other sector), capital flight for South African corporates (Mohamed, 2010) and non-developmental welfarism was implemented on a massive scale to quell popular unrest (Khan, 2013). Welfare grants increased from 3 million beneficiaries in 2000/2001 to 16 million in 2011/2012 – from below 1% of GDP to over 3.5% of GDP in a decade! (Khan, 2013).

This was the context for the recomposition of state apparatuses at the centre of the polity: with exceptions here and there (e.g. some housing, rural development and job creation projects), instead of focusing on building new developmentally oriented state institutions to drive a non-MEC employment-creating 'developmental welfarism' (Khan, 2013:258), the focus was on replacing white with black officials as part of a state-driven 'new racial nationalism' agenda (Glaser, 2011). This strategy, coupled to debt-financed consumerism, stabilized the multi-racial middle class base of the post-apartheid polity but at the expense of what Khan calls the 'bioeconomy' – the bodies of the poor black majority and the resources of nature (Khan, 2013).

South Africa's environmental trajectory

Section 24(b) of South Africa's Constitution states that South Africa is committed to 'secure ecologically sustainable development'. The rationale for a South African ST that would achieve this was elaborated in the aforementioned quote from Minister Trevor Manual. He was, of course, responding to a wide range of negative environmental impacts and resource constraints that were starting to generate ad hoc policy responses, often influenced by the insertion of the South African political elite into global policy dialogues such as the World Economic Forum (WEF). It also marked the beginnings of deeper shifts within the polity at the power and paradigm levels (Swilling, Musango and Wakeford, 2016).

Many policy frameworks have been published since 1994: Chapter 5 of the National Development Plan (NPC) refers specifically to the need for a "just transition" (Republic of South Africa. National Planning Commission, 2012); the Green Economy Accord (Seeliger and Turok, 2016); National Strategy for Sustainable Development (Republic of South Africa. Department of Environmental Affairs, 2011); elements of the New Economic Growth Path (Republic of South Africa. Department of Economic Development, 2011) and aspects of the Industrial Policy

Action Plan (Republic of South Africa. Department of Trade and Industry, 2017) and various sectoral plans in the energy (Integrated Resource Plan) (Republic of South Africa. Department of Energy, 2018), water (National Water Resource Strategy) (Republic of South Africa. Department of Water Affairs, 2013), waste (National Waste Management Strategy) (Republic of South Africa. Department of Water Affairs, 2011), various national transports strategies, urban development (Republic of South Africa. Department of Cooperative Governance and Traditional Affairs, 2016) and biodiversity sectors all reveal a policy-level commitment to structural changes that if implemented would go a long way towards catalysing a ST with significant developmental benefits (Swilling and Annecke, 2012: Chapter 8; Swilling, Musango and Wakeford, 2016).

However, unlike in East Asia, the point of departure was not an institutionalized DS with a reasonably strong relatively autonomous developmental bureaucracy (for a useful discussion, see Southall, 2007; Edigheji, 2010; von Holdt, 2010). This institutional weakness was exacerbated by the overt increasingly well-organized looting of public resources during 2009–2017 (Chipkin and Swilling, 2018).

Recognizing the coal-based carbon-intensive nature of the South African economy, the South African Government has committed itself in global fora to playing its part in mitigating global climate change by limiting its GHG emissions (Trollip and Tyler, 2011). The challenge is massive: ESKOM, the state-owned utility, provides 95% of South Africa's electricity (mostly from coal) and has been struggling for over a decade to build an additional 17,000 MW of capacity to meet growing unmet demand by 2018. Since 2010, year-on-year tariff increases were introduced that resulted in South African electricity going from the cheapest in the world (R0.25c/kWh) to over R1.00/kWh by 2018 – a trend that has resulted in the cost of fossil fuels over the life cycle rising to twice the cost of renewables by 2017 as the costs of renewables continued their downward trend (CSIR, 2017).

South Africa is the most carbon-intensive major developing economy in the world apart from Russia. Per capita CO_2 emissions hover between 8 and 12 metric tons, a figure twice as high as China and four to five times higher than Brazil, Indonesia and India and similar to Britain and Germany (Swilling and Annecke, 2012). South Africa's NDC of 2015 recognizes the potentially severe impact of unmitigated climate change on the country:

> South Africa is especially vulnerable to its [climate] impacts, particularly in respect of water and food security, as well as impacts on health, human settlements, and infrastructure and ecosystem services.

In sum, in many ways, the democratic transition of the mid-1990s did not translate into an equitable and ecologically sustainable South Africa.

At the Fifteenth Conference of the Parties (COP15) to the United National Framework Convention on Climate Change (UNFCCC) held in Copenhagen in 2009, South Africa made a *voluntary commitment* to reduce GHG emissions below business-as-usual (BAU) levels by 34% by 2020 and 42% by 2025.[2] South Africa's

Paris Agreement pledge or Nationally Determined Contributions (NDC), submitted to the UNFCCC on 25 September 2015 and ratified on 2 November 2016, *committed* the country to a long-term peak, plateau and decline (PPD) trajectory in which total GHG emissions will be in a range of 398–614 $MtCO_2e$ in the years 2025 and 2030, equivalent to a target range of 20–82% above 1990 levels by 2030 (Republic of South Africa, 2015) (which excludes consideration of land use, land-use change and forestry [LULUCF]). Thus, South Africa has progressed from a vague pledge to reduce emissions relative to BAU to an absolute emissions range for 2025 and 2030 (Altieri, Trollip, Caetano, Hughes, Merven, et al., 2016). However, there is significant uncertainty about the precise target that will be introduced within this large range.

At present, South Africa is a coal-dependent and carbon-intensive economy. According to the Department of Energy (DoE), about one quarter of the country's coal production is exported, regularly placing it in the 'top five' coal exporters in the world (quoted in Parr, Swilling and Henry, 2018). The remaining three quarters are used domestically of which 90% was used for power generation – a share, according to the International Energy Agency, that exceeds by a large margin the global average of 40% (quoted in Parr, Swilling and Henry, 2018). By 2030, the DoE has forecasted 89,500 MW installed power capacity will be added, of which 46% will be coal, 21% renewables, 13% nuclear and the balance from other sources such as gas, pumped storage and hydro-power (quoted in Parr, Swilling and Henry, 2018). In practice, the outcome is very likely going to be different, not least because rising prices have suppressed demand leading to a decoupling of economic growth rates from the annual growth in demand for electricity. Furthermore, the cost of coal has risen more steeply than predicted, nuclear is too expensive and foreign direct investors are clamouring to invest in renewables.

In his first State of the Nation speech as President in 2018, Cyril Ramaphosa stated that the South African Government sees mining (and presumably this means mainly coal mining given the decline in gold) as a "sunrise industry". The South African faith in coal mining seems at odds with international trends. Indeed, internationally many reports suggest that the coal industry is in terminal decline. The IEA's *Coal Information Overview* Report of 2017, for instance, has found that "world coal production declined in 2016 by 458 Mt, which is the largest decline in absolute terms since IEA records began in 1971". The primary reason for this decline, the Report continues, was "electricity generation from coal-fired power plants in OECD countries fell by 6.1% to a new low of 3029 TWh in 2016" as well as declines in China because of growing concerns about climate change (quoted in Parr, Swilling and Henry, 2018).

A revival in coal's fortunes seems highly unlikely. The IEA's *World Energy Outlook* (WEO) Report of 2016 predicts that coal's share in China's and India's power mix over the period 2017–2040 will fall from 75% to 45% and from 75% to 55%, respectively. Coal demand in the European Union and the United States (which together account for around one-sixth of today's global coal use) will fall by over 60% and 40%, respectively, over the same period. Indeed, by 2040 coal use globally could fall back to levels last seen in the mid-1980s, at under 3000 million metric tons of coal equivalent per year. In sum, the IEA explains, 'there is no global upturn in demand in sight for coal' (all information referred to here quoted in Parr, Swilling and Henry, 2018). Forecasts like these have knock-on effects in the investment community: coal is no longer a long-term bet, with some advisory reports recommending divestment (Wright, Calitz, Bischof-Niemz and Mushwana, 2017; Buckley, 2019).

It is commonly assumed that the country has sufficient coal reserves to last at least 200 years. However, the extent of South Africa's remaining coal reserves is a matter of considerable dispute among government officials, industry players and independent researchers. The official figure for coal reserves is approximately 30 gigatons (Gt) (BP, 2013). However, both Rutledge (2011) and Mohr and Evans (2009) using variants of the 'Hubbert curve' technique estimate that remaining recoverable coal reserves in South Africa may be as low as 10 Gt. Hartnady (2010) estimates that there could be 15 Gt of remaining coal reserves and forecasts a peak in domestic coal production by 2020, while Mohr and Evans' 'best guess' is for a peak in 2036 (Mohr and Evans, 2009). Thus, at some point in the not too distant future – possibly soon after 2020 – rising demand (e.g. to feed ESKOM's two – and possibly three – new coal-fired power plants and to meet export growth targets) could intersect with stagnant or falling production of coal and result in substantial increases in coal prices. Recent trends certainly seem to confirm this trend (Wright, Calitz, Bischof-Niemz and Mushwana, 2017).

Furthermore, the global political economy of coal demand is changing rapidly (the data in this paragraph are from Burton and Winkler, 2014). Traditionally, South Africa exported high-grade coal to the 'West' and subsidized the price of the low-grade coal sold to ESKOM. ESKOM then built a fleet of specially designed power stations to use low-grade coal, which is why CO_2 emissions per kWh of electricity are so high in South Africa. This was part of a strategy that began in the 1930s to massively lower the cost of energy for the South African economy (Jaglin and Dubresson, 2016). However, after the turn of millennium, demand for cheaper coal by new trade partners in Asia and Latin America pushed up the price of cheap coal, while the quality of high-quality coal began to go into decline. In short, the apartheid coal-energy economy started to disintegrate resulting in rising coal prices and therefore of electricity. This trend was reinforced during the post-2010 period by ESKOM's strategy to give preferential prices to black-owned coal companies (Swilling, Bhorat, Buthelezi, Chipkin, Duma, et al., 2017).

Reconciling the developmental role and environmental protection of South African water resources has been a key policy priority (van Koppen and Schreiner, 2014). However, how the contamination and degradation of South Africa's scarce water resources – and their supporting ecosystem – could undermine development goals has, in recent years, become a major concern (Godfrey, Oelofse, Phiri, Nahman and Hall, 2007; Oelofse, 2008b, 2008a; Republic of South Africa. Department of Environmental Affairs, 2016). Water pollution includes the massive threat posed by acid mine drainage, as well as eutrophication resulting from the over-use of chemical fertilizers (Godfrey, Oelofse, Phiri, Nahman and Hall, 2007; Turton, Patrick and Rascher, 2008). Land pollution comes in various forms, including the highly visible impacts of open cast mining, the more subtle impact of subsurface mining and the dumping of solid waste (Blottnitz, 2006; Republic of South Africa. Department of Environmental Affairs, 2016).

These various types of pollution are straining the absorptive capacity of South Africa's natural systems. This in turn risks undermining the integrity of ecosystem services. Biodiversity, of which South Africa has such a generous globally significant endowment, is increasingly under threat not only from the effects of pollution but also from the destruction of habitats as a result of land-use practices including extensive farming, mining and urban sprawl (Driver, Sink, Nel, Holness, Van Niekerk, et al., 2011; Maze and Driver, 2016).

Like fossil fuels, minerals and metals are also finite, non-renewable resources subject to depletion. South Africa's gold production history is a poster child for the so-called Hubbert peak model of non-renewable resource depleted (Hartnady, 2009). Having dominated the global gold industry for a century, South Africa's production reached a peak in 1970 and has been on an inexorable decline ever since.

Soil is also an under-appreciated and under-researched natural resource that provides the foundation for agriculture and arguably for society as a whole (Mills and Fey, 2004; Le Roux, 2007). South Africa's soil fertility is being depleted as a result of a fossil fuel–intensive type of agriculture, with up to a third of its 14 Mha of arable land suffering from degradation.

The long-term impact of resource depletion is increasing both material scarcity and costs of extraction to produce the same level of output (Mudd, 2007), leading to higher and increasingly volatile resource prices (McKinsey Global Institute, 2011). This, in turn, raises costs of production and ultimately the prices of many basic goods and services such as energy and food.

Collibration and the just transition in South Africa

In order for the underlying deep transition to also be a just transition, states will need to become the guarantors of this long-term strategic direction. However, given the overall condition of increasing complexity in ICT-based societies, a return to statism to achieve this will be impossible. At the same time, the mushrooming of bottom-up sustainability-oriented experiments from around the world will continue and accelerate. As a result, a form of collibratory governance will be required that ensures directionality without reducing complexity. In practice, this will be the emergent outcome of shifts within the polity that result from a re-alignment of underlying power relations (dimension 1), the adoption of an appropriate sustainability paradigm (dimension 2) and a clear corresponding set of policies (dimension 4) that could be implemented by a state with the necessary institutional and strategic capacity (dimension 3).

South Africa emerged after 1994 with twin developmental and environmental challenges that were never integrated into a coherent developmentally oriented

ST. Instead, the polity bifurcated into two. On the one hand, a dominant coalition emerged around the governing party that reconciled the continued dominance of the MEC, non-developmental welfarism to pacify the poor majority and policies aimed at fostering a new black elite. On the other hand, a weak loosely organized coalition emerged on the margins of the polity to articulate an environmental agenda that emphasized a mix of biodiversity conservation, climate change adaptation, renewable energy, green jobs, resource efficiency and sustainable farming.

During the 'state capture' years from 2009 to 2018 (see Chapter 9), the dominant coalition within the polity was recomposed around a corrupt power elite led by President Jacob Zuma that focused on reorienting public sector procurement spend to build a 'black industrial class' (Chipkin and Swilling, 2018). At the centre of this political project was an agreement with Russia to build a South African fleet of nuclear power plants (Fig, 2018).

In late 2017, Zuma was voted out of office by the governing party and replaced by Cyril Ramaphosa, who succeeded him as President in early 2018. This shift in power within the polity was a response to the massive build-up of opposition against Zuma and 'state capture' within civil society, the media, the trade union movement and business. The Ramaphosa administration immediately scrapped the commitment to nuclear power and signed contracts to build 27 utility-scale renewable energy power plants. This was followed by a Job Summit and an Investment Summit to re-establish the national compact between government and business.

Three processes are underway in South Africa that illustrate the core argument in this chapter, namely how crisis catalyses shifts within the polity that result in new modes of collibratory governance of a developmentally oriented ST. These are the Renewable Energy Independent Power Producers Procurement Programme (REI4P), the Just Transition stakeholder consultation and the Cape Town Water Crisis.

The REIP has resulted in the accelerated expansion of renewable energy in South Africa. Since 2011, total investment in 102 projects (6327 MW) had exceeded R 200 billion by 2017 (Kruger and Eberhard, 2018). The REIPPPP was introduced as a policy mechanism to address the twin challenges of achieving climate change targets and responding to an electricity supply crisis in the late 2000s (Montmasson-Clair and Ryan, 2014). Equally significant, however, is how the procurement framework has taken into consideration a developmental agenda by including a number of economic development targets, within the price-competitive auction scheme (Eberhard and Naude, 2016). The participation of IPPs (Independent Power Producers) in the generation of utility-scale, grid-connected electricity has taken place without displacing the regime of historically centralized energy governance, including the continued dominant role of ESKOM (Bischof-Niemz and Creamer, 2018). At the same time, the implementation of the REIPPPP has resulted in the dispersion of IPPs across the country, breaking with the conventional concentrated geographic location of South Africa's coal-fired power plants, predominantly in the Mpumalanga Province (Jaglin and Dubresson, 2016). At the centre of this success story was a governance unit within the DoE called the IPP Unit. Established with support from the DBSA, it was a small uncorrupt team that managed the implementation of the regulations in an efficient and effective manner. This included the elaborate quarterly reporting by the privately owned IPPs on development impact that every IPP was required to submit.

A combination of niche innovations supported by demands from global investors and realignments within the polity in response to supply shocks has resulted in a shift from a pro-nuclear policy to a policy framework that includes a significant role for renewables in South Africa's energy future (Baker and Wlokas, 2014; Eberhard and Naude, 2016; Energy Research Centre, 2017; Bischof-Niemz and Creamer, 2018). The key influencing dynamics included international climate commitments, World Bank conditionalities to allocate loan finance to renewables linked to a large World Bank loan to fund coal-fired power, investments by DFIs in the early phase of the innovation cycle and the declining price of renewables relative to coal.

South Africa's National Planning Commission (NPC) comprising independent experts has had a profound impact on South Africa's development trajectory (Republic of South Africa. National Planning Commission, 2012). Chapter 5 of the plan refers to the need for a "just transition". Led by Jeff Radebe, Minister in the Presidency responsible for the NPC until his appointment as Minister of Energy after Ramaphosa became President in 2018, the NPC initiated a series of national stakeholder consultations on the Just Transition. Tasneem Essop, a NPC Commissioner, has facilitated these dialogues. With a long history in the environmental movement, Essop has used this process to slowly gel together a loose coalition of forces supportive of a Just Transition. In essence, this would be an inclusive development trajectory powered by renewable energy, including the gradual phasing out of coal-fired power generation. This strategy received a major boost when a stakeholder group was convened in March 2019 by the Development Bank of Southern Africa (DBSA) to consider a paper on South Africa's energy future that provided a financial analysis of the very serious future risk of stranded assets. Funded by the World Bank, French Development Bank and DBSA, this paper argued that South Africa's near R400 billion investment in new coal-fired power could result in stranded assets equal to $120 billion by 2035 as the world decarbonizes and as the costs of renewable energy continue to fall (Huxham, Anwer and Nelson, 2019).

The Just Transition paradigm reflects a deeper underlying reconfiguration of power relations within the polity that is resulting in new policy frameworks and the reorganization of government around an energy sector that could replace the coal sector over the long run. The new power players are the renewable energy companies, environmental movements, the financial institutions that have heavily invested in renewable energy, global DFIs, global climate funding agencies and the local towns and cities that have much to gain from a decentralized and distributed energy system. Unfortunately, the trade unions have been captured by a pro-coal narrative that blames the decline of coal jobs on renewable energy.

Finally, the Cape Town water crisis of 2017/2018 clearly reveals how climate change can rapidly result in a water crisis that threatens the viability of an entire city (this section draws from the doctoral research of Amanda Gcanga, supervised by myself and from Ziervogel, 2019). Changing rain patterns have reduced dam levels in dams supplying Cape Town from an average of 80% in January 2014 to 20% by January 2018. The City of Cape Town responded by announcing a date in early 2018 when normal water supplies would be terminated. Referred to as "Day Zero", this cataclysmic moment galvanized a new coalition of forces that resulted in a dramatic drop in water consumption by 50% over a period of three months. The Economic Development Partnership (EDP)[3] emerged as the key facilitator of the multi-stakeholder dialogues that resulted in new less technocratic and more inclusive water governance approaches. The EDP was established by the Western Cape Provincial Government (WCPG), but with an independent Board. It receives funding from the WCPG and the City of Cape Town. It employs a staff of 25 and their role is exclusively focused on the facilitation of public-public partnerships for strategic joint action (e.g. between local, provincial and national government departments). Wider stakeholders are brought in if the core public-public partnership holds together. This was the modus operandi that brought key stakeholders together to consider alternatives that had hitherto been entirely dominated by the municipal water engineers and their consultants. The end result is that Day Zero kept being postponed, until it was eventually lifted altogether when the rains finally arrived in May/June/July. However, because of climate change, most experts expect a repeat of this crisis to become the 'new normal' in future years.

The lesson from all three of these cases is that a collibratory governance mechanism was key: the IPP Unit (supported by the DBSA) in the case of the REI4P, the NPC in the case of the Just Transition narrative and the EDP in the case of the Cape Town water crisis. All three mechanisms emerged to coordinate stakeholder responses to crisis. And all three advocated a paradigm shift and policy alternatives. In short, these cases provide some evidence as to how realignments within the polity can emerge that shape the future directionality of developmentally oriented STs.

Conclusion

This chapter has addressed the question of how best to understand the relationship between developmental processes and STs in the global South, with special reference to the political dynamics of DSs. This has become an especially important challenge in light of the rise of the 'BRICS-plus' countries, shorthand for quite a large number of rapidly industrializing economies, many of which are resource-based (not only Brazil, Russia, China and India, but also Indonesia, Vietnam, Kenya, Ghana, Ethiopia, Rwanda, Venezuela, Mexico, Turkey, Botswana, Mauritius, and of course South Africa, etc.). These countries want to implement a twentieth-century conception of accelerated development inspired by the East Asian industrializers, but now in a climate and resource-constrained world. Most important of all, after the publication of the UN SDGs in August 2015, the global discourse is shifting from the old 'MDGs-plus-green economy' framework to the globally approved SDG framework that firmly and irrevocably inserts the 'people-planet-prosperity-peace-partnership' paradigm into official definitions of sustainable development at global and national levels.

Both the DS and ST literatures acknowledge that structural transformation is needed, but each with respect to the hitherto separated goals of human well-being and sustainability. Building on the emerging literatures on East Asia and South Africa that attempt to fuse these separate research trajectories, it was argued that an integrated conception of structural transformation will be needed that is driven by a commitment to both the goals of human well-being and sustainability. However, the expected just transition this could give rise to will not happen simply because there is a shared normative commitment, as is now reflected in the adoption of the SDGs. Nor will much progress be made by formulating bland managerialist policy prescriptions that ignore underlying power dynamics and paradigm differences. An adequate fusion of the core body of concepts in the DS and ST literatures will need to make space for an understanding of the political dynamics of the polity.

The polity, it was argued, is a space of policy-related action and engagement by a wide range of actors within and outside the formal political system that operates in four dimensions: power dynamics, paradigm commitments, state organization and policy programmes. The sustainability-oriented effects (and their countervailing tendencies) at landscape, regime and niche levels are played out within the polity, resulting in changes over time in power dynamics, paradigms shifts, state organization adaptations and the adoption of new policies. These dynamics within the polity can be initiated within one or multiple dimensions, either synchronistically or not.

To illustrate the argument several strands of research on post-apartheid South Africa were integrated, showing how different the South African case is to the East Asian context with special reference to how the political settlement in 1994 protected the economic power structure, thus preventing the implementation of a more radical developmental programme. Poverty was addressed via welfarism and fiscal policy. The emergent outcome is an institutionally weak national state that has not fully dislodged the power of the MEC within the polity, has not to date promoted employment-creating investment-led industrialization, and instead has facilitated accelerated financialization, increased shareholder returns and relatively unproductive transfers to black elites. Eventually, this financialized mode of governance resulted in 'state capture' (see Chapter 9).

At the same time, a myriad of environmental and resource challenges have emerged in post-apartheid South Africa, without an adequate paradigmatic framework or integrated policy response. Except for the REI4P, the result is a wide range of seemingly disconnected ad hoc responses to these environmental challenges. By contrast, the East Asian economies entered the new millennium with a strong state to drive a ST in similar ways to how developmental modernization was driven in the twentieth century.

However, three cases were used to demonstrate positive trajectories in response to crisis. New governance arrangements emerged, reflecting the emergence of new paradigms and policy frameworks.

In light of the adoption of the SDGs, it will be necessary to conduct many more case studies of developing economies in the global South where developmental and sustainability goals need to be reconciled in order to achieve a just transition. This chapter has contributed an approach that could guide this kind of future research by bringing into focus the complex dynamics of collibratory governance of the polity. It was shown how realignments of power dynamics, paradigm shifts, reorganization of government and new policy commitments will be required to shift the deep transition into a just transition that fuses developmental and ST goals.

Notes

1 This definition fuses together various strands in development studies, including Sen's capability perspective, the writing on well-being, traditional concerns with structure in development economics and ecological and institutional sustainability thinking.
2 Although what exactly business-as-usual meant was never properly defined – if the baseline was undefined, so was the end-point.
3 www.wcedp.co.za/

References

Aklin, M. and Urpelainen, J. (2018) *Renewables: The Politics of a Global Energy Transition*. Boston, MA: MIT Press.

Altieri, K., Trollip, H., Caetano, T., Hughes, A., Merven, B., et al. (2016) 'Achieving development and mitigation objectives through a decarbonization development pathway in South Africa', *Climate Policy*, 16, pp. S78–S91. doi: 10.1080/14693062.2016.1150250.

Angel, D. and Rock, M. T. (2009) 'Environmental rationalities and the development state in East Asia: Prospects for a sustainability transition', *Technological Forecasting & Social Change*, 76, pp. 229–240.

Avelino, F., Wittmayer, J. M., O'Riordan, A., Haxeltine, A., Weaver, P., et al. (2014) 'The role of game changers in the dynamics of transformation social innovation', in Avelino, F. and Wittmayer, T. (eds.) *Game-Changers and Transformative Social Innovation: Working Paper – TRANSIT Deliverable 2.1 EU SSH.2013.3.2–1 Grant Agreement no. 613169*. Rotterdam: DRIFT.

Bagchi, A. K. (2000) 'The past and the future of the developmental state', *Journal of World-Systems Research*, 1(2), pp. 398–442.

Baker, L. (2015) 'The evolving role of finance in South Africa's renewable energy sector', *GeoForum*, 64, pp. 146–156.

Baker, L., Newell, P. and Philips, J. (2014) 'The political economy of energy transitions: The case of South Africa', *New Political Economy*, 19(6), pp. 791–818.

Baker, L. and Wlokas, H. (2014) *South Africa's Renewable Energy Procurement: A New Frontier*. Sussex, UK: Tyndall Centre for Climate Change Research.

Berkhout, F., Angel, D. and Wieczorek, A. J. (2009) 'Asian development pathways and sustainable socio-technical regimes', *Technological Forecasting and Social Change*, 76, pp. 218–228.

Bischof-Niemz, T. and Creamer, T. (2018) *South Africa's Energy Transition: A Roadmap to a Decarbonized, Low-Cost and Job-Rich Future*. New York and London: Routledge.

Black, A. and Gerwel, H. (2014) 'Shifting the growth path to achieve employment intensive growth in South Africa', *Development Southern Africa*, 31(2), pp. 241–256.

Blottnitz, H. (2006) *Solid Waste: Background Briefing Paper for the National Sustainable Development Strategy*. Cape Town: Department of Chemical Engineering, University of Cape Town: Report Commissioned by the Department of Environmental Affairs and Tourism.

Bogdan, L., Hurduzeu, G., Josan, A. and Vlasceanu, G. (2011) 'The rise of BRIC, the 21st century geopolitics and the future of the consumer society', *Revista Romana de Geografie Politica*, XIII(1), pp. 48–62.

Bond, P. (2002) *Unsustainable South Africa: Environment, Development and Social Protest*. London: Pluto Press.

BP. (2013) *Statistical Review of World Energy 2013*. London: BP.

Buckley, T. (2019) *Over 100 Global Financial Institutions Are Exiting Coal, with More to Come*. Cleveland, OH: Institute for Energy Economics and Financial Analysis.

Burton, J. and Winkler, H. (2014) *South Africa's Planned Coal Infrastructure Expansion: Drivers, Dynamics and Impacts on Greenhouse Gas Emissions*. Cape Town: Energy Research Centre, University of Cape Town.

Castells, M. and Himanen, P. (2014) *Reconceptualizing Development in the Global Information Age*. Oxford: Oxford University Press.

Chang, H. J. (2007) *Bad Samaritans: Rich Nations, Poor Policies and the Threat to the Developing World*. London: Random House.

Chibber, V. (2002) 'Bureaucratic rationality and the developmental state', *The American Journal of Sociology*, 107(4), pp. 951–989.

Chipkin, I. and Swilling, M. (2018) *Shadow State: The Politics of Betrayal*. Johannesburg: Wits University Press.

CSIR. (2017) *Electricity Scenarios for South Africa: Presentations to the Portfolio Committee on Energy, Cape Town, 21 February 2017*. Pretoria: CSIR.

D'Alisa, G., Demaria, F. and Kallis, G. (2015) *Degrowth: A Vocabulary for a New Era*. New York: Routledge.

Death, C. (2014) 'The green economy in South Africa: Global discourses and local politics', *Politikon*, 41(1), pp. 1–22.

Driver, A., Sink, K., Nel, J., Holness, S., Van Niekerk, L., et al. (2011) *National Biodiversity Assessment 2011: An Assessment of South Africa's Biodiversity and Ecosystems*. Pretoria: South African National Biodiversity Institute and Department of Environmental Affairs.

Eberhard, A. and Naude, R. (2016) 'The South African renewable energy independent power producer procurement programme (REIPPPP) – Lessons learned', *Journal of Energy in Southern Africa*, 27(4), pp. 1–14.

Economic Commission for Latin America and the Carribean. (2013) *Natural Resources: Status and Trends Towards a Regional Development Agenda in Latin America and the Caribbean*. Santiago: United Nations.

Edigheji, O. (2010) *Constructing a Democratic Developmental State in South Africa: Potentials and Challenges*. Cape Town: HSRC Press.

Energy Research Centre. (2017) *The Developing Energy Landscape of South Africa*. Cape Town: University of Cape Town.

Escobar, A. (2015) 'Degrowth, postdevelopment, and transitions: A preliminary conversation', *Sustainability Science*, 10(3), pp. 451–462. doi: 10.1007/s11625-015-0297-5.

Evans, P. (1995) *Embedded Autonomy: States and Industrial Transformation*. Princeton, NJ: Princeton University Press.

Evans, P. (2006) 'What will the 21st century developmental state look like? Implications of contemporary development theory for the state's role', in *Conference on 'The Change Role of the Government in Hong Kong' at the Chinese University of Hong Kong*. Hong Kong: Unpublished.

Evans, P. (2010) 'Constructing the 21st century developmental state: Potentialities and pitfalls', in Edighej, O. (ed.) *Constructing a Democratic Developmental State in South Africa: Potentials and Challenges*. Cape Town: HSRC Press, pp. 37–59.

Fig, D. (2018) 'Capital, climate and the politics of nuclear procurement in South Africa', in Satgar, V. (ed.) *The Climate Crisis: South African and Global Democratic Eco-Socialist Alternatives.* Johannesburg: Wits University Press.

Fine, B. and Rustomjee, Z. (1996) *The Political Economy of South Africa: From Minerals-Energy Complex to Industrialization.* Boulder, CO: Westview Press.

Freund, B. and Witt, H. (2010) *Development Dilemmas in Post-Apartheid South Africa.* Durban: University of KwaZulu-Natal Press.

Geels, F. (2013) 'The impact of the financial-economic crisis on sustainability transitions: Financial investment, governance and public discourse', *Environmental Innovation and Societal Transitions*, 6, pp. 67–95.

Geels, F. (2014) 'Regime resistance against low-carbon transitions: Introducing politics into the multi-level perspective', *Theory, Culture and Society*, 31(5), pp. 21–40.

Gelb, S. (2006) 'A South African developmental state: What is possible?', in Harold Wolpe Memorial Trust's Tenth Anniversary Colloquium 'Engaging Silences and Unresolved Issues in the Political Economy of South Africa', 21–23 September. Cape Town: Unpublished.

Glaser, D. (2011) 'The new black/African racial nationalism in South Africa: Towards a liberal-egalitarian critique', *Transformation*, 76, pp. 67–94.

Godfrey, L., Oelofse, S., Phiri, A., Nahman, A. and Hall, J. (2007) *Mineral Waste: The Required Governance Environment to Enable Re-use.* Pretoria: Council for Scientific and Industrial Research. Available at: http://researchspace.csir.co.za/dspace.

Habib, A. and Padayachee, V. (2000) 'Economic policy and power relations in South Africa's transition to democracy', *World Development*, 28(2), pp. 245–263.

Hajer, M. (2009) *Authoritative Governance.* Oxford: Oxford University Press.

Hart, G. (2008) 'The provocations of neoliberalism: Contesting the nation and liberation after apartheid', *Antipode*, 40(4), pp. 678–705.

Hartnady, C. (2009) 'South Africa's gold production and reserves', *South African Journal of Science*, 105, pp. 320–328.

Hartnady, C. (2010) 'South Africa's diminishing coal reserves', *South African Journal of Science*, 106(9/10), pp. 1–5.

Hess, D. (2014) 'Sustainability transitions: A political coalition perspective', *Research Policy*, 43(2), pp. 278–283.

Huxham, M., Anwer, M. and Nelson, D. (2019) *Understanding the Impact of a Low Carbon Transition on South Africa.* London: Climate Policy Initiative.

Jacobsson, S. and Lauber, V. (2006) 'The politics and policy of energy system transformation: Explaining the German diffusion of renewable energy technology', *Energy Policy*, 34, pp. 256–276.

Jaglin, S. and Dubresson, A. (2016) *ESKOM: Electricity and Technopolitics in South Africa.* Cape Town: UCT Press.

Jayasuriya, K. (2001) 'Globalization and the changing architecture of the state: The regulatory state and the politics of negative co-ordination', *Journal of European Public Policy*, 8(1), pp. 101–123.

Jessop, B. (2003) *The Governance of Complexity and the Complexity of Governance: Preliminary Remarks on Some Problems and Limits of Economic Guidance.* Lancaster, UK: Department of Sociology, Lancaster University. Available at: www.comp.lancs.ac.uk/sociology/papers/Jesson-Governance-of-complexity.pdf.

Jessop, B. (2016) *The State: Past Present Future.* Cambridge: Polity Press.

Johnstone, P. and Newell, P. (2017) 'Sustainability transitions and the state', *Environmental Innovation and Societal Transitions.* Elsevier, 27(October 2017), pp. 72–82. doi: 10.1016/j.eist.2017.10.006.

Khan, F. (2008) *Political Economy of Housing Policy in South Africa, 1994–2004.* Stellenbosch: Stellenbosch University.

Khan, F. (2013) 'Poverty, grants, revolution and "real utopias": Society must be defended by any and all means necessary!', *Review of African Political Economy*, 40(138), pp. 572–588. doi: 10.1080/03056244.2013.854035.

Khan, M. (2004) 'State failure in developing countries and institutional reform strategies', in Tungodden Stern, N. and Kolstad, I. B. (eds.) *Toward Pro-Poor Policies: Aid, Institutions and Globalization.* New York: Oxford University Press.

Kohli, A. (2006) *State and Development.* Princeton, NJ: Edward Elgar.

Kooiman, J. (1993) *Modern Governance: New Government-Society Interactions.* London: Sage.

Kruger, W. and Eberhard, A. (2018) 'Renewable energy auctions in sub-Saharan Africa: Comparing the South African, Ugandan, and Zambian Programs', *Energy and Environment*, pp. 1–13. doi: 10.1002/wene.295.

Lawhon, M. and Murphy, J.T. (2011) 'Socio-technical regimes and sustainability transitions: Insights from political ecology', *Progress in Human Geography*. doi:10.1177/0309132511427960.

Le Roux, J. (2007) 'Monitoring soil erosion in South Africa at regional scale: Review and recommendation', *South African Journal of Science*, 103(7–8).

Leftwich, A. (1995) 'Bringing politics back in: Towards a model of the developmental state', *Journal of Development Studies*, 31(3).

Leftwich, A. (2000) *States of Development: On the Primacy of Politics in Development.* Cambridge: Polity Press.

Marais, H. (2011) *South Africa Pushed to the Limit: The Political Economy of Change.* Cape Town: UCT Press.

Maze, K. and Driver, A. (2016) 'Mainstream ecological infrastructure in planning and investment', in Swilling, M., Musango, J. and Wakeford, J. (eds.) *Greening the South African Economy.* Cape Town: Juta.

Mazzucato, M. (2015) 'The green entrepreneurial state', in Scoones, I., Leach, M. and Newell, P. (eds.) *The Politics of Green Transformations.* London and New York: Routledge Earthscan, pp. 133–152.

McKinsey Global Institute. (2011) *Resource Revolution: Meeting the World's Energy, Materials, Food, and Water Needs.* London: McKinsey Global Institute. Available at: www.mckinsey.com/client_service/sustainability.aspx (Accessed: 10 May 2011).

Meadowcroft, J. (2011) 'Engaging with the politics of sustainability transitions', *Environmental Innovation and Societal Transitions*, 1, pp. 70–75.

Mills, A. J. and Fey, M. (2004) 'Declining soil quality in South Africa: Effects of land use on soil organic matter and surface crusting', *South African Journal of Plant and Soil*, 21(5), pp. 388–399.

Mkandawire, T. (2001) 'Thinking about developmental states in Africa', *Cambridge Journal of Economics*, 25, pp. 289–313.

Mohamed, S. (2008) *Financialization and Accumulation in South Africa.* Unpublished Powerpoint presentation, 7 April. Johannesburg: Corporate Strategy and Industrial Development, University of the Witwatersrand.

Mohamed, S. (2010) 'The state of the South African economy', in *New South African Review 1: 2010 – Development or Decline?* Cape Town: HSRC Press.

Mohr, S. H. and Evans, G. M. (2009) 'Forecasting coal production until 2100', *Fuel*, 88, pp. 2059–2067.

Montmasson-Clair, G. and Ryan, G. (2014) 'Lessons from South Africa's renewable energy regulatory and procurement experience', *Journal of Economic and Financial Sciences*, 7(7), pp. 507–526. doi: 10.1093/ajae/aar051.

Msimanga, B. and Sebitosi, A. B. (2014) 'South Africa's non-policy driven options for renewable energy development', *Renewable Energy*, 69, pp. 420–427.

Mudd, G. M. (2007) *Sustainability of Mining in Australia*. Research Report. Clayton, VIC: Monash University.

National Planning Commission. (2011) *Diagnostic Overview*. Pretoria: National Planning Commission, Republic of South Africa.

Nederveen-Pieterse, J. (2000) 'After post-development', *Third World Quarterly*, 21(2), pp. 175–191.

Netshitenzhe, J. (2011) 'The state of the South Africa state', in *10th Harold Wolpe Memorial Lecture*. Cape Town. Available at: www.wolpetrust.org.za.

Noman, A., Botshwey, K., Stein, H. and Stiglitz, J. (2011) *Good Growth and Governance in Africa: Rethinking Development Strategies*. Oxford: Oxford Scholarship Online. doi: 10.1093/acpro f:oso/9780199698561.001.0001.

Oelofse, S. H. H. (2008a) *Mine Water Pollution – Acid Mine Decant, Effluent and Treatment: Consideration of Key Emerging Issues That May Impact on the State of Environment*. Pretoria: Department of Environmental Affairs and Tourism. Available at: http://soer.deat.gov.za/docport.aspx?m=97&d=28.

Oelofse, S. H. H. (2008b) 'Protecting a vulnerable groundwater resource from impacts of waste disposal: A South African waste governance perspective', *Reflections on Water in South Africa*. Patrick, M. J., Rascher, J. and Turton, A. R. (eds.), 24(3), pp. 477–490.

Offe, C. (2009) 'Governance: An "empty signifier"?', *Constellations*, 16(4), pp. 551–562.

Orthofer, A. (2016) *Wealth Inequality in South Africa: Insights from Survey and Tax Data*. REDI3X3 Working Paper 15. Cape Town: University of Cape Town. Available at: http://web.archive.org/web/20180204212928/www.redi3x3.org/sites/default/files/.

Pant, H. V. (2013) 'The BRICS fallacy', *Washington Quarterly*, 36(3), pp. 91–105. doi: 10.1080/0163660X.2013.825552.

Parr, B., Swilling, M. and Henry, D. (2018) *The Paris Agreement and South Africa's Just Transition*. Melbourne: Melbourne Sustainable Society Institute. Available at: http://bit.ly/2GpB1RE.

Pitcher, A., Moran, M. H. and Johnston, M. (2009) 'Rethinking patrimonialism and neopatrimonialism in Africa', *African Studies Review*, 52(01), pp. 125–156. doi: 10.1353/arw.0.0163.

Republic of South Africa. (2015) *South Africa's National Determined Contribution*. Pretoria: Republic of South Africa.

Republic of South Africa. Department of Cooperative Governance and Traditional Affairs. (2016) *Integrated Urban Development Framework*. Pretoria: Republic of South Africa. Department of Cooperative Governance and Traditional Affairs.

Republic of South Africa. Department of Economic Development. (2011) *New Growth Path: Framework*. Pretoria: Department of Economic Development. Available at: www.economic.gov.za/communications/publications/new-growth-path-series (Accessed: 4 September 2018).

Republic of South Africa. Department of Energy. (2018) *Integrated Resource Plan*. Pretoria: Department of Energy. Available at: www.energy.gov.za/IRP/irp-update-draft-report2018/IRP-Update-2018-Draft-for-Comments.pdf (Accessed: 4 September 2018).

Republic of South Africa. Department of Environmental Affairs. (2011) *National Strategy for Sustainable Development and Action Plan, 2011–2014*. Pretoria: Department of Environmental Affairs. Available at: www.environment.gov.za/sites/default/files/docs/sustainabledevelopment_actionplan_strategy.pdf (Accessed: 4 September 2018).

Republic of South Africa. Department of Environmental Affairs. (2016) *State of the Environment*. Pretoria: Department of Environmental Affairs. Available at: http://soer.deat.gov.za/newsDetailPage.aspx?m=66&amid=16320.

Republic of South Africa. Department of Trade and Industry. (2017) *Industrial Policy Action Plan, 2017/18–2019/20*. Pretoria: Department of Trade and Industry. Available at: www.thedti.gov.za/parliament/2017/IPAP_13June2017.pdf (Accessed: 4 September 2018).

Republic of South Africa. Department of Water Affairs. (2011) *National Waste Management Strategy*. Pretoria: Department of Water Affairs. Available at: www.environment.gov.za/sites/default/files/docs/nationalwaste_management_strategy.pdf (Accessed: 4 September 2018).

Republic of South Africa. Department of Water Affairs. (2013) *National Water Resource Strategy: Water for an Equitable and Sustainable Future*. Pretoria: Department of Water Affairs. Available at: www.dwa.gov.za/documents/Other/Strategic Plan/NWRS2-Final-email-version.pdf (Accessed: 4 September 2018).

Republic of South Africa. National Planning Commission. (2012) *National Development Plan*. Pretoria: Republic of South Africa.

Rock, M., Murphy, J. T., Rasiah, R., van Seters, P. and Managi, S. (2009) 'A hard slog, not a leap frog: Globalization and sustainability transitions in developing Asia', *Technological Forecasting & Social Change*, 76, pp. 241–254.

Rodrik, D., Subramanian, A. and Trebbi, F. (2004) 'Institutions rule: The primacy of institutions over geography and integration in economic development', *Journal of Economic Growth*, 9, pp. 131–165.

Rutledge, D. (2011) 'Estimating Long-term world coal production from S-curve fits', *International Journal of Coal Geology*, 85, pp. 23–33.

Schot, J. and Steinmuller, W. (2016) *Framing Innovation Policy for Transformative Change: Innovation Policy 3.0*. Sussex, UK: Science Policy Research Unit (SPRU).

Scoones, I., Leach, M. and Newell, P. (2015) *The Politics of Green Transformations*. London: Routledge Earthscan.

Seeliger, L. and Turok, I. (2016) 'The green economy accord: Launchpad for a green transition?', in Swilling, M. and Musango Wakeford, J. (eds.) *Greening the South African Economy*. Cape Town: Juta.

Sen, A. (1999) *Development as Freedom*. New York: Knopf.

Smith, A., Stirling, A. and Berkhout, F. (2005) 'The governance of sustainable socio-technical transitions', *Research Policy*, 34, pp. 1491–1510.

Southall, R. (2007) 'The ANC state, more dysfunctional than developmental?', in Buhlungu, S., et al. (eds.) *State of the Nation: South Africa 2006–2007*. Cape Town: HSRC Press, Introduction.

Strange, S. (1996) *The Retreat of the State*. Cambridge: Cambridge University Press.

Swilling, M. (2008) 'Tracking South Africa's elusive developmental state', *Administratio Publico*. Cape Town: HSRC Press, 16(1), pp. 1–29.

Swilling, M. and Annecke, E. (2012) *Just Transitions: Explorations of Sustainability in an Unfair World*. Tokyo: United Nations University Press.

Swilling, M., Bhorat, H., Buthelezi, M., Chipkin, I., Duma, S., et al. (2017) *Betrayal of the Promise: How South Africa Is Being Stolen*. Stellenbosch and Johannesburg: Centre for Complex Systems in Transition and Public Affairs Research Institute.

Swilling, M., Khan, F., Van Zyl, A. and Van Breda, J. (2008) 'Contextualising social giving: An analysis of state fiscal expenditure and poverty in South Africa, 1994–2004', in Habib, A. (ed.) *Social Giving in South Africa*. Cape Town: HSRC Press.

Swilling, M., Musango, J. and Wakeford, J. (2015) 'Developmental states and sustainability transitions: Prospects of a just transition in South Africa', *Journal of Environmental Policy and Planning*. doi: 10.1080/1523908X.2015.1107716.

Swilling, M., Musango, J. and Wakeford, J. (2016) *Greening the South African Economy*. Cape Town: Juta.

Thompson, G. (2009) 'Wither the 'Washington consensus', the 'Developmental State' and the 'Seattle Protests': Can 'Managed Free Trade and Investment' become an alternative development model? *RPD*, 34(131).

Trollip, H. and Tyler, E. (2011) *Is South Africa's Economic Policy Aligned with Our National Mitigation Policy Direction and Low Carbon Future: An Examination of the Carbon Tax, Industrial Policy, New Growth Path and Integrated Resource Plan*, Research Paper for the National Planning Commission Second Low Carbon Economy Workshop. Pretoria: National Planning Commission.

Turton, A. R., Patrick, M. J. and Rascher, J. (2008) 'Setting the scene: Hydropolitics and the development of the South African economy', *Reflections on Water in South Africa*. Patrick, M. J., Rascher, J. and Turton, A. R. (eds.), 24(3), pp. 319–323.

Van Agtmael, A. (2012) 'Think again: The BRICS', *Foreign Policy*, (196), pp. 76–79. Available at: http://foreignpolicy.com/2012/10/08/think-again-the-brics/.

van Koppen, B. and Schreiner, B. (2014) 'Moving beyond integrated water resource management: Developmental water management in South Africa', *International Journal of Water Resources Development*, 30(3), pp. 543–558.

von Holdt, K. (2010) 'Nationalism, bureaucracy and the developmental state: The South African case', *South African Review of Sociology*, 41(1), pp. 4–27. doi: 10.1080/21528581003676010.

Wilson, C. A. (2000) 'Policy regimes and policy change', *Journal of Public Policy*, 20(3), pp. 247–274.

Wolpe, H. (1995) 'The uneven transition from apartheid in South Africa', *Transformation*, 27, pp. 88–102.

Wright Mills, C. (1956) *The Power Elite*. Oxford: Oxford University Press.

Wright, J., Calitz, J., Bischof-Niemz, T. and Mushwana, C. (2017) *The Long-Term Viability of Coal for Power Generation in South Africa*. Pretoria: CSIR.

Ziervogel, G. (2019) *Unpacking the Cape Town Drought: Lessons Learned*. Cape Town: African Centre for Cities.

8

GLOBAL ENERGY TRANSITION, ENERGY DEMOCRACY, AND THE COMMONS

Introduction

The global energy transition is underway. Renewable energy (RE) is now more affordable than fossil fuels in nearly 100 countries across the world. Since 2009, investments in RE have exceeded investments in new fossil fuel generation every year despite the drop in oil prices from $140 per barrel before the collapse of Lehman Brothers in 2008 to below $60 per barrel one year later (climbing again to over US$100 until it dropped to US$43 in 2015, and steadily rising since then) (REN21, 2018). Annual investments in RE since the onset of the global financial crisis in 2007 have increased by 20% each year (REN21, 2018). Between 2009 and 2015, costs of wind energy dropped by 50% and the costs of solar PV modules dropped by 80% between 2008 and 2015 (Michael Liebreich, former CEO of Bloomberg New Energy Finance, quoted in Knuth, 2018:7). RE (including hydro) already meets 26% of global energy needs.

Although there are many causes of this phenomenon, what matters is that RE is no longer a niche innovation – as of 2017, it was a major US$280 billion global industry delivering mature energy systems at affordable prices (despite the volatility – but nevertheless overall decline – of oil prices since 2008). However, the energy landscape is changing so rapidly; as a result, reflections on the social and political implications are only just beginning.

This and the next chapter contribute to this emerging discussion, with particular reference to the argument in Chapter 4 about deep and just transitions. In this chapter it is argued that the RE revolution could potentially create the material base for a new kind of progressive politics of the commons.[1] If this happens, it will

This paper was co-authored with Megan Davies.

be argued, this could be decisive in ensuring that the deep transition underway also translates into a just transition. However, if the RE revolution is delivered via large extractive corporates deploying finance secured via global financial markets, the end result may well be decarbonized capitalism that remains financialized and unequal (Baker, 2015). This is not what a just transition looks like.

Whereas this chapter argues optimistically in favour of realizing the potential of a progressive politics of the energy commons, the next chapter explores the consequences of rising authoritarianism as regimes around the world act to defend the old fossil fuel and nuclear regimes.

This chapter draws together threads from previous chapters. It picks up the argument in Chapter 4 that the translation of the next deep transition into a just transition will depend on which direction the RE revolution goes. The radical potential of incrementalism as a theory of change (Chapter 5) and the case studies of local struggles to construct ecocultural commons across many localities in the global South (Chapter 6) provide the contextual background for imagining the potential flourishing of a RE commons across all world regions. However, unlike much of the literature on the commons and in line with the argument in Chapter 7, it will be necessary to build sustainability-oriented developmental states with appropriate polities that adopt and sustain policies that favour the conditions for perpetuating the energy commons (Cumbers, 2015; Routledge, Cumbers and Derickson, 2018).

One of the key conclusions that emerges from this chapter is that the origins of the RE revolution in Danish and German energy cooperatives suggest that there are already proven historical examples of the way the decentralized materiality of RE infrastructures could align with particular forms of locally controlled communal ownership and solidarity. This, in turn, created an open learning environment that made accelerated socio-technical learning possible. In light of the IPCC's Sixth Assessment Report that refers to a 12-year window to substantially decarbonize the global economy, accelerated socio-technical innovation and learning will be critical. However, this will not be possible if this is done via corporate-controlled, patented intellectual property regimes that have been proven to severely constrain innovation and accelerated learning (Standing, 2016). Unfortunately, the otherwise authoritative and influential Global Commission on the Geopolitics of Energy Transformation completely ignores this: it ignores financialization, the need for rapid learning and the vital importance of the commons as means for ensuring inclusive development (The Global Commission on the Geopolitics of Energy Transformation, 2019). In this light, exploiting the potential of RE to foster cooperative endeavour may not only be significant for those interested in social justice, it may actually be key to accelerated open learning and, therefore, the survival of human civilization as we know it.

Dimensions of the global renewable energy revolution

According to the authoritative REN 21 Report for 2018, 18.2% of total global final energy consumption came from RE (including hydro) in 2016, of which

10.4% comprised of 'modern renewables' (i.e. from all technologies excluding hydro). In 2017 total RE capacity grew faster (9% year-on-year) than any previous year. RE accounted for no less than 70% of all new net additions (from RE and non-RE sources) to total final energy consumption in 2017. This had a lot to do with the extraordinary growth in solar PV, which accounted for 55% of all newly installed RE capacity in 2017. Solar PV is now growing faster than any other type of RE, while hydro is relatively stagnant. More solar PV capacity was constructed in 2017 than fossil fuel and nuclear power capacity combined. The cost of solar PV dropped 73% between 2010 and 2017, to US$100 per MWh (ranging from less than US$10c/kWh in China and India to between 10c and 15c in Africa, Central America, Eurasia, Europe, Middle East and North America) (REN21, 2018:119–220). Offshore wind in 2017 was below US$10c/kWh everywhere. These prices are lower per kWh over the life cycle than fossil fuels and nuclear power. As a result, total investment in RE increased from US$274 billion in 2016 to US$279.8 billion in 2017 – well over the US$200 billion per annum that was being invested per annum in the few years after 2007 (REN21, 2018:17–18). As is clear from Table 8.1, 179 countries had set RE targets at various levels of government by 2017, and 57 countries were committed to meeting 100% of their electricity requirements from RE.

Investment in electrification is growing two-thirds faster than total electricity consumption, driven to a large extent by major programmes to connect mass mobility to electricity grids (e.g. electric cars, buses and trains) and deploy heat pumps (for heating/cooling). As a result, 40% of total power sector investment in 2017 was in expanding the distribution and transmission capacity of the electricity networks (compared to the other investments in coal, gas, oil, nuclear [lowest], solar PV [highest], wind and hydropower) (International Energy Agency, 2018:49). The fact that grids are expanding faster than electricity consumption is a key indicator of the shift away from fossil fuels to renewables which are, by definition, dependent on ICT-enabled grids for their expansion.

As is clear from Figure 8.1, investment in RE since 2007 has steadily risen to US$280 billion, with declines in 2011–2013 and 2015–2016 caused by steep price drops rather than a drop in production of new capacity. Total investment in new

TABLE 8.1 Renewable energy indicators, 2016–2017

Investment	2016	2017
New annual investment in power and fuels (billion US$)	274	279.8
Power		
Renewable power capacity (including hydro)	2017GW	2195GW
Renewable power capacity (not including hydro)	922GW	1081GW
Policies		
Countries with RE targets	176	179
Countries with 100% renewable electricity targets	57	57

Source: Adapted from REN21, 2018:19

Global New Investment in Renewable Power and Fuels in Developed, Emerging and Developing Countries, 2007-2017

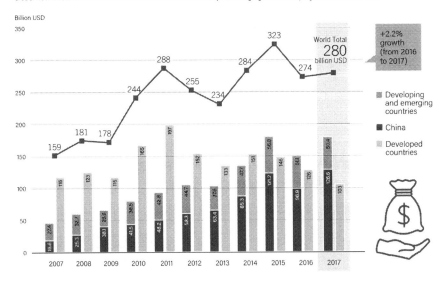

FIGURE 8.1   Global new investment in renewable power and fuels in developed, emerging and developing countries, 2007–2017.

Source: REN21, 2018:140

RE capacity in 2017 was twice the amount invested in new fossil fuel and nuclear capacity combined. However, as RE investments in developed countries declined from 2014, investments have steadily risen in developing countries (except for one year, 2015–2016). China alone accounted for 45% of total investments in RE in 2017 (REN21, 2018:140). Whereas virtually no incentives favouring RE existed in developing countries in the early 2000s, by the second decade of the twenty-first century governments across all major regions in the developing world had adopted policies that favoured RE investments (in particular, the widely used Feed-In Tariff or 'auction' mechanism) (Aklin and Urpelainen, 2018: Kindle Location 4211).

The evidence about investment flows confirms the argument in Chapter 3 about the redirection of investment finance during the post-crisis period into 'green-tech', specifically into RE. Confirming Mazzucato's argument that state support is needed for R&D and risk reduction during the early stages of the innovation cycle (Mazzucato, 2015), up until 2010 most RE firms depended on investments from government agencies of one kind or another (e.g. Development Finance Institutions). However, since 2010, private sector investments in RE have escalated rapidly, most notably drawing funds from a new generation of so-called green bonds (Aklin and Urpelainen, 2018: Kindle Location 4167). Bloomberg New Energy Finance has estimated that cumulative investments in RE between 2014 and 2030 will be US$5.1 trillion, of which more than half (US$3.6 trillion) will be invested in Asia (quoted in Aklin and Urpelainen, 2018: Kindle Location 4179). REN21 tracks six investment flows (as per Table 8.2).

TABLE 8.2 REN21 investment flows

Global research and development	Asset finance	Small-scale distributed capacity investment	Public market investment	Venture capital and private equity	Acquisition activity
US$9.9 billion	US$216 billion	US$49.4 billion	US$5.7 billion	US$1.8 billion	US$114 billion

Source: REN21, 2018:145

The following is the breakdown of these investments for 2017 (REN21, 2018:145).

Global R&D: investments in R&D increased by 6% in 2017 compared to 2016 to US$9.9 billion. However, the government investment remained flat, which means the increase is primarily due to increased investment in R&D by corporates. Interestingly, US investments in R&D outstripped China's despite the total dominance by China of total investments in RE. As explained in subsequent pages, this is because US-based investors are looking for what Bill Gates calls an "energy miracle" led by US tech companies, while the Chinese are content to roll out mature technologies on a global scale.

Asset finance: hitting US$216 billion in 2017, asset financing of utility-scale projects by private and state-owned financial institutions dominated total investment in RE in 2017 (see further analysis of this category of investment in subsequent pages). Chinese investment via state-owned entities in solar alone was US$64.9 billion.

Small-scale distributed capacity: the 15% jump in investments to US$49.4 billion in small-scale solar PV (less than 1 MW) is significant, not least because this suggests where long-term future growth prospects may lie with significant political implications from a commons perspective.

Public market investment: although RE companies raised over US$5 billion via public markets, overall there is a decline in investments from these sources. This reflects the failure of the Silicon Valley-led RE boom of the early 2000s (see subsequent pages), the related decline in the emergence of new companies (seeking to raise funds as IPOs) coupled to a rise in mergers and acquisitions which reflects a degree of consolidation to handle bigger projects.

Venture capital and private equity: there is a general decline in investment from these sources to US$1.8 billion, most probably reflecting the maturing of the technologies in ways that are more appropriate for large-scale corporate investors and dividend-oriented funds rather than the capital gains orientation of VC funds which, in any case, prefer to steer clear of large-scale dividend-oriented investments.

Acquisition activity: although not counted by REN21 as part of the US$280 billion invested in 2017, asset acquisition transactions worth US$114 billion did take place. These included private equity buyouts worth US$11.2 billion. Asset acquisitions and refinancing transactions worth US$87.2 billion also took place. The extent of acquisition activity coupled to a decline in the role of venture capital may reflect the consolidation of the sector characterized by capital-hungry large-scale projects out of reach of venture capital and more appropriate for large-scale companies with balance sheets that give them access to large-scale investment finance (further discussion on these dynamics is given in subsequent pages).

The investment analyses provided by REN21 (2018) and the Global Commission (The Global Commission on the Geopolitics of Energy Transformation, 2019), however, fail to distinguish the investments made by citizens and communities from the other investors (elaborated in subsequent pages). Given that nearly half of these investments were made by citizens/communities in the frontrunner countries – Denmark and Germany – for nearly two decades (Yildiz, 2014; Bauwens, Gotchev and Holstenkamp, 2016), this omission is bizarre. The explanation for this, though, is simple: from a neoliberal perspective, the only investments that matter are those that emanate from the formal financial, industrial, public and utility sectors. It is a perspective, however, that occludes what is most significant about the RE revolution, namely its potential for fostering a new progressive politics of the commons based on the active participation of local communities (discussed further in subsequent pages).

Most RE projects are financed in one of five ways: on-balance-sheet funding from a utility; an independent power producer using debt and equity funding of one sort or another; project finance in the form of bank loans securitized by the project itself; debt and/or equity from a publicly owned DFI; institutional investors (e.g. pension funds) and citizens/communities. In 2017, the large bulk of investments came from utilities (US\$121.5) and project finance (US\$91.2). Investors were able to access funds raised via the issuance of 'green bonds'– a record US\$163.1 billion was raised in 2017, with ROI conditions that tend to stimulate financialized transactions (elaborated further in subsequent pages). These bonds originate from diverse sources, including green bonds issued by sovereign governments, a new generation of non-financial corporations promoting ROI in the 'green-tech' sector and (mainly in the United States) financial institutions trading in asset-backed securities to generate funding for RE investments (related, for example, to the financing of installations of residential solar PV) (REN21, 2018:147).

What really matters is the relative balance between public and private investors. Conventional 'market failure' logic would envisage predominantly public sector investments in upstream R&D (because the returns are public domain knowledge, not privately owned intellectual property) and private sector investment in downstream deployment where risks are lower and returns accrue to the investor. This is not the case in the RE sector (nor was it the case in the IT sector during the last two decades of the twentieth century (Mazzucato, 2011)). As Mazzucato and Semieniuk show, what clearly emerges is that (like the early phases of the IT sector) public sector investment in downstream projects (i.e. not just upstream R&D) has played a crucial role in ensuring that investments in RE have consistently grown by around 20% per annum. In their detailed analysis of asset financing of utility scale RE during the period 2004–2014, Mazzucato and Semieniuk find that

> in spite of widespread energy sector privatization and public sector austerity, public investors are playing an increasingly important role in financing the deployment of RE technologies and are the *only reason* that RE asset finance has experienced any growth at all between the onset of the 2008 financial crisis and 2014.
>
> *(Mazzucato and Semieniuk, 2018:*
> *14 – emphasis added)*

That rising levels of public investment is the *only* reason for rising levels of private sector investments in RE is, indeed, a very strong statement with major implications for climate-related funding flows and the role, in particular, of DFIs.

However, as the aforementioned figures reveal, what matters is not simply the quantities of investments by public and private actors (Figure 8.2) but also what types of investors tend to invest in higher risk projects (Figure 8.3). This factor affects the directionality of innovations in the RE sector: if investments only crowded into low risk investments, the scope for – and spectrum of – innovations over the longer term would be much narrower. The result could then well be the 'locking-in' of lower impact technologies and systems that will only have (relatively marginal) climate effects over a very long period – an outcome that could result in runaway global warming that could destroy the livelihoods of millions. Note,

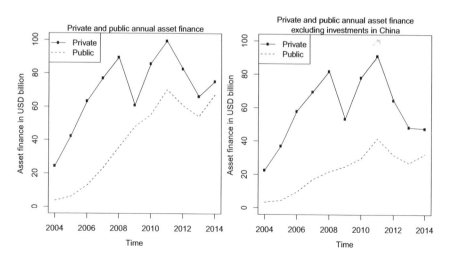

FIGURE 8.2 Volume of annual public and global private asset finance (left panel) and excluding China (right panel).

Source: Mazzucato and Semieniuk, 2018:15

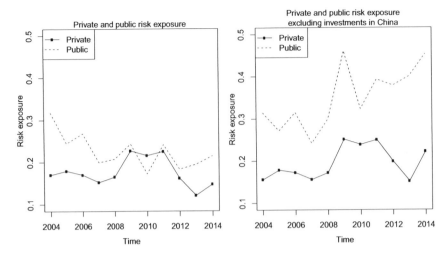

FIGURE 8.3 Exposure to risk of annual public and private asset finance for global investments (left panel) and excluding China (right panel).

Source: Mazzucato and Semieniuk, 2018:15

for example, how the risk profile of PV (CS) has gone from high to low risk in a decade. If investors avoided high-risk investments, the steep learning curve this particular shift implies would not have occurred. To correlate types of investors with a risk assessment of investments, Mazzucato and Semieniuk describe the main technologies according to their risk profile over time as given in Table 8.3.

Mazzucato and Semieniuk then correlate investments by different categories of investors to the technology risk categories. This allows them to conclude that although the overall quantity of public investment in RE was lower than private investment over the 2004–2014 period, public investors tended to invest in higher risk projects than private sector investors. However, as public investment in higher risk projects increased, so private sector investment in higher risk projects also increased (hence, the data referred to here about rising private sector investments in R&D, while public sector investments in R&D tapered off in the 2016–2017 period). This confirms the argument that the passage from niche innovations to mainstreaming depends not only on public investment in upstream R&D but also public investment in actual projects further downstream (Mazzucato and Semieniuk, 2018:16). Following Mazzucato's overall theory of investment (Mazzucato, 2011; Mazzucato and Penna, 2015), the fact that this leverages the private sector into higher risk investments over time confirms that public sector investments have the effect of de-risking the implementation of innovations during the early phase of the innovation cycle. This is certainly what happened in the South African case where sovereign guarantees and investments by DFIs had similar effects. As a general principle, this confirms our understanding of the collibratory governance of non-equilibrium economies going through a deep (and potentially just) long-term

TABLE 8.3 Technology risk classifications, 2004–2014

Technology	Sub-technology	Risk
Wind		
Onshore		Low
Offshore		High
Solar		
Crystalline silicon (PV)		High (2004–2006), medium (2007–2009), low (2010–2014)
Other PV	Thin film PV	High (2004–2009), medium (2010–2014)
	Concentrator PV	High
Concentrated solar power (CSP)		High
Biofuels		
First generation		Low
Second generation		High
Other technologies		
Biomass and waste	Incineration	Low
	Other biomass technologies	Medium
Geothermal		Medium
Marine		High
Small hydro		Low

Source: Mazzucato and Semieniuk, 2018:13

transition: (as argued in Chapter 7) without appropriate interventions by state institutions of various kinds, such a transition would be inconceivable.

Two numbers leap out of the preceding paragraphs that reveal the state of play in the global RE revolution: the growing levels of private sector investment in RE (at both the R&D and downstream deployment levels) and the extent of state-controlled Chinese investment. A key driver of the former trend was the Breakthrough Energy Coalition (BEC) initiated by Bill Gates during the lead-up to the decisive meeting of COP21 in Paris in 2015 which, in turn, gave birth to Breakthrough Energy Ventures – a US$1 billion fund created by a bunch of US-based tech-billionaires. As Knuth observes, the focus of this initiative was on R&D investments in RE 'breakthroughs' intended to change the ballgame – or as Gates himself put it: "We need an energy miracle" (Knuth, 2018:5). Significantly, Gates argued that these funds must be complemented by state investments for the strategy to work (which confirms the argument using different sources in the previous paragraph). In other words, following Knuth's argument, the assumption was that the only way the global economy could be decarbonized is by discovering new 'breakthrough' techno-fixes and state investments needed to complement private sector investments. And the agenda is clear: the promotion of new *US-led* 'breakthrough' innovations to counter the market dominance enjoyed by the Chinese (Knuth, 2018).

Sounds logical enough, except that it ignores the fact that RE is *already* expanding very rapidly using existing mature technologies! While Gates and his friends invest in an 'energy miracle' *that must still happen*, Chinese factories are churning out most of the components of RE energy infrastructures being installed by thousands of businesses all over the world at rates of growth of 20% per annum. In other words, what Gates would like to see happen is already happening – the real 'problem' (for US corporate interests) is that China leads it, not US-based innovators. As Joe Romm wrote in 2016 in an article tellingly entitled "No Bill Gates, We Don't Need 'Energy Miracles' to Solve Climate Change":

> We have seen that aggressive deployment of clean energy technology driven by government policies has – as was predicted – led to precisely the kind of game-changing cost-slashing innovation that Gates mistakenly thinks happens primarily from basic energy research and development (R&D). For six years, Gates has claimed we were wildly under-investing in basic energy R&D. Yet, somehow the very thing Gates says he wanted – huge price drops in key low-carbon technologies (like renewables and efficiency) and key enabling technologies (like batteries for storage) – kept happening. The fact is that accelerated deployment policies around the world created economies of scale and brought technologies rapidly down the learning curve.
>
> *(Romm quoted in Knuth, 2018:6–7)*

The problem is that the 'breakthrough' model was tried and failed dismally in the 2000s. Silicon Valley led a 'clean energy' goldrush with venture capital firms investing US$25 billion between 2006 and 2011 in RE but with very few successes (Knuth, 2018:10). A notable exception was, of course, Elon Musk's Tesla. A key cause of this failure was Chinese competition. Investment in China's RE sector rose from US$2 billion in 2007 to over US$100 billion by 2014 (Knuth, 2018:12), using mature mainly German technologies that Chinese companies were licensed to produce. China's investment levels in R&D have, as a result, been consistently low. China is a manufacturer of proven technologies, not an innovator in search of tech-miracles. To counteract this, besides the Gates initiative, many alternative strategies were pursued in the United States to get around the Chinese market advantage. This includes imposing tariffs on imported Chinese solar products (contested by the Chinese at the WTO), shifting into so-called 'clean tech 2.0' (focusing on distribution of RE by building the 'software' for activating a RE internet of multiple generators-cum-users rather than on the 'hardware' of generation capacity which is a market the Chinese have cornered) and building new financialized business models (Knuth, 2018).

As far as financialization of RE is concerned, two strategies are relevant. Firstly, it means moving away from feed-in tariffs towards auction mechanisms (Leiren and Reimer, 2018). The former is a fixed tariff set by the buyer/regulator for energy fed into the grid (often guided by policies informed by social imperatives) and the latter is a market-oriented price fixed by the seller competing with other sellers bidding to supply energy into the grid. The second is when RE is sold as a service to customers rather than as infrastructure to generate a fixed long-term revenue stream (which can then, in some cases, be refinanced via complex securitization structures (Knuth, 2018)). From a US perspective, combining the so-called financial innovation (which is, in reality, legalized rent seeking) and the 'Internet of Things' to create a new RE economy is, typically, the alternative to the Bill Gates–type 'tech-miracle' approach and the community energy approach, pioneered in Minnesota and discussed later in this chapter (Bolinger, 2005).

In short, if Chinese hegemony is premised on (subsidized) productive superiority in large-scale manufacturing (within and outside China), then the competitive response by coalitions in other countries would inevitably be investments in lowering the transactional costs of RE in the distribution system for the end-user. Usually this means combining 'breakthrough' RE technologies with information and communication technologies aimed at making the acquisition of solar energy as easy as buying a cellphone and airtime. If driven by finance, this would entail high-risk informational and financial innovations aimed at extracting profits from increasingly complex – and therefore fragile – systems that could, ultimately, threaten rather than extend the RE systems that are needed to combat climate change (Knuth, 2018). The alternative to this is, of course, the commons approach that triggered the RE revolution in Denmark and Germany in the first place (more on this in subsequent pages).

Based on in-depth case study work, the authors of the most authoritative recent overview of the political economic history of RE since the 1970s concluded that three future trends can be expected (Aklin and Urpelainen, 2018). Firstly, they expect to see "a massive increase in the deployment of renewables, driven by decreasing costs and higher investments" (Aklin and Urpelainen, 2018:Kindle Location 4211). The frontrunners, Denmark and Germany, will continue to lead because RE has become so entrenched in their respective political economies, there is now a broad cross-party and public consensus that RE is part of the energy futures of these two countries. Many other countries, Aklin and Urpelainen argue, will aim to catch up and even surpass what Denmark and Germany have achieved. What they fail to see is that commons-type initiatives in Denmark and Germany were crucial in achieving this broad public consensus (Debor, 2018). As corporates move into the RE sector as a whole, the driving forces behind RE investments in developing countries are far more top-down, with corporates and/or state institutions playing leading roles. It would be unwise, therefore, to be too optimistic about this expectation.

Secondly, as the RE industries grow in size relative to the rest of the economy, "they acquire political clout" (Aklin and Urpelainen, 2018:Kindle Location 4230–4231). In short, the polity gets recomposed to accommodate these industries as the traditional 'mineral-energy complex' interests lose ground both politically and financially (unless there is a backlash, see Chapter 9). This ensures the consolidation of a political and policy environment that favours the RE sector, albeit not irreversibly. If a backlash is avoided, what matters is not simply the political empowerment of the RE sector per se, but whether this comprises a handful of powerful RE corporations, or whether it comprises a coalition that includes those who operate within the energy commons – the kind of coalitions that were successful in Denmark and Germany in securing policy changes that benefitted the sector. Unfortunately, this distinction between a corporate versus a commons-based coalition is ignored by Aklin and Urpelainen.

Thirdly, they expect to see the acceleration of RE investments in developing countries where, they (incorrectly) assume, there will be very little political opposition (see Chapter 9). For this reason, they expect to see far less political resistance to RE in developing countries than occurred in developed countries. What they ignore are the political consequences of corporate-driven extractive RE programmes in developing countries characterized by extreme inequality and high levels of poverty.

What Aklin and Urpelainen completely miss – as do many others writing about the rise of RE (e.g. Knuth, 2018; Mazzucato and Semieniuk, 2018) – is the significance of the remarkable role played by the commons in driving the energy transitions in Denmark and Germany (Bolinger, 2005; Yildiz, 2014; Moss, Becker and Naumann, 2015; Bauwens, Gotchev and Holstenkamp, 2016). By 2000, 84% of wind capacity in Denmark was owned by communities, which in turn comprised 175,000 households (Bolinger, 2005). By 2010, about half of the RE capacity in Germany was owned by private citizens, while another 40% was owned by strategic investors and public institutions of various kinds. Conventional utilities only owned about 10% of total capacity in Germany by 2010 (Aklin and Urpelainen, 2018:Kindle Location 4349). However, as corporates, utilities and for-profit financial institutions (with access to the funds flowing through the new generation of 'green bonds') have become increasingly involved in the RE sector since 2007, the energy commons has been rolled back by a sophisticated finance-led enclosure movement that is hell-bent on commodifying the entire value chain, including securitization of the annuity flows. This coincides with a shift from Feed-in Tariffs (FiT) to market-oriented price-driven Feed-in Premiums (FiP).

The mainstream model that is emerging from the commodification and financialization of RE seems to be led by two different types of entities: corporates with conventional for-profit financing using familiar securitization mechanisms and Chinese companies which are (partially) state-owned with access to funding from state-owned banks – both these models are antithetical to the commons model. The corporate or China model is what is driving the uptake of RE in developing countries. Alternative community-based models similar to what emerged in Denmark and Germany from the 1980s onwards (with roots in the 1970s) are few and far between in those developing countries where renewables are rapidly expanding.

Aklin and Urpelainen miss completely the political significance of this shift from the commons to commodification. Furthermore, also missing the significance of the commons, the financial analysis provided by Mazzucato and Semieniuk focuses on the period during which non-commons financing was displacing the commons-based financing that launched the energy transition in the frontrunner countries in the first place!

A far more useful set of expectations than the three offered by Aklin and Urpelainen would need to be about anticipating the future directionality of the RE revolution. The first expectation would be that despite the enclosure movement mounted by states and corporates, the nature of RE lends itself to a commons-based approach, especially with regard to the potential of mini-grids (discussed further in subsequent pages). It could spread if a coalition emerged that understood the potential of bottom-up non-financialized and therefore lower risk alternatives.

The second expectation is that China will extend and deepen its dominance of the manufacturing end of the RE value chain by continuing to ensure that it undercuts everyone else on price (using subsidies and relatively cheap labour), thus creating a set of import-cum-installation RE businesses across all world regions that are highly dependent on Chinese manufactured products. The third expectation is that the first and second expectations will be resisted by 'Western' bids by leading tech-capitalists to lead the RE revolution via massive investments in new technology 'breakthroughs' and/or new (inevitably) financialized business models for reducing the transactional costs for the end-user by increasing systemic risk – after all, that is what financialization is all about.

The significance of the emergence of an energy commons (that is now under threat) will be discussed further later. Before that, however, it is necessary in the next section to discuss one of the key drivers of the decline of fossil fuels, namely Energy Return on Investment (EROI). It is also necessary to briefly discuss in the section after that the relationship between energy infrastructures and the dynamics of mass politics, with special reference to coal and oil infrastructures. Without understanding these dynamics, the connection between the RE revolution, global transitions (Chapter 4) and material flows (Chapter 3) will remain opaque.

Energy return on energy invested

In simple terms, the EROI refers to the total amount of energy required to extract a unit of energy, expressed as a ratio – the more energy being extracted, the larger the ratio. In the 1930s, the EROI ratio for oil was 1:100, that is, one barrel of oil made it possible to produce 100 barrels of oil. By the start of the twenty-first century, this ratio had reduced down to between 1:10 and 1:20 (Ayres, 2016:409). Between 1960 and 1980, the EROI declined by more than half, from 35 to 15 even though the total quantity of oil produced increased overall (Ayres, 2016:409–410). Obviously, as the EROI ratio diminishes, so costs per unit of energy goes up. This biophysical reality cannot be conjured away by some clever quantitative trickery.

There are two biophysical drivers of this declining EROI ratio: the depletion of high-quality resources and the consequent increasing dependence on hydrocarbon resources that produce lower quantities of energy that are more expensive to extract (Figure 8.4) (Ahmed, 2017:15).

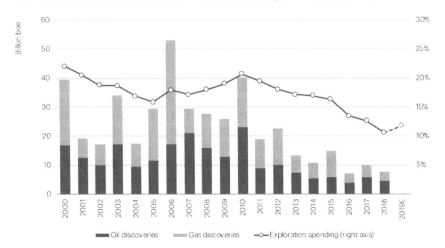

Global conventional resources discoveries and exploration spending as % of total investment

FIGURE 8.4 Global discovered conventional resources and share of exploration in total upstream investment, 2000–2018.

Source: IEA analysis with calculations based on Rystad Energy (2019). IEA (2019), World Energy Investment. All rights reserved.

As a result, only if oil prices remain high (around US$100) will it be possible to profitably extract low-quality hydrocarbons (Ayres, 2016:313). However, high oil prices undermine growth. Without growth, demand contracts and oil prices drop. When oil prices drop, production of unconventional oil (i.e. low EROI resources) is curtailed because it gets less profitable; but growth can be stimulated by the lower oil prices. Growth results in increased demand that pushes up the oil prices, which, in turn, makes unconventional oil extraction possible – for a while. As oil prices rise, growth is undermined, demand contracts, oil prices drop, extraction of unconventional oil becomes unprofitable. And so a global economy addicted to oil bumps along the bottom – volatility becomes the 'new normal'. Add into this mix the increasingly influential disinvestment campaign (that targets the $5 trillion subsidies that the oil industry attracts) and what the 'fracking' bonanza in the United States has done to oil prices, and the gloomy picture facing the oil industry gets clearer.

Figure 8.5 clearly demonstrates how the rate of growth in oil production over time has steadily declined from an average of 7.9% per annum during the period 1965–1974 to 0.1% per annum since the turn of the millennium. The alternatives to conventional oil are tar sands, shale oil and shale gas. However, the EROIs of these resources are between 1.5 and 4, far lower than the current EROI of conventional oil (Ahmed, 2017:18). This is why high prices are needed for these low EROI resources to be profitably exploited.

As far as oil reserves are concerned, former Chief Economist of Royal Dutch Shell Michael Jefferson concluded that the "the standard claim that the world

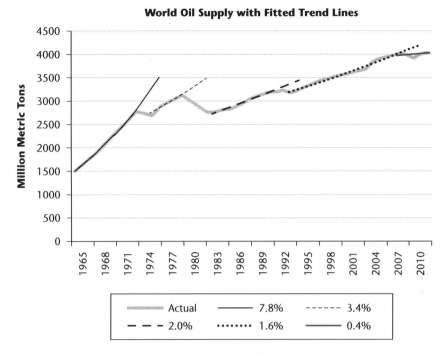

FIGURE 8.5   World crude oil production and fitted growth per cent.

Source: Ahmed, 2017:16

has proved conventional oil reserves of 1.7 trillion barrels is overstated by about 875 billion barrels" (quoted in Ahmed, 2017:19). Nevertheless, the mainstream oil industry remains optimistic that daily oil consumption will – and can – rise from the current 85 million barrels/day to 104 million barrels /day by 2030 (Bradshaw, 2010:277). They concede, however, that this is only possible by significantly increasing output from unconventional oil resources (the same resources with low EROIs). But for this to work, oil prices must remain on or above US$100/barrel – a high risk assumption.

Instead of debating whether we have reached peak oil production or not, it is "more useful to speak of the peak and decline of EROI as a measure of the health of the global energy system" (Ahmed, 2017:20). The EROI of oil and gas peaked in 1999 despite increases in overall production, albeit at ever-declining growth rates (Ahmed, 2017:20). This has major implications for the profitability of oil companies and prospects for future economic growth.

While the six largest oil and gas companies increased investments by 80% between 2007 and 2013, their collective output declined by 6% (Brown, 2015:Kindle Location 351). Major projects have recently been cancelled because oil companies have no certainty that they can expect average oil prices to rise to – and remain at – US$100 per barrel or more (Brown, 2015:Kindle Location 326; see also Ayres,

2016:313–314). Investment advisors and banks are, unsurprisingly, advising their clients against investments in oil stocks, and the world's major financial institutions have announced plans to disinvest from coal. This, in turn, could exacerbate an already dire conundrum that oil companies face.

Oil is the master energy resource of the industrial era. Unsurprisingly, therefore, as reflected in Figures 8.6 and 8.7, economic growth rates have tracked the growth in oil production. As the absolute size of the global economy has expanded (measured in terms of GDP), more energy was consumed (Figure 8.7) at almost the same rate of growth. As the year-on-year rate of growth in oil production has declined, so has the rate of growth in energy use and, of course, GDP. Figure 8.8 shows how economic growth rates have generally tended to decline since 1961.

In short, the data presented in this section clearly reveal the thermodynamics of the interplay between the EROI ratio and the economic growth rate. Conceptually disconnecting the economy from its geophysical context may allow neoclassical economists to assume that economies tend towards equilibrium, but this also makes it impossible for them to explain the average long-term decline in economic growth rates. The clear and compelling evidence of the thermodynamic link between the declining EROI ratio since 1999 and declining growth rates demonstrates how economies – understood as social-ecological systems – tend towards disequilibrium. The declining EROI and the almost zero growth in conventional oil are the clearest indicators that we have hit the socio-metabolic limits of the industrial epoch. Industrial civilization has evolved by assuming oil will always be available to endlessly grow the economy. This assumption no longer holds. It is only a matter of time before this industrial

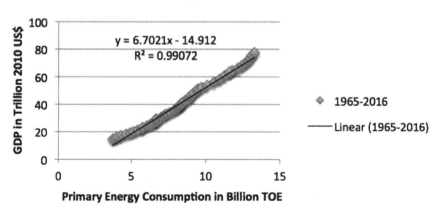

FIGURE 8.6 Energy consumption versus world GDP.

Source: Tverberg, 2018

World Growth in Oil, Energy and GDP

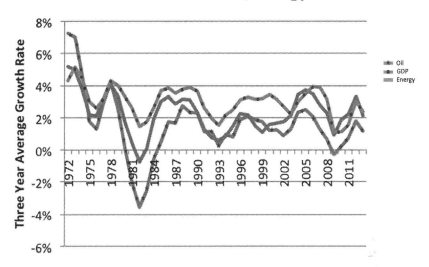

FIGURE 8.7 World growth in oil, energy and GDP.

Source: Tverberg, 2018

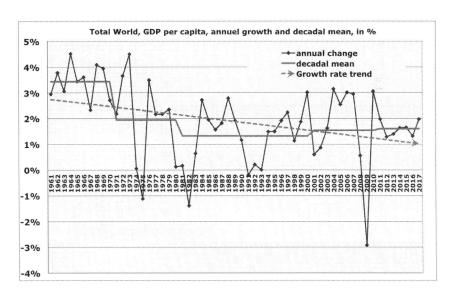

FIGURE 8.8 Declining rate of economic growth.

Source: Ahmed, 2017:28

civilization collapses as it detaches from its biophysical conditions of existence or adapts via a deep and just transition.

Nor are market forces likely to drive the emergence of substitutes that can do what oil can do. As the EROI continues to decline and the global economy bumps along the bottom, the global energy revolution is taking off. This clearly puts in place one of the key conditions for the next long-term development cycle which also correlates with a socio-metabolic transition to a more sustainable socio-metabolic order. In short, what we are witnessing is the commencement of the next deep transition. However, what is the relationship between the decentralized and distributed socio-technical infrastructures that the global energy revolution is putting in place and the political agency required to translate the deep transition into a just transition? To address this question, we need to ask what the relationship was between coal and oil infrastructures and political dynamics. This will help us to explore the politics of RE infrastructures in a subsequent section.

Socio-technical dynamics of energy politics[2]

The specific material configuration of particular energy infrastructures has always shaped the social and political landscape since the onset of the industrial epoch (Mitchell, 2011; The Global Commission on the Geopolitics of Energy Transformation, 2019). This is true for the coal infrastructures built to drive the industrial revolution and colonization in the nineteenth century, social democracy in the twentieth century and the oil infrastructures which became the foundation of globalization and neoliberalism in the second half of the twentieth century/early twenty-first century. As Timothy Mitchell — the most articulate historian of the evolution of 'carbon democracy' as a socio-technical system — puts it:

> Understanding the question of oil and democracy starts with the question of democracy and coal. Modern mass politics was made possible by the development of ways of living that used energy on a new scale. The exploitation of coal provided a thermodynamic force whose supply in the nineteenth century began to increase exponentially. Democracy is sometimes described as a consequence of this change, emerging as the rapid growth of industrial life destroyed older forms of authority and power. The ability to make democratic political claims, however, was not just a by-product of the rise of coal. People forged successful political demands by acquiring a power of action *from within the new energy system*. They assembled themselves into a political machine using its processes of operation. This assembling of political power was later weakened by the transition from a collective life powered with coal to a social and technical world increasingly built upon oil.
>
> *(Mitchell, 2011:12 – emphasis added)*

Mitchell's insight is that the material configurations of energy infrastructures create the necessary – but not sufficient – conditions for particular modes of socio-political organization and mobilization. This notion that social groups acquire "a power of action from within the new energy system" illuminates the linkage between political dynamics and socio-technical systems during different historical phases, starting with the transition to coal-based energy, then to oil-based energy and now to renewables-based energy. As elaborated in subsequent pages, each of these corresponded with very different configurations of energy infrastructures.

The near half century that led up to World War I is often referred to as both the 'age of democratization' and the 'age of empire'. Mitchell demonstrates that both were connected in ways that were profoundly shaped by the material reality of coal infrastructures (Mitchell, 2011). The industrial elites that emerged as the dominant political force from the industrial revolution crafted limited representative democracies to govern the industrial heartlands in North America and Europe, the racist colonial administrations in the different European empires and various 'internal colonialisms' in countries like the United States, Australia and South Africa where the large-scale nineteenth-century migrations of white people and people of colour literally collided in the mines created in these 'new worlds' (Hart and Padayachee, 2013).

In the industrial heartlands, coal was extracted from coal mines concentrated in particular areas and then linked to urban centres via railway networks. Following Mitchel (2011), these mines and railway networks became the material basis for concentrations of large numbers of industrial workers who, in turn, formed militant industrial unions. It was these unions that became the primary opposition to the limited representative democracies in the industrial centres. By the end of World War II, they had become the social basis of the social democratic parties that won the general elections on both sides of the Atlantic after the cessation of hostilities and the conclusion of the agreements reached at the Bretton Woods conference. Fusing together Keynesian thinking with progressive union-based demands for full political and social inclusion, these 'Western' post–World War II governments instigated a so-called golden era of economic growth, rising wages, high taxes, extended social welfare, reduced inequalities and social stability.

The coal-based infrastructure that connected the colonies to the metropoles, however, had a very different socio-political impact within the colonies. The primary purpose of the colonies was to provide raw materials and cheap agricultural produce required by the expanding coal-fired economies of the industrial metropoles. Urbanization was limited, and the majority remained within rural agricultural economies dependent on peasant modes of production. Where unions formed, these were mainly dockworkers employed by the ports that conveyed the outward-bound flow of raw materials or mineworkers where mines existed. Many leaders of dockyard or mining unions in Africa became significant political leaders during the anti-colonial struggles. The current President of South Africa used to be the Secretary-General of the National Union of Mineworkers.

Where colonial authorities encouraged settler enclaves, the disenfranchisement of the majority was justified in racial terms. It was only after World War II that anti-colonial movements started to make progress, culminating in Indian independence in 1947 followed shortly thereafter by independence for most colonies by the early 1960s. The debt and grant funding of the first era of post-colonial 'modernization' was explicitly focused on the construction of energy infrastructures delivered by increasingly powerful multinational corporations. Western banks – and later on the World Bank – played a key role in lending money for these projects. The debt default crisis of the early 1980s hinged to a large extent on problems with energy-related loans.

With the decline of the United Kingdom and the rise of the United States as the Western global hegemon in the context of the Cold War, the United States used the post–World War II Marshall Plan for Europe and its related insistence on the dismantling of the European Empires to promote a transition from coal-based to oil-based energy. Partly related to political battles against union power rooted in coal production within the United States and partly to promote the exports of new-found oil resources and manufactured goods, the promotion of oil-based energy infrastructures only really succeeded from the late 1970s onwards. The oil crises of the 1970s and the peaking of oil production in the United States in the late 1970s forced the United States to realize it needed to secure political and military control of the oil economies of the Middle East. And Maggie Thatcher's first major political move after becoming Prime Minister in 1979 was to smash the mineworkers union and close the coal mines as a precursor for shifting to a dependence on oil within a world dominated by neoliberal economic policies and the deregulation of finance.

Whereas coal-based energy infrastructures in the industrial heartlands enabled the consolidation of powerful industrial trade unionism, oil-based infrastructures resulted in fragmentation and the weakening of labour. Unlike coal, oil was shipped around the world in tankers. These tankers operated outside sovereign jurisdictions and often under 'flags of convenience'. The applications of oil – from lighting to medicines, to plastics and motor vehicles – became increasingly automated: oil (as petroleum or diesel or kerosene) could be piped in and combusted, with no equivalent of a 'stoker' to keep the fire going. As Mitchel sums up,

> In other words, whereas the movement of coal tended to follow dendritic networks, with branches at each end but a single main channel [railways], creating potential choke points at several junctures, oil flowed along networks that often had the properties of a grid, like an electricity network, where there is more than one possible path and the flow of energy can switch to avoid blockages or overcome breakdowns.
>
> *(Mitchell, 2011:38)*

Oil became the master fuel of the age of financialized globalization and neoliberalism. Vast globally connected infrastructures that included oil wells in remote tropical regions or desert regions presided over by authoritarian regimes (in the case of the Middle East) gradually emerged. These comprised dense networks of long-distance pipes, dedicated port terminals in many different countries, massive shipping systems and large-scale centralized refineries. This was not a material basis conducive to organizing a militant unionized labour force with a shared working class culture and political agenda. Unsurprisingly, the post-1970s oil era was a period when labour lost ground and more and more wealth accrued to the increasingly wealthy elites (Stiglitz, 2013; Picketty, 2014). Welfare systems were (partly) dismantled in the developed economies, and in the former colonies the 'resource curse' became a disincentive to diversify economies, thus increasing the dependence of the new elites on global flows of oil-based finance. By the end of the twentieth century, 60% of global energy was generated from oil.

Mitchell's focus is limited to the politics of coal- and oil-based energy infrastructures in the nineteenth and twentieth centuries. He does not explore the political implications of RE. However, using the logic of his analysis, it is clear why his approach will be useful for this chapter's discussion about the political implications of RE infrastructures. The obvious difference between coal/oil infrastructures and RE infrastructures is the fact that RE infrastructures are geographically distributed and decentralized. Already hundreds of thousands of RE plants have been constructed in locales that have never before been significant in industrial or energy terms. And millions more will be constructed. The spatial dynamics of this unique material configuration of energy infrastructures will have social and political impacts of equal magnitude to those catalysed by coal- and oil-based energy infrastructures. Put simply, if coal was the basis for social democracy in mid-twentieth-century industrialized nations and if oil was the basis for globalization and neoliberalism in the late twentieth century, what will/could an emergent progressive politics of RE infrastructures look like? Given that social groups acquire "a power of action from *within* the new energy system" (Mitchell, 2011:12 – emphasis added), what is it about the new decentralized and distributed RE infrastructures that could enable social groups to acquire the power for a particular form of community-based action that makes possible a new progressive politics of the commons? Are there any precedents? And what threatens this form of community-based power?

Towards a progressive politics of the energy commons

As indicated in Chapter 2, the burgeoning literature on the commons has emerged from a search for alternatives to both state-centric and market-led development. What is useful about the notion of 'the commons' is that public value is not automatically associated with state-centric delivery systems, nor is efficiency equated with free markets comprised of individuals who compete to realize their own rationally defined self-interest. Unfortunately, those who often use the term tend to also be anti-statist, that is, they have a limited vision for the role of the state in enabling the commons. As was argued in Chapter 2, it is therefore preferable to refer to the collibratory governance of the commons. As argued in subsequent pages, this becomes the central object of a progressive politics of the energy commons, or what some refer to as a community-oriented version of the energy democracy (ED) movement (Burke and Stephens, 2018).

Historically, the collectivities created by coal-based production-centred industrial systems became the social basis for social democracy and socialism. Both these movements focused on shifting the balance of political power in favour of those who wanted to use the state to achieve social democratic and/or socialist goals (depending on the context and era). To this extent, they were both statist in orientation. The globalized oil-based consumer-oriented economies of the late twentieth century, on the other hand, became the basis for a movement that aimed to roll back the public domain in favour of marketized relations and commodification that created, ultimately, the highly unequal financialized conflictual world we currently live in. For this, neoliberal governments deployed state power in interventionist ways that often contradicted their neoliberal claims (Mitchell and Fazi, 2017).

Drawing on evidence from the frontrunners in the RE revolution (Denmark and Germany) plus isolated examples from other contexts (New Zealand, Central America, Italy), it will be argued that there is sufficient evidence to conclude that the material configurations of the new RE infrastructures create the necessary – but not sufficient – conditions for the emergence of a new progressive politics of the energy commons – or what Debor refers to as a "rising cosmopolitan community" (Debor, 2018: Kindle Location 4650). This overlaps with – but is not identical to – what some refer to as 'energy democracies' (Burke and Stephens, 2018). If this can be substantiated, it helps confirm the argument in Chapter 4 that the translation of the deep transition into a just transition will depend on the directionality of the RE revolution and the modes of governance selected to guide it.

As already mentioned, by 2000 over 80% of wind capacity in Denmark was owned by citizens/collectives of some sort and by 2010 nearly half of the RE capacity (wind and solar) in Germany was owned by similar entities (Nolden, 2013; Moss, Becker and Naumann, 2015; Bauwens, Gotchev and Holstenkamp, 2016). As the expanding research on these initiatives reveal, what the commons literature would call 'commoning' within a supportive collibratory environment was what led to the emergence of a significant RE sector in Denmark and Germany during the 1980s and 1990s.[3] These were unique conditions for incrementalist contextually rooted innovations involving manufacturers, users, collectives and policymakers that rapidly (less than two decades) created a technologically, institutionally, financially and socially viable socio-technical alternative to combustion-based energy – an alternative that is now being applied worldwide. How did these emerge from *within* the way RE systems are configured? And does this provide sufficient evidence to conclude that the material configurations of RE infrastructures create, in practice, the necessary but insufficient conditions for the flourishing of ED as a new progressive politics of the commons?

The story of the ED movement in Denmark and Germany is instructive. It reveals the extent and maturity of a commons-oriented movement that triggered the global energy revolution, including a favourable policy and regulatory environment that best exemplifies the kind of collibratory governance that is needed. In his summary of the state of play of wind energy in 2000, Bolinger (2005) shows that over 80% of all wind-based RE in Denmark and Germany could be defined as "community-owned" (see Table 8.4). By 2012, 48% of all RE in Germany was

TABLE 8.4 Community wind power development in selected European countries (2000)

	Total wind capacity (MW)	Community-owned wind capacity (MW)	Percentage of community-owned (%)	Number of household investors
Germany	6,161	−5,400	88	−100,000
Denmark	2,268	−1,900	84	−175,000
Sweden	240	−30	13	−15,000
The United Kingdom	414	−3	1	−2,000
Total	9,083	7,333	81	292,000

Source: Bolinger, 2005:559

owned by "citizens" via cooperatives and partnerships known as Closed-End Funds (CEFs) (Yildiz, 2014:678). The predominant form of ownership in Denmark was cooperatives.

Bolinger argues that the remarkable growth in socially owned RE was only possible because of the existence of an enabling policy and regulatory framework (see elements of this framework listed in Table 8.5). He shows that Denmark and Germany had in common feed-in tariffs, standardized interconnection protocols and a wind turbine industry that worked collaboratively with the cooperatives to drive innovations. In Denmark, additional incentives included tax exemption for income from RE, a subsidy for reducing CO_2 and, most significantly – and uniquely – restrictions on private ownership of RE generation. German RE plants also enjoyed the advantages of accelerated depreciation. Neither Denmark nor Germany suffered from the permitting denials that obstructed the progress of RE in the United Kingdom. The cooperative nature of RE growth in Denmark and Germany ensured broad-based support across a wide range of locales, which is why permitting denials was not an issue during the 1980s and 1990s. However, after both Germany and Denmark dismantled the collibratory governance that had supported the growth of ED before 2000 in the case of Denmark and before 2014 in the case of Germany in favour of policies that enabled corporate capture of RE, anti-RE environmental movements emerged causing a rise in permitting denials. In other words, by taking spatial relations for granted, the corporates began burning their fingers. Neoliberal governance could do little to counter this trend, other than encourage locals to buy shares in RE businesses – a far cry from the collectivist spirit and energy that inspired the energy cooperative movement.

Reinforcing the collibratory governance approach, Denmark and Germany both had long cooperative traditions stretching back into the early nineteenth century (partly to facilitate rural energy provision) and well-developed anti-nuclear activist movements that had successfully opposed the construction of nuclear power plants during the 1960s and 1970s (Bauwens, Gotchev and Holstenkamp, 2016).

TABLE 8.5 Historical drivers of community wind power development

	Denmark	Sweden	Germany	The United Kingdom	The United States
Feed-in laws	√	√	√		
Standardized interconnection	√	√	√		
Tax-free production income	√	√			
Energy/ CO_2 tax refund	√	√		√	
Flow-through depreciation			√		
Wind turbine manufacturing industry	√		√		
Ownership restrictions	√				
Permitting denials				√	

Source: Bolinger, 2005:559)

By the late 1970s, the Danish 'community energy' movement managed to secure an enabling policy that included the following elements (Mey and Diesendorf, 2018:117):

- all farmers and rural households could install a wind turbine on their own land;
- local residents could become members of local cooperatives in their municipalities or neighbouring municipalities;
- exclusive local ownership was a condition for operating permits – electricity utilities could only build large wind farms in agreement with the government and if they did not violate the wishes of farmers and local residents;
- private individuals could only own shares in wind turbines corresponding to the household's private consumption (6,000 kWh per year, extended to 9,000 kWh and to 30,000 kWh per person over 18 living in the household).

The ban on corporate ownership of wind turbines is particularly striking.

Significantly, Vestas – currently a leading global wind turbine producer – started off as a company producing low-tech turbines. It worked collaboratively with the Danish wind cooperatives in order to test innovations and to learn from practice what worked and what did not work. This was effectively a knowledge commons that enabled rapid innovation and learning –something that would not have been possible in a patented IP environment. Over time, Vestas emerged as a global leader because it effectively commercialized the IP. Unfortunately, as is usual practice, this also entailed patenting. It would not be surprising if this curtails the rate of innovation.

By 2012, following research by Yildiz (2014) and my own engagements with the sector via the Westphalian Churches during the course of 2016, there were 754 energy cooperatives in Germany, of which 431 managed solar energy generation, 47 onshore wind, while the remainder were in bioenergy or provided energy services. These 754 cooperatives had 136,000 members, 125,000 of whom were classified as citizens (i.e. not businesses). Together, these 754 cooperatives raised 416 million Euros in equity and over 800 million Euros in debt, resulting in a total investment by these cooperatives of 1.2 billion Euros in RE by 2012. Half the debt capital was derived from cooperative banks (e.g. Triodos, GLS and the various Church-owned banks that are prominent in Germany) and subsidized loans from DFIs (e.g. KfW). By 31 December 2015, there were 1,055 registered energy cooperatives (Debor, 2018: Kindle Location 4664), of which 933 were involved in RE (Debor, 2018:4766). Significantly, registrations of German cooperatives declined drastically from 2014 onwards as collibratory governance gave way to market-orientated neoliberal policies (see subsequent pages). The German energy cooperative movement was a victim of its own success.

Significantly, German cooperatives are not autonomous forces initiated by communities – they were clearly products of collibratory governance. Local governments of various kinds were involved in (co-)founding 38% of all cooperatives registered by 2014 and were on the Boards of 26%. Cooperative banks helped (co-) found 36% and were on the Boards of 16%. Municipal energy providers helped (co-)found 11% of the registered cooperatives and were on the Boards of 6%. Other entities involved in the founding and governance of cooperatives included other energy cooperatives, RE project management and consulting companies, RE energy plant providers, RE investment companies, RE providers, regional associations and institutions, regional climate institutions, municipal institutions, research institutions and national operating companies (Debor, 2018: Kindle Location 5034–5079). Communities, cooperative banks and municipal institutions were the three most significant partners of German energy cooperatives within a clearly defined collibratory governance framework.

In addition to the cooperatives, there were also the CEFs for facilitating investments in RE. These were effectively partnerships (known as GmbH & Co, KG) comprising two groups: limited liability partners whose sole role is to invest and a management partner that manages the business. This model is best suited for those who do not want to participate in decision-making and require specialist management skills to manage higher risk ventures – they can feel like cooperatives, without the sweat equity investment by members in return for a say in decision-making as pertains in the cooperatives. By 2012, CEFs had raised 723.2 million Euros in equity, with an average of about 15,000 investors per fund. While most of these funds were active in the RE sector, a minority were involved in non-RE investments (Yildiz, 2014).

In short, total investment in Germany by 2012 by cooperatives and CEFs in RE was nearly 2 billion Euros plus the unknown amount of debt raised by the CEFs. This represents a remarkable investment in a knowledge and resource commons that was left unprotected from a corporate enclosure movement that followed.

It is puzzling that in their analysis of public and private investments in RE, Mazzucato and Semieniuk do not refer to any of these investments by the cooperative sector (2018). Instead of noticing that it was neither public nor private investment that drove RE innovations in Denmark and Germany in the 1980s and 1990s,

Mazzucato and Semieniuk focus exclusively on investment patterns in the 2000s when public sector investment enabled an increase in private sector investments. It was this entry of the increasingly large-scale private sector investments that began to reverse the socio-institutional and cultural conditions that made RE innovations possible in the first place. This calls into question whether Mazzucato's overall approach to investment in RE innovation (Mazzucato, 2015) is actually appropriate to the RE sector. Mazzucato's blindness to commons-oriented investments stems, ultimately, from her Keynesian focus on the interaction between public and private investment – a focus that may have been appropriate for analysing the rise of the internet but is too narrow for understanding the early phases of the RE innovation cycle.

Rising public investments in RE in the 2000s were aimed at de-risking rising private sector investments in the RE sector, notably enabled by pro-corporate policies adopted by a newly elected right-wing Government in Denmark and the German Government after 2014 (Bauwens, Gotchev and Holstenkamp, 2016). The outcome would have been very different if the dominant coalition within the polities of these two countries was supportive of the energy cooperatives and CEFs.

What is now clear is that the rapid growth of RE infrastructures across all world regions has inspired the imaginaries that have started to emerge from the ED movement. What unites the ED movement is the vision of a decentralized and distributed RE system that is counterposed to a centralized vision of the future of RE as propagated by energy corporates (Strachan, Cowell, Ellis, Sherry-Brennan and Toke, 2015), many energy utilities, large-scale investors, DFIs, policy elites and the technical professions engaged in building RE infrastructures around the world. The centralized mode of governance is a continuation of the energy systems built up during the coal and oil eras but now applied to RE. Significantly, the technologies emerged first within the ED movement (especially in Denmark and Germany), but since 2007 the centralized model powered by large-scale investment funding has been extracting the technological know-how from the ED movement and deploying these innovations within a conventional policy-enabled top-down corporate format via a new generation of mega-RE projects (e.g. Desertec in the Sahara to supply Europe with RE, the giant windfarms off the coast of Denmark and the utility-scale RE plants in South Africa).

The emerging ED movement envisages a future comprising a vast multitude of decentralized and geographically distributed local/regional energy democracies that derive all their energy from RE sources within an ecologically sustainable and socially equitable economic system (Strachan, Cowell, Ellis, Sherry-Brennan and Toke, 2015; Burke and Stephens, 2018). Unfortunately, the ED literature has not engaged with the CBPP literature, in particular the institutional mechanisms for collective knowledge platforms coupled to entrepreneurial forms. This could result in the marginalization of ED to a few democratic experiments. To go to scale, the CBPP-type operating system will be required. Imagine an intelligent smart grid that linked together 'prosumers' generating/consuming energy and trading via a blockchain currency, with a knowledge commons at the centre. The learning platform could log continuous improvements, but it could also become the economic core of a makers guild of some kind, from local vegetable production to new apps for similar settlements elsewhere in the world.

Significantly, as reflected in Table 8.6 from Stephens and Burke, the ED's perspective on the role of the state is not made explicit. While supportive of 'community

TABLE 8.6 Two strategic frameworks for advancing renewable energy futures

Topic	Centralized model of renewable energy	Decentralized model of renewable energy
Analysis of the crisis	The climate crisis is separate from the economic crisis. This implies that the climate crisis can be resolved without addressing the economic crisis, and vice versa	The economic and climate crises are inextricably linked – an integrated crisis reflecting the collision of globalized capitalism with the Earth's ecological limits
Solution to the crisis	The solution to the climate crisis is to replace fossil fuel energy with RE in order to transition to a decarbonized capitalism. The solution to the economic crisis is seen as a separate matter	Replace the globalized capitalist system and its inherent growth dynamic with sustainable economic development based on RE to meet the needs of human beings rather than the needs of capital accumulation
Structural aim	Decarbonize the current economic system without fundamentally changing it	Transition to a new, decarbonized, ecologically sound, life-sustaining economic system that can serve the needs of the world's peoples
Programmatic approach	Reduce GHG emissions – mainly through market mechanisms and new technology, but within the current structure of corporate economic and political power	Create an alternative, equitable social and economic order based on democratic principles and an energy platform that seeks to replace the corporate energy establishment with alternative institutions
Socio-economic change agents	Those who have benefited most from the current globalized capitalist system: corporations and supporting states	Those most impacted by globalized capitalism: workers, low-income communities, and communities of colour
View of energy	Energy is a commodity, the basic enabler of capital accumulation and an expanding growth economy, all of which increases the contradictions of the existing economic and political system	Energy is a resource, a basic enabler of economic life – to be democratized and harnessed to meet human needs and transition the world to an ecologically sustainable economic future

Source: Burke and Stephens, 2018:84

ownership' of RE infrastructures (especially during the early stages of the struggle against the centralized corporate system), the ED movement realizes that to go to scale fast enough (in light of the urgency of climate change), public utilities of one kind or another will have to ultimately play a major role in creating a new decommodified RE infrastructure that supports the proliferation of locally controlled RE generators. These locally controlled RE generators can take the form of municipal entities ('remunicipalization' movement) or non-state collective structures like cooperatives (predominant in Denmark and Germany) or social enterprises (Strachan, Cowell, Ellis, Sherry-Brennan and Toke, 2015), or, of course, a CBPP-type formation.

This is why the ED literature can be said to embrace such a wide range of emergent forms (even though not all actors – nor those who analyse them – use the 'ED' label to self-identify their practices), from remunicipalization (Becker, Beveridge and Naumann, 2015; Moss, Becker and Naumann, 2015; Becker and Naumann, 2017; Routledge, Cumbers and Derickson, 2018) to cooperatives (Bolinger, 2005; Nolden, 2013; Yildiz, 2014; Bauwens, Gotchev and Holstenkamp, 2016; Kalkbrenner and Roosen, 2016; Van Der Schoor, Van Lente, Scholtens and Peine, 2016; Debor, 2018; Mey and Diesendorf, 2018), to community energy in various parts of the world (Seyfang, Park and Smith, 2013; Strachan, Cowell, Ellis, Sherry-Brennan and Toke, 2015; Magnani and Osti, 2016; MacArthur and Matthewman, 2018; Madriz-Vargas, Bruce and Watt, 2018) and partnerships of various kinds (Bolinger, 2005; Davies, Swilling and Wlokas, 2017; Wlokas, Westoby and Soal, 2017).

As discussed further later, influenced by regimes designed to make mining more socially responsible, South Africa has a somewhat exceptional – albeit bizarre – hybrid regime of corporate-delivered RE infrastructures, coupled to funding flows for community-based development projects delivered in quite a paternalistic way (that are not necessarily related to energy) (Davies, Swilling and Wlokas, 2017).

So what is it about the material configuration of RE infrastructures that helps explain the emergence of the ED movement? The three most obvious are their modular nature (from 1 kW to 30 MW), their geographical dispersal across a large number of locales and how rapidly they can be installed (Burke and Stephens, 2018:83). Together, this means specific citizens/groups within particular contexts (drawing on their unique relational character and cultural capabilities, for example, a history of membership of cooperatives like in Denmark and Germany) can activate installations ranging from as small as the cost of a solar home system to multi-million dollar investments in utility-scale renewables. Even if the ROIs for renewables in the 1980s and 1990s were attractive to corporates, the transaction costs for a corporate of negotiating the complexities of access to each locale would have been prohibitive, and it still is. When locally led initiatives bubble up from below these transaction costs get taken care of by the non-financial 'sweat equity' of the participants who invest in collaborating, raising funds and securing local support within a de facto knowledge commons. Indeed, large-scale decentralization without the mobilization of this local 'sweat equity' in CBPP-type knowledge commons is inconceivable for this reason.

Needless to say, the ROI for RE in the 1980s and 1990s was 4% or less, making it an unattractive proposition for corporates (before transaction costs of a multi-site business context are taken into account). Since then, net metering (i.e. the two-way flow of energy onto and off the grid) has redefined the grid from being a one-way transmission belt for large energy utilities without much choice for end-users to a kind of energy commons – a meshwork with multiple 'pro-sumers' that could, in theory, be equated to the internet commons. "For decades", Stephens and Burke note,

> observers have declared a variety of benefits of the decentralized renewable energy model beyond electricity output. Small- and medium-scale renewable systems, deployed at the scale of urban neighborhoods or rural villages, are expected to reduce overhead including capital and administrative costs, reduce energy costs, reduce transmission and distribution losses, increase grid reliability, and reduce incidence of blackouts. Smaller operations reduce the distance between generation and point of use, and allow users to generate and sell energy. Community-scale projects require smaller land areas, minimizing the need for costly transmission and distribution lines and use of eminent domain. Optimal economies of scale are realized at relatively modest sizes for wind and solar facilities, making mid-size projects more cost effective than larger projects. Distributed generation is also expected to significantly reduce financial risk and allow deployment of renewables at a faster pace.
>
> *(Burke and Stephens, 2018:84)*

In recent years, distributed energy generation has been reinforced by the emergence of sophisticated low-cost internet-based mini-grid technologies which effectively enable 'prosumers' to organize themselves into an 'energy commons', where energy is traded between themselves, with surpluses sold onto the grid. A recent review of the most technically advanced mini-grids in The Netherlands commissioned by that country's Ministry of Economic Affairs argued that a new generation of mini-grids is emerging in Europe that they refer to as a SIDE (Smart Integrated Decentralised Energy) (de Graaf, no date). The report predicts that by 2050, half of all households in the European Union will generate RE, and a third of these will be part of local 'energy communities'. A SIDE is a cooperatively owned mini-grid that is almost totally self-sufficient and highly flexible, with maximum integration of a multiplicity of components (not just solar panels and hot water systems). Specifically, SIDE is an acronym that refers to these local systems as follows:

- Smart: this means they are managed via an ICT-enabled local energy management system (LEMS) that automates most transactions, including a local trading currency;
- Integrated: using a sophisticated LEMS, synergies between all the components are maximized via continuous learning;
- Decentralized: they are local systems with a clear system boundary;
- Energy: multiple energy flows, in particular heat/cooling, power and food.

A SIDE usually includes the following technologies, each of which has evolved separately via their own specific innovation pathways but now integrated: solar PV panels, solar thermal panels, heat pumps, combined heat and power systems, wind turbines, electric vehicles, district heating and cooling, batteries, hot water tanks, hydrogen fuel cells, electric boilers, wood stoves, electric heating, biodigesters, seasonal thermal aquifer storage, cooling systems and heat recovery systems. The LEMS ensures these are aligned and integrated and also interconnected with the grid so that surpluses can be fed back into the grid (Figure 8.9) (de Graaf, no date).

Not only are SIDEs the cutting edge in RE innovation globally, they clearly provide a material configuration that is well suited to those who want to (a) cease depending on the grid and (b) become a self-governing community – in short, a resource-cum-knowledge-cum-social commons as envisaged in the literature on CBPP alternatives. Although the grid could in theory be cut up into mini-grids/SIDEs and concessioned off to corporates to run a future decarbonized energy

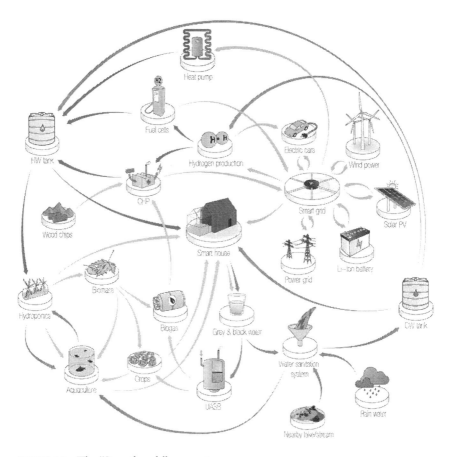

FIGURE 8.9 The "Smarthoods" concept.

Source: F. de Graad, florijn@spectral.energy

system along neoliberal lines, the state clearly could prevent this by incentivizing cooperatively owned SIDEs which perfectly complement the building of the commons. Germany and Denmark in the 1980s/1990s is a clear precedent for this type of intervention. The social benefits of solidarity and trust-building to make a SIDE work (including 'sweat equity' in the maintenance and ongoing innovation) far outweigh the apparent 'efficiencies' of a market-based alternative. Reinforcing the social potential of cooperatively owned SIDEs would be a perfect project of collibratory governance of a new twenty-first century energy commons. This is what should be the core focus of the ED movement. If it goes to scale, that could be truly transformative.

Despite the remarkable achievements in Denmark and Germany, conditions shifted in favour of the corporates by the early 2000s. By the 2000s, the hardware and software had evolved in ways that made corporate-owned utility-scale RE power plants financially viable: larger windmills, cheaper solar PV panels, better storage, more efficient long-distance transmission and, of course, the evolution of internet-based mechanisms for centralizing control of large numbers of decentralized systems in a complex system. Gone are the days when engineers need centralized 'brick-and-mortar' infrastructures with physical dials and meters connected to flows a few hundred meters away to give them real-time information to ensure maximum control of resource flows like electricity, sewage, water and waste. Today, internet-based systems create information flows that make it possible to algorithmically control a vast multitude of installations distributed across vast territories (locally, regionally, nationally, globally). All these features combined, plus the de-risking made possible by rising levels of public investment of various kinds, has made it possible for RE to become an attractive proposition for conventional large-scale financial and corporate investors (including the old Utilities such as ENEL in Italy that has decided to shut down half its coal-fired power stations and become a global leader in RE). This makes possible the opposite of what is envisaged by the ED movement (as captured by Burke and Stephens), namely "distributed and decentralized energy [a]s ... the best opportunity to reassert democratic control of energy sources and renewable energy development" (2018:83). The emergence of a new generation of algorithmically coordinated large utility-scale RE power plants across all world regions is evidence of an enclosure movement that is reversing the gains made by the ED movement during a time when corporates had no interest in technologies that hold the key to fighting climate change. The Global Commission (The Global Commission on the Geopolitics of Energy Transformation, 2019) ignores the anti-democratic and non-inclusionary consequences of this enclosure movement.

Despite the threat of this enclosure movement, Bauwens et al. observe a "double movement" within the cooperative sector from about 2010 onwards (Bauwens, Gotchev and Holstenkamp, 2016:144): the policy reversals (movement one) of the early 2000s in Denmark and Germany generated greater regional collaboration between local-level organizations to cope with a more market-oriented environment (movement two). The combined impact of EU pressures to deregulate the energy market and the electoral victory of a conservative government in 2001 in Denmark resulted in the dismantling of the collibratory governance of the energy commons that enabled the evolution of the world-leading Danish wind energy sector (movement one). Overnight, private sector investors were granted permission to invest in wind turbines, the FiT was replaced with the market-oriented FiP, the geographical location of turbines was restricted, incentives for collective ownership were abolished and in 2004 purchase guarantees for RE generators were scrapped. As Mey and Diesendorf conclude,

> With a lack of new projects and existing cooperatives consolidating or dissolving, CRE [Community Renewable Energy] activities consequently plummeted.
>
> *(Mey and Diesendorf, 2018:111)*

Even the collection of data by government statistical agencies about ownership of wind turbines ceased, signalling how this sector had ceased overnight to even be regarded as worth thinking about.

However, after the hostile years of 2003–2008, the conservative Government in Denmark realized that the subversion of the commons had undermined Danish leadership in the global RE market (not least because Vestas got cut off from the source of innovation – the cooperatives). This, plus the revival of environmental awareness around the hosting in Copenhagen of the 2009 UNFCCC conference on climate change, led to the reinstatement of support for community-based RE (movement two).

The Promotion of Renewable Energy Act 2009 in Denmark provided for a fixed premium for RE which made community-based projects financially viable again, including a requirement that 20% of the shares need to be owned by the local community as a way of reducing local resistance to onshore windfarms (Mey and Diesendorf, 2018:111).[4] This is part of a wider Danish commitment to become fossil-free by 2050 by promoting large-scale offshore wind farms funded, owned and operated by the private sector. In other words, from leading RE innovation via support for community-based RE in the 1980s and 1990s, Denmark aims to become the world leader in market-driven RE systems that will, inevitably, be large-scale and offshore, and therefore free from the socio-spatial constraints that haunt onshore RE generators from those who now feel excluded.

In Germany, the revolutionary idea of a FiT was abandoned in 2014 in favour of the more market-oriented FiP scheme, including reduction in the subsidy that the FiT made possible (movement one). This has tended to favour the growth in CEFs relative to cooperatives. However, 'movement two' in Germany has two different trends: the first is towards remunicipalization of energy delivery (particularly in Berlin and Hamburg), with a focus on RE (Becker, Beveridge and Naumann, 2015); the second is the formation of Federations of RE cooperatives to create economies of scale and to increase political leverage within the polity (Bauwens, Gotchev and Holstenkamp, 2016).

The rise, decline and re-emergence of the ED movement in Denmark and Germany has major implications for the implementation of RE systems in the developing world. The policy shifts in Denmark in 2001 and Germany in 2014 were about creating enabling environments for massively expanding market-driven corporate investments in RE, legitimized by global narratives calling on the private sector to respond to climate change (Newell, 2015) and the inevitability of the 'energy transformation' (The Global Commission on the Geopolitics of Energy Transformation, 2019). This was preceded by a decisive shift in power within the polities of these two countries, culminating in the consolidation of well-publicized public-private partnerships to rapidly decarbonize the economies of these two countries. Angela Merkel's famous response to the Fukushima disaster in 2011 by announcing the closure of nuclear plants represented the ultimate irony: it was the final victory of the anti-nuclear protests of the 1960s but also the start of the justification for the neoliberal enclosure of the energy commons that the anti-nuclear movement gave birth to. As the commons-oriented origins of the RE revolution in Denmark and Germany get eclipsed, governments in developing countries consider investments in large-scale RE projects delivered by corporations (Baker and Wlokas, 2014; Baker, Newell and Philips, 2014; Newell and Phillips, 2016; Davies, Swilling and Wlokas, 2017). In other words, what has been 'globalized' since the 2010s is not the 'collibratory governance of the energy commons' alternative to combustion that emerged from the unique material configurations of RE infrastructures in the 'frontrunner' countries (Denmark and Germany) but rather the commodification alternative that has been 'normalized' in influential analyses (such as those provided by Brown, 2015; Aklin and Urpelainen, 2018; Mazzucato and Semieniuk, 2018; The Global Commission on the Geopolitics of Energy Transformation, 2019). It does not have to be this way, especially if we want to see the deep transition translated into a just transition.

Conclusion

This chapter's conclusions are pretty straightforward. Firstly, it is indisputable that the RE industry has grown rapidly and will continue to do so into the foreseeable future. This was driven by public, private and community-based investors. Secondly, a key driver of the rise in RE investments has been the declining EROI ratio. The issue is not whether there is enough oil or not, but whether it is affordable to get it out of the ground and to market if the oil price remains volatile and/or below $100 per barrel. Fracking and the rise of renewables help keep the prices down, and low prices means fewer profits for large oil companies increasingly dependent on ever more costly oil and subsidies. Thirdly, the mode of combustion affects the nature of politics. Coal essentially resulted in the industrial unions. The union movement was the backbone of the social democratic parties that won elections after World War I and initiated the post–World War II long-term development cycle that ended in 2009. Oil was the basis of neoliberal globalization that, in turn, resulted in financialization and, ultimately, the onset of the global financial crisis in 2007. If the politics of coal was social democracy and the politics of oil neoliberalism, what is the politics of RE? The answer lies partly in the fourth conclusion: the material configurations of RE became the basis for the emergence of a significant energy cooperative movement in Denmark and Germany, aided and abetted by an exemplary collibratory governance framework. During the 1980s and 1990s, communities, cooperatives, cooperative banks and state institutions were the most important collaborators during the innovation phase of the RE industry. However, as the costs of RE came down and new technologies emerged, a wide range of public and private investors moved into the RE sector. The collibratory governance framework that supported the energy commons was dismantled, and public sector investments helped to de-risk private sector investments which have, as a consequence, grown rapidly. However, the growth of mini-grids/SIDEs could provide the material basis for the re-establishment of a new form of energy commons.

Returning to the overall storyline of the book, it was argued in Chapter 4 that the directionality of the deep transition will depend on the directionality of the RE revolution. As is clear from this chapter, there is good news and bad news. The good news is that the material configurations of RE do seem to create the necessary but not sufficient conditions for a progressive politics of the commons. In Denmark and Germany, a pro-commons collibratory governance framework plus a history of cooperative organization and anti-nuclear activism put in place the sufficient conditions for the emergence of a significant energy cooperative movement. However, the extent of its dependence on a supportive policy framework became clear when conditions changed in Denmark after 2001 and in Germany after 2014. If the large utility-scale RE power plant built, owned and operated by the private sector grows to become the global norm across all regions, the deep transition has very little chance of translating into a just transition. This trajectory, however, need not be inevitable.

Obviously, the environmental movements need to campaign vigorously to ensure that government policies do not reinforce this trajectory. Instead, government policies must aggressively support the growth of the energy commons. The DFIs can play a key role here, as KfW has already shown in the German context. The cooperative banks must continue to hold the line rather than focusing only on safe investments in decarbonization, thus ignoring the commons potential that lies in funding cooperatives. Remunicipalization is going to become more significant. Finding a large number of sites for big utility-scale RE plants will become increasingly difficult for private investors in a full world where spaces are always contested in one way or another. Municipalities might be best placed to build political coalitions supportive of RE investments. This could take the form of municipal ownership or cooperative ownership, or some kind of 'public-commons' hybrid. Enlightened investors sensitive to the threat of climate change and the need for a more just transition will also emerge to play a role.

Notes

1 This is similar to, but significantly different, to what some are referring to as 'energy democracy'(Burke and Stephens, 2018).
2 This section draws heavily from the book *Carbon Democracy* by Timothy Mitchell.
3 It is worth noting that the ED literature and the literature on community energy do not draw on the commons literature to make sense of more commons-oriented modes of social organization that emerged in various parts of the world.
4 The South African REI4P has a similar provision.

References

Ahmed, N. M. (2017) *Failing States, Collapsing Systems: Biophysical Triggers of Political Violence.* Cham, Switzerland: Springer.

Aklin, M. and Urpelainen, J. (2018) *Renewables: The Politics of a Global Energy Transition.* Boston, MA: MIT Press.

Ayres, B. (2016) *Energy, Complexity and Wealth Maximization.* Berlin: Springer.

Baker, L. (2015) 'The evolving role of finance in South Africa's renewable energy sector', *GeoForum*, 64, pp. 146–156.

Baker, L., Newell, P. and Philips, J. (2014) 'The political economy of energy transitions: The case of South Africa', *New Political Economy*, 19(6), pp. 791–818.

Baker, L. and Wlokas, H. (2014) *South Africa's Renewable Energy Procurement: A New Frontier.* Sussex, UK: Tyndall Centre for Climate Change Research.

Bauwens, T., Gotchev, B. and Holstenkamp, L. (2016) 'What drives the development of community energy in Europe? The case of wind power cooperatives', *Energy Research & Social Science*, 13, pp. 136–147. doi: 10.1016/j.erss.2015.12.016.

Becker, S., Beveridge, R. and Naumann, M. (2015) 'Remunicipalization in German cities: Contesting neo-liberalism and reimagining urban governance?', *Space and Polity*, 19(1), pp. 76–90. doi: 10.1080/13562576.2014.991119.

Becker, S. and Naumann, M. (2017) 'Energy democracy: Mapping the debate on energy alternatives', *Geography Compass*, 11(8), pp. 1–13. doi: 10.1111/gec3.12321.

Bolinger, M. (2005) 'Making European-style community wind power development work in the US', *Renewable & Sustainable Energy Reviews*, 9(6), pp. 556–575.

Bradshaw, M. (2010) 'Global energy dilemmas: A geographical perspective', *The Geographical Journal*, 176(4), pp. 275–290.

Brown, L. (2015) *The Great Transition: Shifting from Fossil Fuels to Solar and Wind Energy.* New York and London: W.W. Norton & Co.

Burke, M. J. and Stephens, J. C. (2018) 'Political power and renewable energy futures: A critical review', *Energy Research and Social Science*, 35, pp. 78–93.

Cumbers, A. (2015) 'Constructing a global commons in, against and beyond the state', *Space and Polity*, 19(1), pp. 62–75. doi: 10.1080/13562576.2014.995465.

Davies, M., Swilling, M. and Wlokas, H. L. (2017) 'Towards new configurations of urban energy governance in South Africa's renewable energy procurement programme', *Energy Research and Social Science*. doi: 10.1016/j.erss.2017.11.010.

de Graaf, F. (no date) *New Strategies for Smart Integrated Decentralised Energy Systems.* Amsterdam: Metabolic. Available at: www.metabolic.nl/wp-content/uploads/2018/08/SIDE_SystemsReport-1.pdf.

Debor, S. (2018) *Multiplying Mighty Davids: The Influence of Energy Cooperatives on Germany's Energy Transition.* New York: Springer.

Hart, K. and Padayachee, V. (2013) 'A history of South African capitalism in national and global perspective', *Transformation: Critical Perspectives on Southern Africa*, 81(1), pp. 55–85. doi: 10.1353/trn.2013.0004.

International Energy Agency. (2018) *World Energy Investment 2018*. International Energy Agency. Available at: www.iea.org (Accessed: 14 April 2019).

Kalkbrenner, B. and Roosen, J. (2016) 'Citizens' willingness to participate in local renewable energy projects: The role of community and trust in Germany', *Energy Research and Social Science*, 13, pp. 60–70. doi: 10.1016/j.erss.2015.12.006.

Knuth, S. (2018) '"Breakthroughs" for a green economy? Financialization and clean energy transition', *Energy Research & Social Science*, 41, pp. 220–229. doi: 10.1016/j. erss.2018.04.024.

Leiren, M. and Reimer, I. (2018) 'Historical institutionalist perspective on the shift from feed-in tariffs towards auctioning in German renewable energy policy', *Energy Research and Social Science*, 43, pp. 33–40. doi: 10.1016/j.erss.2018.05.022.

MacArthur, J. and Matthewman, S. (2018) 'Populist resistance and alternative transitions: Indigenous ownership of energy infrastructure in Aotearoa New Zealand', *Energy Research and Social Science*. Elsevier, 43, pp. 16–24. doi: 10.1016/j.erss.2018.05.009.

Madriz-Vargas, R., Bruce, A. and Watt, M. (2018) 'The future of Community Renewable Energy for electricity access in rural Central America', *Energy Research and Social Science*, 35, pp. 118–131. doi: 10.1016/j.erss.2017.10.015.

Magnani, N. and Osti, G. (2016) 'Does civil society matter? Challenges and strategies of grassroots initiatives in Italy's energy transition', *Energy Research & Social Science*, 13, pp. 148–157. doi: 10.1016/j.erss.2015.12.012.

Mazzucato, M. (2011) *The Entrepreneurial State*. London: Demos.

Mazzucato, M. (2015) 'The green entrepreneurial state', in Scoones, I., Leach, M. and Newell, P. (eds.) *The Politics of Green Transformations*. London and New York: Routledge Earthscan, pp. 133–152.

Mazzucato, M. and Penna, M. (2015) *Mission-Oriented Finance for Innovation: New Ideas for Investment-Led Growth*. London and New York: Rowman & Littlefield International.

Mazzucato, M. and Semieniuk, G. (2018) 'Financing renewable energy: Who is financing what and why it matters', *Technological Forecasting & Social Change*, 127, pp. 8–22.

Mey, F. and Diesendorf, M. (2018) 'Who owns an energy transition? Strategic action fields and community wind energy in Denmark', *Energy Research and Social Science*, 35(October 2017), pp. 108–117. doi: 10.1016/j.erss.2017.10.044.

Mitchell, T. (2011) *Carbon Democracy: Political Power in the Age of Oil*. London and New York: Verso.

Mitchell, W. and Fazi, T. (2017) *Reclaiming the State: A Progressive Vision of Sovereignty for a Post-Neoliberal World*. London: Pluto Press.

Moss, T., Becker, S. and Naumann, M. (2015) 'Whose energy transition is it, anyway? Organisation and ownership of the Energiewende in villages, cities and regions', *Local Environment*, 20(12), pp. 1547–1563. doi: 10.1080/13549839.2014.915799.

Newell, P. (2015) 'Civil society, corporate accountability', *Global Environmental Politics*, 8(3), pp. 122–153.

Newell, P. and Phillips, J. (2016) 'Neoliberal energy transitions in the South: Kenyan experiences', *Geoforum*, 74, pp. 39–48. doi: 10.1016/j.geoforum.2016.05.009.

Nolden, C. (2013) 'Governing community energy-feed-in tariffs and the development of community wind energy schemes in the United Kingdom and Germany', *Energy Policy*, 63, pp. 543–552. doi: 10.1016/j.enpol.2013.08.050.

Picketty, T. (2014) *Capital in the Twenty-First Century*. Boston, MA: Belknap Press.

REN21. (2018) *Renewables 2018: Global Status Report*. Paris: REN21 Secretariat.

Routledge, P., Cumbers, A. and Derickson, K. D. (2018) 'States of just transition: Realising climate justice through and against the state', *Geoforum*, 88, pp. 78–86. doi: 10.1016/j. geoforum.2017.11.015.

Seyfang, G., Park, J. J. and Smith, A. (2013) 'A thousand flowers blooming? An examination of community energy in the UK', *Energy Policy*, 61, pp. 977–989. doi: 10.1016/j. enpol.2013.06.030.

Standing, G. (2016) *The Corruption of Capitalism: Why Rentiers Thrive and World Does Not Pay*. London: Biteback Publishing.

Stiglitz, J. (2013) *The Price of Inequality*. New York: W.W. Norton & Co.

Strachan, P., Cowell, R., Ellis, G., Sherry-Brennan, F. and Toke, D. (2015) 'Promoting Community Renewable Energy in a Corporate Energy World', *Sustainable Development*, 23(2), pp. 96–109. doi: 10.1002/sd.1576.

The Global Commission on the Geopolitics of Energy Transformation. (2019) *A New World: The Geopolitics of the Energy Transformation*. Abu Dhabi: International Renewable Energy Agency.

Tverberg, G. (2018) 'How the economy works as it reaches energy limits – An introduction for actuaries and others', *Our Finite World*, 11 May. Available at: https://ourfiniteworld. com/2018/05/11/how-the-economy-works-as-it-reaches-energy-limits-an-introduc tion-for-actuaries-and-others/.

Van Der Schoor, T., Van Lente, H., Scholtens, B. and Peine, A. (2016) 'Challenging obduracy: How local communities transform the energy system', *Energy Research and Social Science*, 13, pp. 94–105. doi: 10.1016/j.erss.2015.12.009.

Wlokas, H., Westoby, P. and Soal, S. (2017) 'Learning from the literature on community development for the implementation of community renewables in South Africa', *Journal of Energy in Southern Africa*, 28(1), pp. 35–44.

Yildiz, Ö. (2014) 'Financing renewable energy infrastructures via financial citizen participation – The case of Germany', *Renewable Energy*, 68, pp. 677–685. doi: 10.1016/j. renene.2014.02.038.

9

RESISTING TRANSITION: AUTHORITARIANISM, ENERGY DOMINANCE AND ELECTRO-MASCULINITY[1]

Introduction

President Donald Trump on energy: four decades of US energy policy premised on the doctrine of "energy interdependence" with other nations has been replaced by the policy of securing "American energy dominance". To give effect to this, Trump's Interior Secretary issued Secretarial Order 3351 which states:

> Achieving American energy dominance begins with recognizing that we have vast untapped domestic energy reserves [read: fossil fuels]. For too long America has been held back by burdensome [read: environmental] regulations on our energy industry. The Department is committed to an America-first energy strategy.
>
> *(Juhasz, 2018)*

"Trump", Juhasz continues, "has unleashed a massive, untethered expansion of oil, natural gas and coal production, designed to make this country the world's foremost dirty energy powerhouse" (Juhasz, 2018). The *New York Times* has calculated that since Trump's election, 78 environmental laws have been targeted, with 47 "rollbacks" already achieved by December 2018 and 31 still in process (Popovich, Albeck-Ripka and Pierre-Louis, 2018).

President Donald Trump on women: responding to former Hewlett-Packard boss Carly Fiorina's intention to run for the Republican Presidential nomination, *Rolling Stone* quoted Trump as saying: "Look at that face. Would anyone vote for that? Can you imagine that, the face of our next President? I mean, she's a woman ... really, folks, come on. Are we serious?" (Cohen, 2017)

Bolsonaro on environment: within a few hours after he was sworn in as President of Brazil on 2 January 2019, the newly elected right-winger Jair Bolsonaro issued an executive order transferring the Amazon's indigenous reserves to the Ministry of Agriculture as the prelude to an assault on these environmentally sensitive and protected areas, including the indigenous people who live within them. The Ministry of Agriculture serves the interests of large-scale agribusinesses. He also intends auctioning off to global oil companies the right to extract 15 billion barrels of oil from the so-called pre-salt deposits located many miles below the sea floor off Brazil's Atlantic coast.

Bolsonaro on women: during a TV interview in 2003 when he was a member of the Brazilian Congress, he responded to Congresswomen Maria do Rosario's accusation that he is violent and a rapist: "I would never rape you because you don't deserve it" (Kaiser, 2018).

Duerte on energy: in November 2018, Philippine President Rodrigo Duerte concluded an historic deal with the Chinese Premier Xi Jinping that envisages the joint exploitation of oil and gas reserves in the much-disputed South China Sea. No reference was made to climate change in any of the reports on this much-publicized meeting; instead, the reports were about Duerte's decision to break from the Philippines' traditional US-centred strategic alliances.

Duerte on women: he responded to the rape and killing of an Australian missionary in Davoa City (where he was once Mayor before becoming President) by saying: "They raped her, they lined up. I was angry because … she was so beautiful, the mayor should have been first. What a waste" (Freedland, 2018).

Zuma on nuclear energy: in November 2018, South Africa's Public Enterprises Minister Pravin Gordhan testified at a Judicial Commission of Inquiry into State Capture that former President Zuma had made the procurement of 9,600 MW of nuclear power from Russian energy company Rosatom a top policy priority. He said he was fired from his position as Finance Minister in March 2017 because he opposed the deal. As did his successor, Nhlanhla Nene, who presented a case against the deal to the Cabinet on the morning of 9 December 2016 and was fired that afternoon. The governing party, the ANC, never approved these decisions. Both Finance Ministers said the deal was unaffordable, and Zuma appointed six different Ministers of Energy during his eight-year term. The Russians have celebrated the deal and have remained committed to seeing it implemented.

Zuma on women: in a highly publicized trial in 2006, Jacob Zuma (64) was tried and acquitted for raping the daughter of a family friend, a woman half his age – gender activists dubbed her "Khwezi" (star). He never denied having sex with her. While in court he flaunted his credentials as a Zulu traditionalist (to pander to his political base), and declared in Zulu that he is no rapist: "Angisona isishimane mina" ('I don't struggle to have liaisons with women/I am not a sissy'). He also argued that it was against Zulu culture for a man to leave a sexually aroused woman unsatisfied – an interpretation that was contested by people knowledgeable about Zulu culture. Outside the court, large numbers of women supporters chanted: "burn the bitch".

These quotes reflect how a new generation of authoritarian political leaders from many different world regions have fused together a resurgent toxic masculinity with a non-renewable energy boosterism in ways that offer certainty in an increasingly insecure and uncertain world – a world in transition. Or, in the words used in the Introduction to this book, a reassertion and defence of what being a human in a racist misogynistic world without resource limits should mean. Why is this happening? It is not good enough to simply say "conservative 'strong men' like that always hate women and the environment". This may be true, but surely we need to explain the way misogynistic masculinities have been coupled to authoritarian anti-ecological political projects just as we enter into the transition to the sustainability age.

Four quotes capture this historical juncture – from Yanis Varoufakis, former Minister of Finance in the Greek Government; from Bernie Saunders, US Senator; from Guardian columnist Jonathan Freedland; and from Equality Now's report on rape as a "global epidemic":

> Our era will be remembered for the triumphant march of a globally unifying rightwing – a Nationalist International – that sprang out of the cesspool of financialised capitalism. Whether it will also be remembered for a successful humanist challenge to this menace depends on the willingness of progressives in the United States, the European Union, the United Kingdom as well as countries like Mexico, India and South Africa, to forge a coherent Progressive International.
>
> *(Varoufakis, 2018)*

> At a time of massive wealth and income inequality, when the world's top 1% now owns more wealth than the bottom 99%, we are seeing the rise of a new authoritarian axis.
>
> *(Sanders, 2018)*

> Trump's appointment of Brett Kavanaugh to the Supreme Court despite credible evidence that he raped Christine Blasey Ford "shone a light on a phenomenon that Trump both feeds and exemplifies: a sense of male entitlement so extreme it resents any restraint. It is a swaggering machismo that believes rules are for limp-wristed wimps; that in its most radical form places itself above the law. This phenomenon stretches beyond the partisan battles of Washington DC, beyond even the battlefield of sexual harassment: it is instead a core, if underplayed, aspect of the populist wave currently upending the politics of Asia, continental Europe and Britain.
>
> *(Freedland, 2018)*

> Around the world, rape and sexual abuse are everyday violent occurrences – affecting close to a billion women and girls over their lifetimes. However, despite the pervasiveness of these crimes, laws are insufficient, inconsistent, not systematically enforced and, sometimes, promote violence.
>
> *(Equality Now, 2018)*

Can this coupling of authoritarianism, anti-environment and toxic masculinity be interpreted as an attempt to resist and push back the age of sustainability by violently suppressing the most vibrant and vociferous forces of inspiration from around the world? And if so, is this a more useful (or at least important complementary) explanation than the more common explanations of resistance? – namely, the addiction to consumption, finance-driven short-termism (quarterly reporting, etc.), cognitive dissonance ('death of the old narrative, absence of a new one'), climate denialism of all kinds, fear of loss and the moral superiority of 'greed is good'. Indeed, can this be interpreted as a backlash against the emergence of the post-racial, post-misogynistic, post-ecological relational humanism that underpins the age of sustainability?

I will try to address this question by re-examining a context I know best but for the first time from a Critical Men's Studies (CMS) perspective. I want to rethink what in South Africa is called 'state capture' but this time by factoring in the role played by a resurgent toxic masculinity within a society with one of the highest rape rates in the world and where intimacy has become increasingly decoupled from the institution of marriage.[2] Although South Africa has many unique features that have resulted in particular forms of mineral-energy extraction and modes of masculinity, how the post-1994 democratic transition was hijacked may provide insights into what is clearly a global trend away from a relational humanism and the 'transformed world' envisaged in the Preamble to the Sustainable Development Goals (SDGs).

Authoritarianism, declining EROI and petro-masculinities

As argued in Chapter 4, 2009 marked the end of the post–World War II long-term development cycle. The collapse of Lehman Brothers in October 2008 brought an end to the illusion that the financialization of the global economy was economically and fiscally sustainable. The so-called bailouts and nationalizations transformed private debt into massive sovereign debt burdens. This, coupled to a sustained commitment to austerity economics (reduced welfare spending, quantitative easing, more debt and low interest rates to protect the banks from inflation), has ensured that economic growth levels have not returned to pre-2007 levels. Instead, inequalities have widened as the super-rich – the '0.1%' – have got richer and the middle class has continued to shrink in developed economies, while in developing countries the growth of the middle has slowed down (Stiglitz, 2013; Neate, 2017). According to Oxfam, in 2018, 27 individuals owned wealth equal to the poorest 3.8 billion people.

As argued in Chapter 8 and confirmed by recent research by renowned economists (Brandt, 2017; Courst and Fizaine, 2017), although EROI peaked in the 1960s, since then the declining EROI has driven the gradual decline in economic growth rates. There is now a substantial body of literature that argues that resource depletion in general will persistently drive low growth – and involuntary degrowth – for the foreseeable future (Daly, 1996; Heinberg, 2011; Sorman and Giampietro, 2013;

Fagnart and Germain, 2016; Bonaiuti, 2017; Kallis, 2017). Under these conditions – and if all else remains equal – increased prosperity and well-being for the majority under present conditions is unlikely, while worsening inequalities can be expected. In short, the onset of the global financial crisis is not merely about the limits of particular (neoliberal) economic policies but marks the end of a 250-year epoch of continuous economic growth over the long term. As argued in Chapter 4, this is what establishes the objective conditions for a 'deep transition' which, in turn, could catalyse a 'just transition' (depending, as argued in Chapter 8, on the directionality of the energy transition).

As argued by Heinberg and Crownshaw (2018) and Ahmed (2016), declining EROI and the end of growth could result in the rise of authoritarianism and increased violence as resource wars of various kinds spread. These are not simply wars between nations over access to oil (e.g. the US invasion of Iraq) or minerals (Western Sahara) but also intra-national resource conflicts over water (Syria), oil (Sudan), land (Zimbabwe, South Africa) and minerals (DRC) and violent local conflicts (e.g. India's 'building sand wars').

Incontrovertibly, low growth and a shrinking middle class have stoked the fires of reactionary anger, especially in the developed world but also with cultural variations in the developing world (e.g. Brazil, Turkey) (Mishra, 2018). Underneath this simmering rage lies the "global anxiety epidemic", with 300 million people worldwide estimated to be suffering from anxiety and depression (ATKearny, 2018).

To artificially prop up growth rates to mitigate these disruptive trends, debt levels have been allowed to rise above 2008 levels, creating – according to the IMF – the conditions for another financial crash but this time with more severe consequences because state capacities to bail out private banks are virtually non-existent (Inman, 2018). Indeed, this suggests that declining EROI could continue to drive rising debt levels and, therefore, more instability, more anxiety and thus increased support for right-wing populist movements and regimes who promise the opposite. As Varoufakis so elegantly suggests with reference to the early twentieth century:

> Fascists did not come to power in the mid-war period by promising violence, war or concentration camps. They came to power by addressing good people who, following a severe capitalist crisis, had been treated for too long like livestock that had lost its market value. Instead of treating them like "deplorables", fascists looked at them in the eye and promised to restore their pride, offered their friendship, gave them a sense that they belonged to a larger ideal, allowed them to think of themselves as something more than sovereign consumers.
>
> That injection of self-esteem was accompanied by warnings against the lurking "alien" who threatened their revived hope. The politics of "us versus them" took over, bleached of social class characteristics and defined solely in terms of identities. The fear of losing status turned into tolerance of human rights abuses first against the suspect "others" and then against any and all dissent. Soon, as the establishment's control over politics waned under the

weight of the economic crisis it had caused, the progressives ended up mar-
ginalised or in prison. By then it was all over.

(Varoufakis, 2018)

All true, but what most commentary about the rise of right-wing populism as a
response to crisis under-emphasizes is its appeal to men, and in particular in the
global North context, white heterosexual men suffering from what Daggett calls
"hypermasculinity". This, she argues, "arises when agents of hegemonic masculinity
feel threatened or undermined, thereby needing to inflate, exaggerate, or otherwise
distort their traditional masculinity" (Daggett, 2018). For Daggett, the most serious
threats arise not only from the likes of the #MeToo movement but also from calls
to mitigate climate change and the related potential decline of the "petrocultures"
that "secure cultural meaning and political subjectivities".[3] "In other words", she
continues,

> fossil fuels matter to new authoritarian movements in the West because of
> profits and consumer lifestyles, but also because privileged subjectivities are
> oil-soaked and coal-dusted. It is no coincidence that white, conservative
> American men – regardless of class – appear to be among the most vocifer-
> ous climate deniers, as well as leading fossil fuel proponents in the West.
>
> *(Daggett, 2018)*

To grasp this nexus between hypermasculinity and climate change denial (which
serves to justify expanding oil and coal industries), Daggett proposes the notion
of "petro-masculinity". This concept, Dagget argues, makes sense of the fact that
misogyny and climate denial are "mutually constituted, with gender anxiety slith-
ering alongside climate anxiety, and misogynist violence sometimes exploding as
fossil violence" (Daggett, 2018). Given that misogyny is essentially about the use of
violence by men and given that extraction of fossil fuels requires the use of violence
to manage the locales of extraction and the disruptions caused by climate change
(including mass migration), Daggett concludes that the "wilful continuation of fossil fuel regimes" is "a misogynist practice" (Daggett, 2018). Indeed, this does seem an
appropriate depiction of the symbolism and impacts of the new doctrine of "American energy domination". It certainly accords with the language used by miners to
describe mining as 'extraction', and the generally accepted notion that it is 'mother
earth' that must get mechanically penetrated and forced open to yield 'her wealth'.

However, Daggett makes clear that petro-masculinity is not the only masculinist
response to climate change. Ecomodernism recognizes and accepts climate science
and the notion that the environment is under threat (Hajer, 1995). However, the
changes envisaged are adjustments to the status quo, not a fundamental change.
Nevertheless, the ecomodern masculinities (that incorporate 'care and compassion'
into a techno-fix narrative) associated with this eco-masculinized position (think
Elon Musk, Arnold Schwarzenegger) are very different to the aggressive misogynistic climate denialism associated with petro-masculinities that get expressed in

movements like the Proud Boys, Rollin' Coal and the self-righteous racism of the right-wing Christian evangelical Churches (Daggett, 2018). Indeed, taken to its logical conclusion, this kind of advanced ecomodernist masculinist techno-wizardry has found a respectable place in the new literature on 'transhumanism' that co-opts relationality into a seamless algorithmic utopia that must deny the politics of inequality (Savulescu and Bostrom, 2009). The alternative, of course, to both is an inclusive pro-feminist ecocultural 'post-humanist' masculinity that enables men to work cooperatively with men and women in expanding the creative potentials and relational practices of the commons (see Chapter 6).

Although Daggett's notion of petro-masculinity is helpful in addressing her primary concern which is the convergence of hypermasculinity and the reactive reassertive promotion of fossil fuels, it is not helpful in addressing nuclear power as an equally powerful denial of renewable energy. Indeed, Daggett's critique of the US movement for "American energy dominance" (by exploiting untapped US resources) as a misogynistic project can equally be applied to Russia's strategy to secure global energy dominance via nuclear energy (as discussed in subsequent pages). It may be preferable, therefore, to refer to *electro-masculinity* if we want to widen the lens to incorporate nuclear power. Indeed, Putin argues that Russia's nuclear agency Rosatom could build fleets of nuclear power plants in Russia and around the world that are even more centralized, controllable and powerful than fossil fuel infrastructures (discussed further in subsequent pages). The projection of power this makes possible feeds directly into the masculinist revival of conservative Russian nationalist identities at a time when Russia's economy is going from bad to worse.

South African case

At first glance, the narrative thus far may appear at odds with the South African reality. After all, South Africa's history of class, race and gender relations is markedly different to most of those that have been the locus of right-wing movements since 2007 (with the possible notable exception of Brazil). Nor does it have a mainstream ideological history of the majority (who are obviously black) that includes popular support for right-wing nationalism. Organized white racist cultural formations remained active after 1994 but are confined to certain Churches, marginal extreme (mostly armed) political groups, a few scattered racially exclusive self-organized communities and various clubs and associations (bikers, hunters, fight clubs, music groups, special bars). Nevertheless, over 90% of assets are still owned by 10% of the population (Orthofer, 2016), most of whom are white and, it can safely be assumed, the majority of the actual asset owners are male – thus, a society where racism abounds and which remains profoundly patriarchal and deeply misogynistic.

During the decade that started with the election of Jacob Zuma as President of the African National Congress in 2007 culminating in his Presidency of the country in 2009, South Africa experienced a silent coup that led to what is generally referred to as 'state capture' (Chipkin and Swilling, 2018). By the time he was elected, Zuma still faced 786 charges of fraud and had been questionably acquitted of rape in 2007. He proceeded to lead state capture, which was a political project initiated by a power elite coordinated by an alliance of families, which included amongst others the Zuma family (the 'patrons') and the Gupta family (the 'brokers')– the latter consisting of a group of three brothers who moved to South Africa from India in the late 1990s. At the centre of this silent coup was a determined effort to implement what came to be referred to as the 'nuclear deal' that was struck between Zuma and Russia's President Putin in 2014.

To date, the feminist critique of the so-called 'Zunami' (Hunter, 2011; Motsei, 2007), on the one hand, and the political economy of state capture on the other (von Holdt, 2013; Chipkin and Swilling, 2018; Madonsela, 2018) have not been connected in the literature. Instead, the resurgent traditionalist misogynistic patriarchy that Zuma promoted has been treated as a distinct and separate phenomenon from the Zuma-led state capture project. If a link is made in the popular narratives, it is a moral one ('Zuma is a sexist and corrupt'), with in some cases suspect racist undertones ('that is what African leaders are like'). This conceptual separation but moral conflation masks the deeply entangled nature of these phenomena within the complex logics of South Africa's gendered political economy of race and class, thus impoverishing our understanding of South African politics.

I intend to argue that by doing the nuclear deal with Putin, Zuma agreed to become a key element of Putin's global nuclear-based energy strategy, which aimed to break Russia's dependence on fossil fuels but was also a bid to mount a global alternative to renewable energy. For Zuma, this R1 trillion project was the crown jewel in his policy of 'radical economic transformation' aimed at creating a black industrial class that would displace the white capitalist elite. However, in order to implement the agreement, extra-legal means were required which, in turn, depended on state capture and the consolidation of a shadow state for managing the required transactions. This, in turn, was dependent on a wider political project that depended on two means of violent control: the systemic use of violence to manage political dynamics (von Holdt, 2013) and the reinforcement of misogynistic masculinities (Gqola, 2007; Kopano, 2008; Motsei, 2007). Drawing on a rich tradition of writing about masculinity (Posel, 2005; Walker, 2005; Gqola, 2007; Morrell, 2007; Kopano, 2008; Hunter, 2011; Morrell, Jewkes and Lindegger, 2012; Ratele, 2014), I will argue that South Africa's version of *electro-masculinity* is thus deeply rooted in its specific racial, class and gender history and its peripheral dependence on global hegemons like Russia.

South African context

An overview of South Africa's socio-economic context was provided in Chapter 7, in particular the extraordinary degree of asset-based inequality. By 2009, 15 years after democratization, the contradictions of the post-1994 era had come to a head. As elaborated in detail elsewhere (Swilling, Bhorat, Buthelezi, Chipkin, Duma, et al., 2017; Chipkin and Swilling, 2018), a combination of economic policies influenced by neoclassical economics had failed to foster increases in productive investment. Instead, large-scale net transfers of wealth to traditional shareholders ('shareholder value movement'), legal and illegally exported flows of potentially productive capital, rent transfers to the new black elites ('black empowerment') and increased indebtedness of the increasingly multi-racial middle class ('financialization') resulted in the failure to direct large-scale investment into massive increases in employment for the expanding pool of unemployed people. As manufacturing went into decline, the traditional core of the South African economy – the mineral-energy complex – was reinforced (Mohamed, 2010; Zalk, 2011), including the well-documented associated masculinities at all levels (white male shareholders, mining engineers, managers and black male migrant workers and supervisors) associated with extractivism (Delius, 2014).

By 2009, power dynamics within the polity began to shift as trade unions became increasingly critical of government for not being interventionist enough to halt job losses; black business leaders became dissatisfied with the continued dominance of white capital; investment confidence amongst white business declined reinforcing non-investment (the famous 'capital strike') and the growth of liquid funds; various political elites became dissatisfied with the centralization of power in the hands of President Mbeki. A new coalition of forces coalesced around Jacob Zuma who became President in 2009.

Zuma initiated a policy shift that focused on the expenditures of state-owned enterprises ('our industries') as a key strategy for building a new 'black industrial class' as the vanguard of 'radical economic transformation', including the prioritization of the nuclear deal (Chipkin and Swilling, 2018). While recognizing the endemic nature of corruption in the pre-democratic polity (Hyslop, 2005; Van Vuuren, 2006), this potentially productive strategy (with many positive precedents elsewhere in the world, especially the 'Asian Tigers') was corrupted, as the resultant flow of rents got captured by an increasingly confident power elite clustered around President Zuma. This power elite captured the polity and repurposed state institutions by appointing loyalists to the Boards and Executives of key state-owned enterprises (e.g. ESKOM, Transnet, DENEL and many others) and strategic institutions (e.g. South African Revenue Service and security agencies). The result was the consolidation of neo-patrimonial governance (Pitcher, Moran and Johnston, 2009; Lodge, 2014), with the constitutional state manipulated by what we called the 'shadow state' (Chipkin and Swilling, 2018).

The construction of a shadow state is effectively the archetypal neo-patrimonial form of collibratory governance. It entails the restructuring of the polity: the constitutional state remains intact, albeit hollowed out, and interacts symbiotically with a shadow state comprised of a powerful network of powerbrokers who manage extra-legal transactions sanctioned by the patron-in-chief (Zuma). None of this would have been possible without the deployment of violent practices which, in turn, limited the rights promised by democracy. von Holdt has analysed how the use of violence during elections, violent settling of disputes within organizations, political assassinations and the violent subversion of the rule of law by democratic institutions generate and shape violence, while providing a certain level of constraint on violence (von Holdt, 2013). The result was the emergence of an institutionalized "violent democracy" (von Holdt, 2013) after 1994 that abetted the construction of the increasingly authoritarian neo-patrimonial mode of collibratory governance after 2009.

At the centre of Zuma's political project was the nuclear deal that he entered into with Russian President Vladimir Putin in 2014, with high-level discussions between the two leaders going back a few years. Over the course of 13 cabinet reshuffles between 2009 and 2017, Zuma appointed six different Ministers of Energy, each with a renewed mandate to accelerate the nuclear build programme and, after 2014, to execute the nuclear deal (Fig, 2018).

But in the shadows, in May 2010, a Gupta-owned company, Oakbay Resources and Energy, which included Zuma's son Duduzane Zuma as a shareholder via his company Mabengela Investments, quietly bought the Toronto-listed Uranium One's Dominion mine in Klerksdorp, located in the Northwest Province. The state-owned investment fund, the Public Investment Corporation, was hauled in to partly fund this transaction on suspect terms. The rationale was obvious: based on the assumption that an expansion of the nuclear energy sector was inevitable, the demand for uranium was set to skyrocket. The (largely illusory) foundation of a new supposedly black-owned mineral-nuclear-energy complex was being put in place by the power elite managing the symbiotic relationship between the shadow and constitutional state.

Under the leadership of two women anti-nuclear activists, Liz McDade and Makoma Lekalakala,[4] two NGOs, Earthlife Africa and the South African Faith Communities Environmental Institute (SAFCEI), instigated a court case in 2015 against the President, Minister of Energy and other government bodies challenging the legality of the signed Inter-Governmental Agreement (IGA) between Russia and South Africa (Chutel, 2018). To everyone's surprise, the court ruled in favour of the applicants in 2017 and the nuclear deal was effectively scrapped.

Significantly, the court noted that the IGA included wide-ranging binding provisions that did not appear in any of Russia's IGAs with other countries. Specifically,

> the IGA provides for a strategic partnership, which would focus on the development of a comprehensive nuclear new-build programme, including the design, creation and decommissioning of nuclear plants; use of the Russian Water-Water Energetic Reactor (VVER) technology for a total capacity of 9.6 GW; collaborating on implementing two units of 2.4 GW at specifically stated sites and with additional IGAs to be signed on how this would be done, with joint committees to oversee this; favourable tax regimes and other incentives provided for Russia; and with South Africa incurring all liability as a result of any nuclear incidents.
>
> *(Prins and Davies, 2018:16)*

The court ruled that the wide-ranging nature of these provisions were such that Parliament was legally obliged to "approve" the IGA – instead, ignoring the advice of the State Law Advisor, the Minister of Energy at the time only tabled the IGA in Parliament for "noting" (Prins and Davies, 2018). The National Treasury's criticism of the financial provisions of the deal was also ignored, including public statements to the effect that South Africa's national sovereignty was at risk.

The only way to understand why the authors of the (originally secret[5]) IGA had mistakenly overreached their legal mandate is to recognize how Zuma had effectively consented that South Africa would become a pawn in Russia's new post-2007 grand strategy to become a global nuclear power. In return for subverting the much-lauded South African Constitution, there is evidence the Russians funded the ANC's election campaigns and inserted key intelligence operatives into the Presidency to bring the deal to closure.

To prepare for his global strategy, in 2007 Putin signed a decree that provided for the restructuring of the Russian nuclear industry that effectively integrated the entire value-chain into a single unified state-owned entity (Merdan, 2018). Building on Russia's long Cold War history of nuclear armament (including the accumulation of vast technical know-how, large-scale nuclear infrastructures and uranium mining capabilities), this strategy to build a civilian nuclear energy sector made strategic sense. Rosatom was tasked with increasing nuclear energy in Russia's energy mix to 25% by 2030, and in parallel it concluded agreements to build nuclear plants (including floating nuclear plants) in Egypt, Jordan, Algeria, UAE, Kuwait, India, Turkey, Thailand, Indonesia and a number of Eastern European countries (Merdan, 2018). Reports on Russian nuclear plants built outside Russia suggest they are a kind of hybrid between a military base and an embassy, underwritten by state guarantees that effectively give Russia extensive control of these country's energy economies. Russia's 2010 Energy Strategy document clearly stated that Russia's commitment to sustainable development means reducing dependence on fossil fuels by increasing the supply of nuclear energy (Safronova, 2010). Later drafts[6] reinforced the commitment to nuclear-based sustainable development. Nevertheless, as Eberhard and Lovins conclude, buying a Russian nuclear plant is an inadvisable high-risk venture:

> Russia is facing economic challenges, sliding in terms of rated world economies to number 15, below Mexico. Sovereign debt is a real concern and low oil prices and Western sanctions in response to Russia's aggression towards the Ukraine and other areas are making matters worse. . . . Russia wants to build nuclear power plants and needs huge amounts of capital to finance its nuclear commitments around the world. It hardly appears to be a stable financial partner.
>
> *(Eberhard and Lovins, 2018)*

After nearly eight years in power, a broad-based coalition of civil society, business, faith and human rights organizations coalesced around a demand for Zuma's resignation and a return to constitutionalism. In December 2017, the governing party voted to replace President Zuma with Cyril Ramaphosa as President of the ANC. After Zuma resisted pressures to resign to buy time to finalize the nuclear deal (Swilling, 2019), he eventually 'resigned' as President of the country in February 2019 and Ramaphosa replaced him. One of the first acts of Ramaphosa's new Minister of Energy, Jeff Radebe, was to sign 27 renewable energy contracts that the previous President put on hold to prepare the way for the finalization of the nuclear deal. Ramaphosa informed the World Economic Forum (WEF) in January 2018 that nuclear was not affordable and there were rumours that Putin tried to strong-arm him into respecting the terms of the IGA. Ramaphosa, however, may support renewable energy (with evidence that his personal investment fund has renewable energy investments), but given his history (as Secretary-General of the National Union of Mineworkers) he is also a strong supporter of the coal industry, describing it in his maiden speech as a "sunrise industry" without referring to renewables despite the remarkable growth of renewables in South Africa since 2011. However, in his 2019 State of the Nation address he corrected this, referring to climate change and renewable energy investments. His Minister of Finance reinforced this in his budget speech a few weeks later.

In sum, it was the Zuma/Gupta-centred power elite within the polity that committed South Africa to a particularly noxious energy choice. This included colluding with the largely illusory Russian bid to mount a global nuclear strategy that would have given Russia control of South Africa's economy by way of the state guarantee that Minister Nene was pressured to sign. It was this choice that contributed significantly to the decomposition of democratic governance (including the firing of two Ministers of Finance and five Ministers of Energy) and nearly resulted in South Africa becoming another Russian-controlled failed state, inclusive of the endemic violence and fear required to maintain political loyalty. It is a scenario that will repeat itself in many developing countries, no matter which global hegemon is pulling the strings.

Gender and masculinity in transition

The rising significance of political violence since 1994 and especially since the commencement of Zuma's presidency in 2009 (von Holdt, 2013) cannot be separated from the 'crisis of masculinity' that South African men experienced after the introduction of 'constitutional sexuality' (Walker, 2005) in 1994. This crisis, as Motsei so brilliantly reveals in her profound reflections on Zuma's rape trial, has increasingly been 'resolved' by way of intensified forms of intimate violence directed against large swathes of South Africa's women (Motsei, 2007). Most importantly, this 'crisis of masculinity' is prevalent across all types of men – white urban males, urbanized black men from poor and middle class areas, marginalized right-wing Afrikaner men rooted in a romanticized 'boer' identity and men still rooted in traditional rural African cultures and contexts.

It is a mistake to equate this crisis of masculinity and the associated sexual violence with poor urban black youth on the rampage, as so often happens in the popular press and everyday conversations. This ignores a wide spectrum of violence that is not on the streets but is in the homes of many respected middle-class families (both black and white). However, there is also plenty of evidence of the emergence of positive pro-feminist masculinities amongst men from all races and classes more aligned with the Constitution (Morrell, 2001; Walker, 2005; Ratele, 2014) and more consistent with indigenous cultural norms (Motsei, 2007). Indeed, NGOs like Sisonke Gender Justice have large programmes working with men in poorer communities who want to find alternative ways of being men – and fathers – that reconnect them to those they love most. This is not because Sisonke Gender Justice thinks that these are the only men who are problematic, but rather they are men who tend to lack access to support services to help with what it means to be a different kind of man. I am part of a men's group that meets weekly as part of the wider movement called The Mankind Project – this group comprises middle-class black and white men, as well as younger black men from poor communities in Cape Town. We all wrestle with the same challenges, feeling in our everyday experience what this 'crisis of masculinity' actually means. A misogynistic culture does not only destroy women's lives, it destroys men's lives too – but of course, in a completely different way.

Over half the South African population (52%) are women and mostly youthful (a third are below 15 years). By 1994, 40% of all 'family units' were headed by women and rising since 1994. The mean age of marriage for women is 28 years, whereas the majority of women have their first child before the age of 21. What this suggests is that a relatively high number of couples have children before they are formally married (Morrell, Jewkes and Lindegger, 2012). Indeed, by 2001, only 30% of couples were married, a drop from 38% in 1991, 42% in 1980 and 57% in 1960. "Indeed", Hunter suggests, "marriage in South Africa has faced perhaps one of the sharpest reductions in the world . . . wedlock is virtually a middle-class institution today" (Hunter, 2011). Again, this is not a normative statement about the 'collapse of family values' but rather a reference to the increasing precarity of a way of cohabiting described by the notion of 'family life'.

One of the reasons for this is the unaffordability of *lobola* ('bridewealth'): as conditions become more insecure for women, concerned parents demand a higher *lobola* as evidence that the man has the financial means to support a family; but as the *lobola* increases, fewer men - especially the expanding number of unemployed young men - can afford it which results in more and more couples living out of wedlock. This affects both poorer and better off couples. This leaves women in increasingly precarious positions when it comes to child care and other related obligations if the man leaves, especially if he has been a breadwinner and shirks his maintenance responsibilities. Fathers play little or no role in bringing up their children – a challenge that some NGOs such as Sisonke Gender Justice have made a major focus of their work.[7] In 1993, 36% of children had absent fathers, and by 2002 it was 46% (Morrell, Jewkes and Lindegger, 2012). Of the total births registered in 2016, 62% contained no information about the father and nearly half of all mothers were single (Lehohla, 2019). Only 31% of mothers in 2016 were legally married, while 60% of all fathers were legally married (Lehohla, 2019). Unsurprisingly, the poorest families are those with one breadwinner, more so if that person is a woman given that she is likely to earn less. These figures reveal South Africa's extreme gendered precarity.

Endemic violence exacerbates these gender dynamics. Although the murder rate dropped by 42% between 1994 and 2008, at 38.6 murders per 100,000 people, South Africa's murder rate is still four times the global average. Almost all of the men that were murdered during this period were killed by other men. By contrast, the female homicide rate is six times the global average and half the women were murdered by their intimate partners (Morrell, Jewkes and Lindegger, 2012). Half of all South African women report experiencing one form or another of domestic violence (Walker, 2005). The rape rate in 1998 was the highest in the world – 115.6 cases per year for every 100,000 people. Taking into account unreported rapes, it has been estimated that as many as 1 million rapes occur annually in South Africa (Walker, 2005). Infant and child rape has also increased: between January and December 2000, 13,540 children under the age of 17 were raped, of whom 7,899 were under the age of 11 (Walker, 2005). At 5.2 million people, South Africa has the highest number of people living with HIV. The majority are women, and women who have experienced intimate partner violence are more likely to get HIV, while men who perpetrate partner violence are more likely to be HIV infected (Morrell, Jewkes and Lindegger, 2012). There has also been a rise in homophobic violence as homosexuality has become more visible since 1994.

By contrast, the 'non-sexism' provided for in the South African Constitution and a raft of related legislation envisage a very different kind of masculinity than what is reflected in these cold but horrifying statistics. Formally, on paper, the legal position of women has changed profoundly. Women make up 30% of the legislature, marital rape is a crime, domestic violence is a serious offense, court orders can be obtained by women to ensure men pay maintenance, labour legislation obliges organizations to employ women and pay them at the same level as men, women have access to abortion on demand without parental or partner consent, contraception is free, user fees for maternity services were removed, maternity rights are mandated, sexual harassment outlawed and gay and lesbian people/couples were given entrenched rights (including the right to marry) (Walker, 2005). Women's position in society, Hassim argues, has been "transformed from one of presence to power" (Walker, 2005). And yet, Walker argues,

> the very liberal version of "constitutional sexuality" does not speak to many masculinities of the past. Those masculinities, steeped in violence and authoritarianism, are anathema to the "gender equality" prescribed by the Constitution and the battery of policies and laws, which have been written in its wake. The ideal South African man in this frame *is one who is nonviolent, a good father and husband, employed and able to provide for his family.*
>
> *(Walker, 2005 – emphasis added)*

Indeed, this is most likely the 'ideal man' envisaged by parents who set the *lobola* for their daughters and is widely regarded across all race groups as the normalized building block of the new South African nation. However, it assumes the existence of a job-creating economy and safe living environments for all. Unsurprisingly, therefore, it is an ideal that has not been realized. As Posel concludes, rampant sexual violence not only shattered this conception of masculinity, sexual violence – and baby rape in particular – also triggered a widely recognized moral crisis in the early 2000s about the entire viability of the nation-building project:

> From this standpoint, therefore, the country's moral crisis was fundamentally a crisis of manhood: if men failed to don the mantle of responsible father-hood, they jeopardized the possibility of responsible nationhood.
>
> *(Posel, 2005:249)*

These cold statistics can never really communicate the pervasive everyday horrors that South African woman experience, either consciously or not. Nor do they capture the "scandal of manhood" triggered by rampant sexual violence. Ironically, as baby rape reports increased, old sexist and misogynistic stereotypes of rape victims and rapists began to shift – after all the common assumption that 'she asked for it by the way she dressed' and 'he was tempted' cannot be applicable to a baby even in the mind of the most extreme sexist! (Posel, 2005). In her searing critique of the huge gap between 'constitutional sexuality' and the realities women experience, Pumla Gqola echoes many others – for example, the contributors to *Writing What We Like* (Qunta, 2016) – when she argues,

> We know that today women do not feel safe in the streets and homes of South Africa, that women's bodies are seen as accessible for consumption – touching, raping, kidnapping, commenting on, grabbing, twisting, beating, burning, maiming – and control, that women are denied the very freedom that "empowerment" suggests, the very freedom the Constitution protects. And the problem is often made women's responsibility.
>
> *(Gqola, 2007)*

This contrast between the gendered reality of all South Africans and 'constitutional sexuality' provides the complex context for contested hegemonic masculinities during the post-1994 period (for a case study of masculinity as experienced by men in a Cape Town soccer club, see Abreu, 2016). Up until this point, the preceding evidence and argumentation implicates *all* South African men across race and class, no matter what the reader may be imagining from her/his vantage point. But, building on Morrel's work, it can be argued that there are essentially three hegemonic masculinities in South Africa, shaped in turn by South Africa's unique historical patterns of social development (Morrell, 2001): "white masculinity", "African masculinity" and "black urban masculinity". Gqola is surely right when she argues they *all* share – to a lesser or greater degree – a predilection for a "cult of femininity" – the view (which Gqola suggests is also shared by many women) "that women must adhere to very limiting notions of femininity... [they must] prove that they exhibit traditionally feminine traits" (Gqola, 2007). But, as Gqola makes clear, there is a rich feminist tradition of African writers like her self, Motsei and many others who have challenged this hegemonic view for decades.

"White masculinity" is associated with the economic dominance of a white asset-owning male class that has morphed from the overt rigid puritanical racist authoritarian man constructed by Calvinistic Afrikaner culture (as reflected in the name and conduct of the 'Broederbond'[8]) into a more metropolitan, sauve, race-savvy, clean-shaven, shareholder/CEO, manipulative, (often) ultra-rich 'stay-out-the-limelight-but-keep-control' masculinity. It is a masculinity that likes to distance itself from what these men regard as the distasteful riffraff – remnants of a lower class white racist bearded masculinity that persists in publicly less overt forms in certain known bars, sports clubs, shooting clubs and biker gangs. While, at the same time, deep down they might still harbour exactly what they love to distance themselves from: racist and sexist assumptions, often reflected in everyday conduct in office politics, the night club, braai parties and in the way they fund and control their extramarital affairs and sugar-daddy arrangements with younger women (of all races). The proliferation of press reports about seemingly reputable white men at the peak of their professions having to resign in the face of accusations of sexual harassment or the men who shoot their partners and children before taking their own lives speaks to this reality. Unfortunately, the large bulk of the literature on South African masculinities does not reflect deeply enough about the crisis of 'white masculinity' per se (even in the seminal text by Morrell, 2001). And yet, within movements like The Mankind Project, white men wrestle all the time with the contradictions between what they are projected to be by a gendered society in crisis and how they really feel (for an example of this pattern of reflection, see Chalklen, 2018).

"African masculinity" reflects a rurally based masculinity reproduced via indigenous institutions/practices (or what Ratele (2014) prefers to call "tradition") such as chiefship, communal land tenure, customary law, *lobola*, cattle wealth, polygamy (in some areas) and the unique mix of ancestral and Christian beliefs reproduced in many of the African Churches. It is a masculinity that is not place-specific and is profoundly shaped by migrancy, that is, it is dominant in rural areas but found in urban areas (especially informal settlements). The ties to land, identity and having a 'job' become essential ingredients of this version of masculinity.

"Black urban masculinity" has evolved through urbanization and the construction of South Africa's "townships" where black people were located under apartheid and still mostly live. Increasingly disconnected from their "African masculinities", these black urban masculinities have bifurcated into those reproduced by an increasingly successful debt-financed consumption-oriented black middle class that aspires to 'move into the suburbs' with their nuclear families (while often retaining their 'township girlfriends') and the masculinities of an urban precariat that barely survives above the poverty line and often on the margins of legality

(Walker, 2005; Hunter, 2011). Between them lies the established unionized working class with stable jobs, bonded homes, local Churches, dart clubs and favourite watering holes.

Many NGOs emerged after 1994 to promote a pro-feminist grassroots masculinity amongst boys and men (Walker, 2005) and white CEOs like Woolworth's Colin Hall ran workshops to promote a new style of inclusive leadership that proselytized the bleached race/gender values of the new post-1994 business elite. However, as Kopano makes clear, "ruling masculinities" are closely associated with prevailing political leadership styles (Kopano, 2008). Kopano and Morrell et al. contrast masculinities associated with Nelson Mandela and Zuma. Following Unterhalter (2000), Morrel et al. argue that Mandela projected and represented a "heroic" struggle-oriented masculinity that challenged the violent and authoritarian behaviours of apartheid's white male political leaders, the violent 'camp culture' of the liberation movement and the traditionalist African masculinities promoted by the Bantustan leaders and Chiefs. Although hotly contested around the time Winnie Mandela died in 2017, Mandela did, Morrell et al. argue, project "a new, more egalitarian masculinity to South Africa". (Morrell, Jewkes and Lindegger, 2012) Zuma, by contrast, "epitomized a rejection of more thoughtful, egalitarian masculinities, rather asserting in the name of 'tradition', a masculinity that was heterosexist, patriarchal, implicitly violent and that glorified ideas of male sexual entitlement, notably polygamy, and conspicuous sexual success with women" (Morrell, Jewkes and Lindegger, 2012). His rape trial became a show trial of South African misogyny. And yet, he attracted substantial support from women, exemplified by the women who chanted "Burn the bitch" outside the court.

Based on a case study in the heart of Zuma-supporting KwaZulu Natal, Hunter found that women supported Zuma because he represented the opposite of their experience of men (at a time of "profoundly gendered . . . increasing class inequalities"): Zuma, after all, actually *married* all 'his women' and generously supported all of them and their children. And at his rape trial, speaking in isiZulu he said he offered to pay *ilobolo* to marry Kwezi – a gesture that was derided by the liberal press. The Zulu press, however, presented it differently: they understood the gravitas of this gesture and what *ilobolo* meant at a time when marriage is rare and *ilobolo* increasingly unaffordable. As Hunter concludes,

> *Ilobolo* long connected work and family, house and home, production and reproduction. Zuma therefore drew on the high status of marriage and fertility to society to position himself as a respectable patriarch, an *umnumzana*, and not a rapist.
>
> *(Hunter, 2011)*

As Motsei's extraordinary book reveals, Zuma's strategy was immensely successful, and it explains his continued popularity in a deeply misogynistic patriarchal society.

The message that emerged from his trial, she argues, reinforced the worst of about such a society:

> It is also clear from the reportage on the trial that a female rape victim who doesn't fight back is perceived as a willing participant. If she fights back, however, she runs the risk of injury or death. If she chooses not to speak out, she will die inside. If she speaks out, she is a devil and deserves to burn in hell. Either way there is a possibility of death. Therefore, it is better to speak out and, like Biko, die for an ideal that will live on.
>
> *(Motsei, 2007:34–35)*

Zuma, and Julius Malema (former ANC Youth League leader and subsequent leader of opposition party the Economic Freedom Fighters), effectively mounted a backlash against gender equality, presenting it as anti-African and equating it with modernity, white middle-class values and a threat to male economic advancement (Kopano, 2008). Consistent with his well-known dislike of the Constitution in general, Zuma in his conduct and rhetoric undermined the 'constitutional sexuality' that the women's movement had struggled to achieve for so long. This further reinforced a particularly misogynistic version of the 'cult of femininity' that Gqola so eloquently describes (Gqola, 2007). Rather than resolving the 'crisis of masculinity' in post-apartheid South Africa, Zuma's symbolic gender politics reinforced the underlying causes of the problem, thus deepening rather than dissolving the misogynistic 'cult of femininity'.

Converging in electro-masculinity

This section brings the analysis of state capture-as-energy choice into conversation with the analysis of the changing nature of 'ruling masculinities'. The meeting point lies in an understanding of a particular mode of collibratory governance, namely authoritarian neo-patrimonialism, and the central constitutive role that energy generation plays in this political project (and, indeed, all previous South African political projects).

Following Daggett, what matters is not simply the political economy of the mineral-energy complex but also the "psycho-effective dimensions" of energy generation (Daggett, 2018). This includes the collective desires of those whose identities are tied to the elaborate system of employment, migration, earnings, dependence and rural/urban livelihoods that extractivism and coal-based energy production in particular have historically made possible since the late 1800s (Delius, 2014). By connecting the delivery of a new black-led mineral energy sector based on coal/nuclear power and his role as a 'respectable patriarch' who could build the economy so that other men could do what he did (earn good money, pay *lobola*, establish a family, buy a house), Zuma effectively fused these two narratives into what could be referred to as 'electro-masculinity'.

Electro-masculinity can take many forms, but in essence captures the fusion between two deep patriarchically defined male desires: the desire for individual economic freedom in ever-expanding economies powered by fossil fuels and/or nuclear energy, and the desire for a restored masculinity within a functional patriarchal order that provides certainty. These fuse in the notion of *electro-masculinity*. It is a notion that seems more appropriate to the South African context than Daggett's notion of 'petro-masculinity' which, by definition, has a limited reference to fossil fuels. *Electro* is derived from the word *electrum* which etymologically refers to two substances: an amber-coloured natural alloy of gold and silver used to make the first coins in seventh century BC Greece and amber itself whose electrostatic properties gave rise to the modern English words for electron and electricity. In other words, the etymological roots of 'electro' refer to mineral wealth and energy flows, the two key historic drivers of the South African economy and the linked economic activities that were the basis for providing millions of working men with the archetypal South African job – mining (Delius, 2014).

Working on the hardness of the rockface and releasing the flows of energy to build the economy became the oft-reproduced symbolic images of the all-pervasive masculinist culture that united 'mining men' from CEO to rock-driller. Initially profoundly racist, these narratives were steadily de-racialized from the 1980s onwards, thanks mainly to the role played by the National Union of Mineworkers which was founded, of course, by the man who replaced Zuma as President, Cyril Ramaphosa. As the extractivist origin and foundation of the modern South African economy, it is a culture that has always permeated what it meant to 'work', and who was expected to 'work' (Delius, 2014) – a theme popularized in Hugh Masekela's famous song *Stimela*. The desperate desire shared by everyone to 'work' in post-1994 South Africa was inseparable from this profound sense that the economic priority was to ensure that 'men had jobs' – in particular, black men. Unsurprisingly, therefore, one of Zuma's key strategies was to capture the mining sector (via Zwane, his politically incompetent bombastic Minister of Mines), including support for the infamous 'Tegeta deal' – the archetypal manoeuvre engineered by the CEO of ESKOM, Brian Molefe, to force out an old white mining company and replace it with a 'black empowered' company (notwithstanding the fact that being Gupta-owned does not qualify as 'black empowerment') (Chipkin and Swilling, 2018).

The clue to the psycho-political fusion between extractivism/energy and masculinity lies in an unreported YouTube recording of a speech Zuma gave in Zulu in July 2016:

> If it were up to me, and I made the rules, I would ask for six months as a dictator. You would see wonders, South Africa would be straight. That's why, if you give me six months, and allow Zuma to be a dictator, you would be amazed. Absolutely. Everything would be straight. Right now to make a decision you need to consult. You need a resolution, decision, collective petition. Yoh! It's a lot of work!
>
> *(Quoted in Swilling, Bhorat, Buthelezi, Chipkin,*
> *Duma, et al., 2017:21)*

By revealing his deep yearning for an authoritarian order as his way to get things 'straight', Zuma confirmed his frustrations with the constitutional state (see also Zweni, 2018), thus clarifying why he pursued the establishment of the shadow state where acting like a dictator through loyal 'brokers' was made possible (Chipkin and Swilling, 2018). And for this he continues to enjoy support in the ANC, even after his demise as President. This authoritarian impulse strikes a chord in a society – and in the governing party – that is deeply frustrated with the persistence of poverty, the slow pace of change and continued economic dominance of the white elite. His alliance with Putin added considerably to his prestige as a 'man of power'. His message was clear: if given the chance (including the disappearance of fraud charges and constitutional constraints), he would deliver economic freedom and hence the foundation for a respectable patriarchal masculinity based on a specific interpretation of African 'tradition'. He effectively used the fact that he lacked this freedom to present himself as a victim of circumstance who could not, therefore, be blamed for the lack of change.

The well-established European literature on authoritarianism (going back to Adorno and the Frankfurt School, but also to Erich Fromm and Michelle Foucault) has always tried to understand what it is about authoritarianism that explains why it is able to mobilize popular support (see Morelock, 2018). The answer lay in the conclusion that the authoritarian impulse resides "within us all", as Foucault put it,

> in our heads and in our everyday behaviour. . . [it] causes us to love power, to desire the very thing that dominates and exploits us.
>
> *(Daggett, 2018)*

In short, during uncertain times, the 'authoritarian personality' within us all colludes with the certainties promised by the 'strong man' who embodies a collective memory of a particular past, present and future. In the writing of Frantz Fanon, we find a similar concern. In the famous chapter in *The Wretched of the Earth* on postcolonial politics entitled "The Pitfalls of National Consciousness", Fanon writes:

> The leader pacifies the people. For years on end after independence has been won, we see him, incapable of urging the people to a concrete task, unable really to open the future to them or of flinging them into the path of national reconstruction . . .; we see him reassessing the history of independence and recalling the sacred unity of the struggle for liberation. . . . The leader, seen objectively, *brings the people to a halt* and persists in either expelling them from history or preventing them from taking root in it. During the struggle for liberation the leader awakened the people and promised them a forward march, heroic and unmitigated. Today, he *uses every means to put them to sleep*, and three or four times a year asks them to remember the colonial period and to look back on the long way they have comes since then.
>
> *(Fanon, 1963:168–169 – emphasis added)*

But, as Daggett argues, the authoritarian political project that "brings the people to a halt" always includes an appeal to men, exploiting their hypermasculinity and deep-seated rage with deteriorating economic conditions that threaten their inviolable self-identities as the primary breadwinners – a sense of impotence if they cannot be the primary breadwinners (Daggett, 2018). Without this 'right to be real men', that rage can catalyse the threatening spectre of political violence, or even worse: the exploitation of men's fears by ruthless populist leaders bent on activating the male predilection for violence against 'the other'. Reassertion of a stricter purer hierarchical patriarchal order – including the glorification of misogyny as reflected in the quotes at the start of this chapter – is proclaimed as a precondition for achieving wider political and economic certainty. By implication, breakdown of this order is depicted as the *cause* of moral decay and the reason why it is *women* (in the classic misogynistic framing) who are 'out of control' (politically, sexually, culturally, etc.), not men. Beneath the authoritarian call for restoration of a culturally specific patriarchal ideal (often with reference to 'tradition'), there is

> an underlying fear of the social fragility of masculinity, as well as a shared sense among members of each having personally fallen short of that ideal. Capitalist crises, such as the worldwide depression of the 1930s or the 2008 financial crisis, do not help; they only make it more difficult for many to achieve that *essential emblem of modern masculinity: a breadwinner job* .... [These] men ... "show deep-seated fears of weakness" in themselves. The meaning of weakness to these men seems to be tied up with intense fears of nonmasculinity. To escape these fears they try to bolster themselves up by various antiweakness or pseudomasculinity defences, where pseudomasculinity means "boastfulness about such traits as determination, energy, industry, independence, decisiveness, and will power".
>
> *(Daggett, 2018)*

All these traits are apparent in the South African imbroglio: an economic context that leaves large swathes of men feeling 'weak' because they are unemployable and without the means to secure a wife and family; rising frustration in the face of an increasingly conspicuous rich white male predatory elite that gets defined as 'the other' – the obstacle to upward mobility and wealth creation (what Zuma referred to as 'white monopoly capital'); a sexual violence epidemic perpetrated by men (from all classes and races) caught in the gendered contradictions of class and race, with poorer men often left feeling disempowered by what they assume women expect/reject; widespread use of violence by men from all races and classes to settle political disputes; a pseudomasculinity that draws on many reference points – for some it is interpretations of 'African traditions' for narratives of pride and assertiveness, while for others the same is achieved in the new 'brotherhoods' of the biker gang, or the expensive 'extreme sports' preferred by the sons of the old white elite; and finally a patriarchal 'ruling masculinity' that promises a degree of certainty to men and women at a time of plummeting marriage rates and pervasive sexual violence.

The failure of South Africa's economy to deliver to the majority what it clearly delivered to the white minority before 1994 exacerbated a profound sense of collective impotence amongst many different categories of men. This dovetailed with norms about masculinity reproduced by schooling systems attended by men from all classes: that weakness is to be attacked, to feel pain is to be weak, to feel vulnerable is dangerous, that sexual power is about conquest, and that misogyny is normal. It was this that Zuma tapped into when he said "Angisona isishimane mina" and promised 'radical economic transformation' that would replace 'white monopoly capital' with a new 'black industrial class' in command of a new 'big and shiny' uranium-nuclear energy complex. All he needed to 'straighten out' this vaguely defined electro-masculinity, he said, was "six months as a dictator".

Electro-masculinity, however, is not simply an elite project, nor is it always misogynistic. It is also expressed in legitimate organized political and cultural practices that reinforce the collective desires of those whose identities are tied materially and symbolically to particular forms of energy generation.[9] On 28 February 2017, for example, thousands of (mainly male black and some white) truck drivers (who drive the trucks that transport coal) mounted a highway blockage and a march on the capitol, Pretoria, in protest against the continued expansion of the renewable energy sector. They argued that renewable energy would result in the loss of 50,000 coal mining jobs and threaten the future of ESKOM. Speaking on national radio, acting CEO of ESKOM at the time, the now disgraced 'state capture' collaborator Matshela Koko, exploited the fears workers and truck drivers have of renewable energy by blaming ESKOM's reduced demand for coal on the expansion of the renewable energy sector. Although factually incorrect, this unprecedented joint action between workers, truck drivers and ESKOM management reinforced a general patriotic perception that renewable energy is a threat to the very foundation of the South African economy – coal mining and ESKOM's role as South Africa's energy generator.

Another example from the cultural sphere is provided by the black biker gangs. They are significant because they have been taken over by groups of financially successful professionals and businessmen. The upwardly mobile urban black masculinity they express is best reflected in the names of this new generation of biker gangs: Elite Bikerz, Big Fellas, Commandos, Real Kings, Suspecs, Convics, African Gladiators, Vikings and Smoking Guns. These exclusive male associations, where members call each other 'brother', merge a sense of charity/community support with a swaggering machismo identity – leathered 'patch-wearing bikers' on their speedy, dangerous, increasingly expensive shiny 'super-bikes' (Tucker, 2017). The black bikers and white bikers go an annual "ubuntu" tour, but remain largely separate. A common site is the screaming roar of an expensive high-performance combustion engine as the rider pulls off a 360 degree 'circle wheelie' in front of a large township audience. At highway filling stations, pumped up black biker gangs have been seen parading their power by revving up their engines in an ear-splitting display of metal, rubber and leather, partly to just attract the envy and admiration that all bikers crave but it also deliciously 'pisses off the whites'. During the Zuma years, it became increasingly common practice for large groups of bikers riding in semi-military formation to rev up their engines into a crescendo of combustive noise power as they circled slowly around the stadium at the start of political rallies as a kind of honorary 'guard of honour' ahead of the political leaders who later address the crowd. These spectacles are about boastful male power (big men in leathers), financial success (expensive bikes) and black leadership ('bikers are no longer only white'). But subliminally, they also connect the combustive power of high-performance petrol engines with the symbolic icons of a financially successful upwardly mobile socially aware black masculinity.

Conclusion

The purpose of this chapter was to reflect on the deeper intertwined dynamics of political and personal power that are shaping resistances to change across many world regions, including developed and developing countries. Contested energy cultures are at the centre of these dynamics. Returning to themes addressed in Chapters 1 and 2, the old racialized, sexualized and naturalized identities need to give way to *ukama* – the Shona word for 'relatedness' to all beings, species and things. Consistent with broad swathes of Sub-Saharan African conceptions of the relational self (Murove, 2009a, 2009b; Coetzee, 2017), *ukama* is more appropriate for the requirements of the next epoch – the age of sustainability. This relational sensibility is in some way a reaction – a kind of responsive spiritual advanced guard – to the combined threats of femicide, ecocide and genocide that may well be the shadow legacy of industrial modernity if nothing changes. It is also an emergent property of the unfolding wider metatheory emerging from Western scholarship referred to in Chapter 2 as *complex integral realism* which, of course, accords with the deeper and longer Sub-Saharan African axiological traditions.

It is obvious that the emergence of a relational self and the ascendance of a relational metatheory for making epistemological and ontological sense of the world will, almost by definition, be resisted. However, although this resistance may well be articulated explicitly in certain academic texts, the real resistance will emerge amongst those who fear loss of the pre-existing certainties that industrial fossil- and nuclear-based modernity made possible.

This chapter uses the South African context to reveal how the struggle over energy futures gets interwoven with the complex struggles to redefine old certainties as the industrial age comes to an end. The authoritarian defence of an outdated unaffordable twentieth-century 'modern' energy future got fused with a violent reassertion of a toxic masculinity that reinforced a pandemic of sexual violence. Electro-masculinity became a kind of leitmotif of the Zuma years. Zuma embodied a fusion of a particular kind of misogyny that drew on 'tradition' for legitimation with a vision of an energy future that ran contrary to the kind of renewable energy future that is rapidly becoming the energy foundation of the sustainability age at a global level (see Chapter 8). How toxic masculinity and unsustainable energy futures get fused into an electro-masculinity that will be contextually specific to each country must become the focus of attention of activists and analysts interested in the politics of transition to an age of sustainability. Drawing on the new energy humanities literature, this chapter has provided one detailed case study of how this fusion occurred. The age of sustainability requires that we challenge outdated conceptions of what it means to be human in order to create space and resources for viable, generative alternatives to emerge. A relational post-racist, post-sexist ecologically connected 'self'/'we' is more likely to find meaning and joy in the collaborations required to build the commons that will need to underpin the age of sustainability.

Notes

1 This chapter would not have been possible without the many conversations with Megan Davies and Amanda Gcanga, PhD researchers in the CST. However, I take responsibility for the argument and conclusions.
2 I do not want to depict the plummeting marriage rate as 'the problem' as if marriage is some sacrosanct norm that reflects the 'health' of society – a perspective that is rooted in the cultural construct of the nuclear family that was consciously promoted to support the twentieth-century consumerism and presumes heterosexuality as the norm. Indeed, the breakdown of marriage might in fact reflect the aspirations of an emerging generation of increasingly economically independent women who resist the oppressive constraints of the conventionally accepted norms about marriage that have not always been to their emotional, physical, economic and psychological advantage.
3 This notion that fossil fuels are not merely a source of energy but the basis for entire cultural assemblages has emerged from the energy humanities literature, a sub-theme of the burgeoning environmental humanities literature (see Mitchell, 2011; LeMenager, 2016; Wilson, Carlson and Szeman, 2017)
4 In recognition of their efforts, they won the Goldman Environmental Prize in 2018.
5 Vladimir Slivyak, head of Ecodefence, a Moscow-based NGO, found the 'secret' IGA lodged on the website of the Russian foreign ministry and he conveyed it to Earthlife Africa in Johannesburg, which made the court application possible.
6 https://policy.asiapacificenergy.org/node/1240
7 See www.genderjustice.og.za
8 The Broederbond – literally meaning 'association of brothers' – brought together the most powerful Afrikaner men during the apartheid years to coordinate strategies for dominance and advancing the interests of the Afrikaner 'nation'.
9 This line of thinking draws directly from the energy humanities literature cited earlier.

References

Abreu, D. (2016) 'It's Not for Fun Anymore': Performance of Masculinity as a Source of Stress Amongst Professional Soccer Players in South Africa. Master's Thesis, University of Cape Town.
Ahmed, N. M. (2016) Failing States, Collapsing Systems: BioPhysical Triggers of Political Violence. doi: 10.1007/978-3-319-47816-6.
ATKearny. (2018) Year-Ahead Predictions 2019. Seoul: AT Kearney.
Bonaiuti, M. (2017) 'Are we entering the age of involuntary degrowth? Promethean technologies and declining returns of innovation', Journal of Cleaner Production. doi: 10.1016/j.jclepro.2017.02.196.

Brandt, A. (2017) 'How does energy resource depletion affect prosperity? Mathematics of a minimum energy return on investment', *BioPhysical Economics and Resource Quality*, 2(2). doi: 10.1007/s41247-017-0019-y.

Chalklen, W. (2018) *White Privilege and the Road to Building a United South Africa*. Available at: https://warrenchalklen.com/2015/01/07/white-privilege-and-the-road-to-building-a-united-south-africa/?fbclid=IwAR29tisdS9qYfz52U8mI4HJ5WFEfBwC6f7OW01OIRiNdZfvRVLz10EHDEkY (Accessed: 14 November 2018).

Chipkin, I. and Swilling, M. (2018) *Shadow State: The Politics of Betrayal*. Johannesburg: Wits University Press.

Chutel, L. (2018) 'How two South African women stopped Zuma and Putin's $76 billion Russian nuclear deal', *Quartz Africa*, 25 April.

Coetzee, A. (2017) *African Feminism as a Decolonising Force: A Philosophical Exploration of the Work of Oyeronke Oyewumi*. Stellenbosch: Stellenbosch University.

Cohen, C. (2017) 'Donald Trump sexism tracker: Every offensive comment in one place', *The Telegraph*, 14 July.

Courst, V. and Fizaine, F. (2017) 'Long-term estimates of the energy-return-on-investment (EROI) of coal, oil, and gas global productions', *Ecological Economics*, 138, pp. 145–159.

Daggett, C. (2018) 'Petro-masculinity: Fossil fuels and authoritarian desire', *Millennium: Journal of International Studies*, (0130). doi: 10.1177/0305829818775817.

Daly, H. E. (1996) *Beyond Growth: The Economics of Sustainable Development*. Boston, MA: Beacon Press.

Delius, P. (2014) 'The making and changing of migrant workers' worlds, 1800–2014', *African Studies*, 73(3), pp. 313–322. Available at: doi: 10.1080/00020184.2014.969479.

Eberhard, A. and Lovins, A. (2018) 'Five facts that drive South Africa's nuclear power plan should die', *The Conversation*, 31 January.

Equality Now. (2018) *The World's Shame: The Global Rape Epidemic*. New York, Nairobi, and London: Equality Now. Available at: https://d3n8a8pro7vhmx.cloudfront.net/equalitynow/pages/308/attachments/original/1527599090/EqualityNowRapeLawReport2017_Single_Pages_0.pdf?1527599090.

Fagnart, J. and Germain, M. (2016) 'Net energy ratio, EROEI and the macroeconomy structural change', *Journal of Economic Dynamics and Control*, 37, pp. 121–126.

Fanon, F. (1963) *The Wretched of the Earth*. New York: Presence Africaine.

Fig, D. (2018) 'Capital, climate and the politics of nuclear procurement in South Africa', in Satgar, V. (ed.) *The Climate Crisis: South African and Global Democratic Eco-Socialist Alternatives*. Johannesburg: Wits University Press.

Freedland, J. (2018) 'Kavanaugh has revealed the insidious force in global politics: toxic masculinity', *The Guardian*, 29 September.

Gqola, P. (2007) 'How the "cult of femininity" and violent masculinities support endemic gender based violence in contemporary South Africa', *African Identities*, 5(1), pp. 111–124. doi: 10.1080/14725840701253894.

Hajer, M. (1995) *The Politics of Environmental Discourse: Ecological Modernization and the Policy Process*. New York: Oxford University Press.

Heinberg, R. (2011) *The End of Growth: Adapting to Our New Economic Reality*. Gabriola Island, CN: New Society Publishers.

Heinberg, R. and Crownshaw, T. (2018) 'Energy decline and authoritarianism', *BioPhysical Economics and Resource Quality*, 3(3), p. 8. doi: 10.1007/s41247-018-0042-7.

Hunter, M. (2011) 'Beneath the "Zunami": Jacob Zuma and the gendered politics of social reproduction in South Africa', *Antipode*, 43(4), pp. 1102–1126. doi: 10.1111/j.1467-8330.2010.00847.x.

Hyslop, J. (2005) 'Political corruption: Before and after apartheid', *Journal of Southern African Studies*, 31(4), pp. 773–789. doi: 10.1080/03057070500370555.

Inman, P. (2018) 'World economy at risk of another financial crash, says IMF', *The Guardian*, 3 October.

Juhasz, A. (2018) 'Trump's pursuit of "American energy dominance" threatens the entire plane', *Los Angeles Times*, 9 December.

Kaiser, A. (2018) 'Woman who Bolsonaro insulted: "Our president-elect encourages rape"', *The Guardian*, 23 December.

Kallis, G. (2017) 'Radical dematerialisation and degrowth', *Philosophical Transactions of the Royal Society A*. doi: 10.1098/rsta.2016.0383.

Kopano, R. (2008) 'Ruling masculinities in post-apartheid South Africa', in Cornwall, A., Correa, S. and Jolly, S. (eds.) *Development with a Body: Sexuality, Human Rights and Development*. London and New York: Zed Books, pp. 121–135.

Lehohla, P. (2019) 'Talk by Pali Lehohla, former statistician general of South Africa', in *Talked Presented at Workshop on Towards a Socially Just and Sustainable Economy: Deepening the Dialogue*, 18 March. Cape Town. Available at: www.markswilling.co.za/2019/03/towards-a-socially-just-and-sustainable-economy-deepening-the-dialogue-co-hosted-with-thuli-madonsela/.

LeMenager, S. (2016) *Living Oil: Petroleum Culture in the American Century*. New York: Oxford University Press.

Lodge, T. (2014) 'Neo-patrimonial politics in the ANC', *African Affairs*, 113(450), pp. 1–23. doi: 10.1093/afraf/adt069.

Madonsela, S. (2018) 'Critical reflections on state capture in South Africa', *Insight on Africa*, pp. 1–18. doi: 10.1177/0975087818805888.

Merdan, E. (2018) 'New energy export strategy: Russia's shift to nuclear energy', *Energy*, 9 April. Available at: www.aa.com.tr/en/energy/analysis/new-energy-export-strategy-russia-s-shift-to-nuclear-energy/19565.

Mishra, P. (2018) *Age of Anger: A History of the Present*. London: Penguin Random House.

Mitchell, T. (2011) *Carbon Democracy: Political Power in the Age of Oil*. London and New York: Verso.

Mohamed, S. (2010) 'The state of the South African economy', in *New South African Review 1: 2010 – Development or Decline?* Cape Town: HSRC Press.

Morelock, J. (2018) *Critical Theory and Authoritarian Populism*. London: University of Westminster Press. Available at: https://doi.org/10.16997/book30.

Morrell, R. (2001) *Changing Men in Southern Africa*. Pietermaritzburg/London: University of Natal Press/Zed Books.

Morrell, R. (2007) 'Men, masculinities and gender politics in South Africa: A reply to Macleod', *Psychology in Society*, 35(1), pp. 15–26. Available at: www.pins.org.za/pins/pins35/pins35_article04_Morrell.pdf.

Morrell, R., Jewkes, R. and Lindegger, G. (2012) 'Hegemonic masculinity/masculinities in South Africa: Culture, power, and gender politics', *Men and Masculinities*, 15(1), pp. 11–30. doi: 10.1177/1097184X12438001.

Motsei, M. (2007) *The Kanga and the Kangaroo Court*. Johannesburg: Jacana.

Murove, M. (ed.) (2009a) *African Ethics: An Anthology of Comparative and Applied Ethics*. Durban: University of KwaZulu-Natal Press.

Murove, M. (2009b) 'An African environmental ethic based on the concepts of ukama and ubuntu', in Murove, M. (ed.) *African Ethics: An Anthology of Comparative and Applied Ethics*. Durban: University of KwaZulu-Natal Press.

Neate, R. (2017) 'Richest 1% own half the world's wealth, study finds', *The Guardian*, 14 November.

Orthofer, A. (2016) *Wealth Inequality in South Africa: Insights from Survey and Tax Data*. REDI3X3 Working Paper 15. Cape Town: University of Cape Town. Available at: http://web.archive.org/web/20180204212928/www.redi3x3.org/sites/default/files/.

Pitcher, A., Moran, M. H. and Johnston, M. (2009) 'Rethinking patrimonialism and neopatrimonialism in Africa', *African Studies Review*, 52(01), pp. 125–156. doi: 10.1353/arw.0.0163.

Popovich, N., Albeck-Ripka, L. and Pierre-Louis, K. (2018) '78 environmental rules on the way out under Trump', *New York Times*, 28 December.

Posel, D. (2005) 'The scandal of manhood: "Baby rape" and the politicization of sexual violence in post-apartheid South Africa', *Culture, Health and Sexuality*, 7(3), pp. 239–252. doi: 10.1080/13691050412331293467.

Prins, N. and Davies, E. (2018) *South Africa's Nuclear New-Build Programme: Who Are the Players and What Are the Potential Strategies for Pushing the Nuclear New-Build Programme?* Cape Town: World Wide Fund for Nature (WWF).

Qunta, Y. (2016) *Writing What We Like: A New Generation Speaks*. Cape Town: Tafelberg.

Ratele, K. (2014) 'Currents against gender transformation of South African men: Relocating marginality to the centre of research and theory of masculinities', *Norma*. Taylor & Francis, 9(1), pp. 30–44. doi: 10.1080/18902138.2014.892285.

Safronova, G. (2010) *Energy Strategy of Russia for the Period up to 2030*. Moscow: Ministry of Energy of the Russian Federation.

Sanders, B. (2018) 'A new authoritarian axis demands an international progressive front', *The Guardian*, 13 September.

Savulescu, J. and Bostrom, N. (2009) *Human Enhancement*. Oxford: Oxford University Press.

Sorman, A. and Giampietro, M. (2013) 'The energetic metabolism of societies and the degrowth paradigm: Analyzing biophysical constraints and realities', *Journal of Cleaner Production*, 38, pp. 80–93.

Stiglitz, J. (2013) *The Price of Inequality*. New York: W.W. Norton & Co.

Swilling, M. (2019) 'Zexit, a nuclear deal and Russia's strong-arm persuasion', *Daily Maverick*, 8 January.

Swilling, M., Bhorat, H., Buthelezi, M., Chipkin, I., Duma, S., et al. (2017) *Betrayal of the Promise: How South Africa Is Being Stolen*. Stellenbosch and Johannesburg: Centre for Complex Systems in Transition and Public Affairs Research Institute.

Tucker, W. (2017) 'The rise and rise of black biking', *Mail & Guardian*, 29 September.

Unterhalter, E. (2000) 'The work of the nation: Heroic adventures and masculinity in South African autobiographical writing of the anti-apartheid struggle', *The European Journal of Development Research*, 12, pp. 157–178.

Van Vuuren, H. (2006) *Apartheid Grand Corruption: Assessing the Scale of Crimes of Profit in South Africa from 1976 to 1994, Security Studies*. Pretoria: Institute for Security Studies. Available at: http://dspace.africaportal.org/jspui/bitstream/123456789/31239/1/APARTHEIDGRANDC2.pdf?1.

Varoufakis, F. (2018) 'Our new international movement will fight rising fascism and globalists', *The Guardian*, 13 September.

von Holdt, K. (2013) 'South Africa: The transition to violent democracy', *Review of African Political Economy*. Taylor & Francis, 40(138), pp. 589–604. doi: 10.1080/03056244.2013.854040.

Walker, L. (2005) 'Men behaving differently: South African men since 1994', *Culture, Health and Sexuality*, 7(3), pp. 225–238. doi: 10.1080/13691050410001713215.

Wilson, S., Carlson, A. and Szeman, I. (eds.) (2017) *Petrocultures: Oil, Politics, Culture*. Montreal: McGill-Queen's University Press.

Zalk, N. (2011) 'South African post-apartheid policies towards industrialization: Tentative implications for other African countries', in Norman, A., Botchwey, K., Stein, H. and Stiglitz, J. E. (eds.) *Good Growth and Governance in Africa: Rethinking Development Strategies*. Oxford: Oxford Scholarship Online, pp. 345–371. doi: 10.1093/acprof:oso/9780199698561.001.0001.

Zweni, Z. (2018) 'Zuma lectures students on state capture', *Sowetan*, 13 September.

PART IV

Transdisciplinary knowing

10

TOWARDS AN EVOLUTIONARY PEDAGOGY OF THE PRESENT

Introduction

After all is said and done, paradigm shifts have to be learnt by others. The extraordinary human capacity to learn from the previous generation is what creates the collective memory that makes inter-generational cultures possible. Babies born today may know as little as they knew a thousand years ago, but within a few short years they imbibe and learn what they need to know to be and live within the unique cultural context of their present existence, no matter how complex this may be. And yet, at certain points in our lives, we need to learn, unlearn and relearn anew as we are forced to confront realities that our learning to date never prepared us for.

The span of knowledge and experience captured in this book cannot remain confined to the covers of this book. It needs to be shared. However, this must happen via a learning process that transcends traditional cognitive learning. These ideas need to be embodied if they are to be fully appreciated, and for that cognitive learning needs to be fused together with experiential place-based learning. This is what I have endeavoured to do over the past nearly two decades, working with an amazingly creative group of South Africans at the SI in Stellenbosch, South Africa. Influenced by her training as a Montessori teacher and experience at Schumacher College in the UK, Eve Annecke played a particularly significant role in crafting this evolutionary pedagogy of the present (see Annecke, 2013; Freeth and Annecke, 2016). Graduates from our programme, who now run the SI under the capable leadership of Jess Schulschenk, continue to evolve this pedagogic legacy in their own ways.

With hindsight, I can see that over the last two decades what we were figuring out is a pedagogy that prepares people for uncertain futures by rooting them in an evolutionary experience of the present in a uniquely configured context. The context I am referring to here is, of course, the SI and its location in the Lynedoch EcoVillage (as discussed in Chapter 1). But it is the teaching and learning approach that emerged within this context that really matters here.

By being in a space where we could experiment with a particular teaching and learning approach, we also realized it was necessary to translate this into a research methodology that equips students to be activist researchers. Here I am referring to the emergence of a way of knowing that has been formalized in a body of knowledge we now call 'transdisciplinary knowing' (see Van Breda, 2008; Muhar, Visser and van Breda, 2013; Swilling, 2014; van Breda and Swilling, 2018; van Breda, 2019).

Both the teaching and research approaches that emerged intertwined within each other and over time they have been analytically disentangled, so they could each be elaborated in their own right and reconnected into our everyday practice. The end result was the establishment of a new research centre in 2016 within Stellenbosch University called the Centre for Complex Systems in Transition, of which I was the founding Co-Director. Working closely with the SI, the CST houses the doctoral and postdoctoral research that further elaborates and applies what has come to be called transdisciplinary research.

This chapter[1] reflects on this still unfolding pedagogical-cum-research helix that has emerged over the past decade and a half. To condense this enormously rich experience into a digestible story is essentially impossible. However, a start will be made by describing the teaching and learning approach that has evolved at the SI over the years. To communicate the atmosphere and feel of this, a description of the content and process of the first module of the master's programme as it was delivered in 2019 will be provided. This will be followed by a discussion of how we evolved our transdisciplinary research approach. The penultimate section reflects on these interwoven learning and researching practices from the perspective of the arguments elaborated in the preceding chapters. By depicting (with hindsight) what has emerged as an *evolutionary pedagogy of the present*, I want to explore how we have evolved a pedagogy that is appropriate for the interdisciplinary themes that permeated the preceding chapters: an *Ukamian* perspective on the relational self, an integral complex realism, asynchronous transitional dynamics, building the ecocultural commons, incrementalism, why context matters and dealing with misogynistic authoritarian forms of resistance to change.

Place-based experiential learning

Late January in Stellenbosch is crisp, dry and not too hot. Cooled by the south-easterly winds coming off the cold Atlantic Ocean currents, it precedes February when temperatures climb up to the lower- to mid-30s as the winds seem to drop and the sky seems to enlarge into a deep brilliant cloudless blue. The majestic presence of the Helderberg mountain range embraces the patchwork of vineyards, farmsteads and settlements that spread out across the valley floor towards the foothills of the mountains in the distance. In the morning hours, the sun glints off the countless leaves of the large trees and shrubs that now envelope the SI. Everything seems so alive, a shimmering dance of gold and green against the scent of rich composted soils. Unlike when we arrived in 2003 at this windswept dusty unkempt site, now there are vast choruses of birds and beetles that celebrate the new summer day, drowning out the ever-present sound of distant traffic from the passing road. The two ancient giant ficus trees just outside the entrance of the SI push their magnificent limbs into the air, supported by huge trunks that can only be fully embraced by five people. The ficus trees inspired the iconic image of the SI – a labyrinth that symbolizes – like in many cultures – the inner mystical journey, represented in the shape of the ficus trees. The ficus trees also symbolize generosity – they provide an immense shelter for all species, nutritious food and meeting spaces for so many activities and celebrations.

I have often wondered what it would be like to tell the story of Lynedoch from the perspective of the two ficus trees in conversation with one another. Maybe it would start with something like this: "Hey, wake up, check out these new humans who've arrived. They stand and gawk at us, even hug our trunks". "Yeah, you're right – and look, now they are planting another one of us right at the front entrance of their Institute, and many more tree friends. I wonder why they're doing this". "I dunno, but it's fun – I really like it in the mornings when they come out below us in all shapes and colours to stretch and do what they call the 'Lynedoch Shout'. They have such a great young energy".

For so many years now, I've stood at the entrance to the SI on the morning we expect the students to arrive on the first day of orientation: I take in the surrounding sights, scents and sounds; feel the cool gentle south-easterly breeze against my skin; and I imagine in my mind's eye the arrival of the students. Who will pitch up? What stories will they bring? What will they expect? What are their dreams and hopes? How will they fold into the collective memory of those that preceded them? How will they react to a space that looks very different to what they imagined Stellenbosch University looks like?

After going through the formal university registration procedure, one late January morning in 2019, all 60 master's students were asked to gather in the main hall. Surrounded by the golden yellow of the unfired clay brick walls with the sound of young children in their classrooms on the other side of the walls, the group was seated in a circle. All the teaching and administrative staff were also in the circle, including SI staff involved in child care, farming and education projects. Our introductory greeting, after a poem by Ben Okri, included passing a stone around: holding the stone, they were asked to say their name and to offer a single word into the circle that captured their expectations. The rest of the two-day orientation was taken up with introductions to the modules and our teaching and learning approach. Jess Schulschenk introduced her amazing team – cooks, teachers, gardeners, administrators, artists and a dedicated 'learning architect'.

The most important part of orientation was when they were each given a piece of clay and asked to mould something that represented their expectations, which was then placed on a piece of paper, with a sentence each wrote that talked to the symbol. These 60 sentences were later assembled into a word-cloud and fed back into the group at the end of the introductory module, and the images were captured in poster form by Fiona, our resident artist and creative coordinator.

During the course of the following two weeks, students were introduced to a rich tapestry of ideas that contextualized the global commitment to sustainable development embodied in the Sustainable Development Goals (SDGs) within an interpretation of an Africa-centred sense of deep time and contemporary economic history. Before continuing with the story of this module as a kind of hologram of our entire programme, see Box 10.1 that describes the structure and logic of our teaching programme.

BOX 10.1

Students register for an Honours-level degree called the Postgraduate Diploma in Sustainable Development (PGdip-SD). This comprises eight one-week modules, followed by six weeks of essay writing. To do this in a year is demanding and must be done on a full-time basis or part-time over two years. Students select their eight modules from the following selection of modules (formal module names are followed by a brief sense of what the course is about):

- Sustainable Development: introductory overview of themes covered in the entire programme
- Complexity Theory and Systems Thinking: complex adaptive systems as taught by a philosopher
- Learning Transitions and Environmental Ethics: other ways of knowing and being in this world
- Food Security and Globalized Agriculture: why the global food system is fundamentally flawed
- Food System Transitions: what the alternatives are to the global food system
- Transdisciplinary Design for Transformation: design thinking tools for designing and implementing sustainable alternatives
- Renewable Energy Policy: overview of the global renewable energy revolution and the policy implications from an energy democracy (ED) perspective
- Systems Dynamics Modelling: computer-based systems dynamics tools for participatory assessment and planning of sustainable alternatives
- Biodiversity and Ecosystem Services: ecosystem approach to sustainable development, the dynamics of biodiversity and why biodiversity matters

- Globalization, Governance and Development: why the global economy is in crisis and an introduction to heterodox and ecological economic theories that provide alternatives to neoliberal economics
- Applied Economics: political economy of South Africa, with special reference to the mineral-energy complex, post-1994 governance and policy, economic theories and policies that address the real challenges of inequality, poverty and unemployment
- Renewable Energy Financing: how renewable energy projects can be structured to attract investments
- Sustainable Cities: why the global sustainable development challenge is being addressed in the most interesting ways in cities, with a special focus on African cities
- Corporate Governance and Sustainable Enterprise: how corporates address the sustainability challenge, and how a new generation of social enterprises is emerging as an alternative

After completing the PGdip-SD, students are eligible for admission into the MPhil in Sustainable Development (MPhil-SD). The PGdip-SD is the only accepted entry qualification into the MPhil-SD. The primary output of the MPhil-SD is a research-based master's thesis of around 60,000 words. With some exceptions, students are required to develop a conceptual framework, methodology and empirical research strategy. Over the years, reflecting our own learning curve, more and more students have used a transdisciplinary research methodology for their master's and doctoral research.

The place-based discussion learning continues to evolve at the SI within the wider context of the Lynedoch EcoVillage. In 2019, the intake into the PGdip-SD increased to 60 and 35 were admitted to the research-based MPhil-SD. Starting in 2018, the SI in partnership with SU launched a four-year vocational undergraduate degree in sustainable development and entrepreneurship. With an annual intake of 25, by year 4 (2021) there will be 100 young overwhelmingly black students enrolled in this programme. This means at any one time, there will be around 200 students studying at the SI at various levels. This excludes the doctoral and postdoctoral students based at the CST, some of whom have emerged as teachers in the undergraduate, PGdip and the MPhil programmes. When they all meet under the ficus trees in the mornings to listen to poetry and stretch, the ordinariness of this everyday routine can easily mask the extraordinary conjoining of a beautiful transformative space with a unique learning approach.

What is most remarkable about the master's programme is that it is located within the wider context of the Lynedoch EcoVillage. Over nearly 20 years, this vibrant experiment in living and learning has evolved into a de facto knowledge commons that connects a diverse range of collective learning processes. With SI acting as the animator of a relational web of interconnected but institutionally separate activities, the following are worth noting:

- in partnership with Stellenbosch University, SI's learning programmes for nearly 200 university students, including the new four-year undergraduate degree in entrepreneurship and sustainability and the master's programme;
- the primary school, which will be a fully fledged Montessori primary school for 300 students as from 2020;
- the Montessori preschool, including a baby-care centre;
- the Home Owners Association and its ongoing management of the site and a diverse community;
- an agro-ecological training academy to train young farmers via a three-year programme; and
- a Further Education and Training College for running various training programmes.

Various NGOs and social enterprises (e.g. WWF and iShack) rent spaces on site, thus contributing to the richness of the local economy that has been created. By May 2019, the SI website listed 61 members of the 'people of the institute', most of whom were full-time staff while the rest were research associates or Stellenbosch University staff.

Interdisciplinary learning

Although there has been a paradigm swing in teaching practice from the traditional 'download' (in the form of an 'expert-centred' lecture) to facilitated experiential discussion learning (sometimes referred to as the 'flipped classroom'), it is a mistake to treat these as binary opposites. Appropriately balanced, they can reinforce each other, especially when you have the advantage of being embedded within an actual learning laboratory for sustainable living.

At the start of the first day of formal teaching, students were asked to think of two members of their family whom they have physically touched: the oldest and the youngest. Using the birth dates of the oldest and the dates when the youngest is most likely to die of old age (assuming a lifespan of 75 years), students discover they have physically touched the equivalent of 200 years of lived experience. From here they are introduced to the SDGs, and most had not heard of them. They were asked to fill in a form stating which SDG meant most to them, and the two they think are most supportive of achieving the selected primary goal. In 2019, the socio-economic goals meant most to students – ending poverty, decent work and economic growth, reduced inequalities and responsible production and consumption.

What followed was a lecture on the seven documents that I regard as shifting a key aspect of how we understand our world (drawn from Chapter 2 of Swilling and Annecke, 2012):

- 2003 UN Habitat Report entitled *The Challenge of Slums* which demonstrated that one in three urbanites live in slums – this dismantled the assumption that urban equates with the end of poverty;
- 1998 Human Development Report which focused on inequality during the heyday of neoliberalism and which was responsible for producing the oft-quoted figure that 20% of the global population are responsible for 86% of consumption expenditure;
- 2007 Fourth Assessment Report of the IPCC that firmly established the anthropogenic causes of climate change but more importantly demonstrated that the poor will suffer first and most even though they have contributed least to the problem;
- 2005 Millennium Ecosystem Assessment which demonstrated that 15 out of the 24 ecosystems we depend on are degraded or used unsustainably;
- 2008 World Energy Outlook published by the International Energy Agency that demonstrated that the "era of cheap oil is over";
- 2008 report entitled Agriculture at a Crossroads: International Assessment of Agricultural Knowledge, Science and Technology for Development which essentially argued for the incorporation of ecosystem science into the new agro-ecological science aimed at making agriculture more sustainable; and
- the 2011 report by the International Resource Panel (IRP) entitled *Decoupling Natural Resource Use and Environmental Impacts from Economic Growth* demonstrated the total quantity of materials the global economy depends on and why economic activity needs to be decoupled from the rising resource use.

Each of the above culminates in one or a cluster of key SDGs by 2015. By joining the dots, students come to grasp how different bodies of knowledge have fused together since the 1970s in ways that changed our perception of the world over the course of a generation. Later that afternoon, they watched the documentary *End of Suburbia*. Cognitive shifts and the limits of an all-pervasive material reality are connected in this way.

On the morning of the second day (Tuesday), more in-depth discussion of climate change, ecosystems and resource flows takes place. In the afternoon, a pictorial history of the last 2.5 million years of pre-/present human evolution is presented, with iconic images used to illustrate the 12 revolutions that resulted in the world we live in today – with at every stage the inclusion of references to African history. The story was told in a way that illustrated Harari's argument that the defining feature of *homo sapiens* is our dual capacity to organize in large numbers and live in imaginary worlds (Harari, 2011). The 12 revolutions are as follows: 1 Tool-making Revolution, 2.5 m ybp; 2 Second Human Revolution – the great expansion of homo sapiens from 100 ybp; 3 Agriculture Revolution from 13,000 ybp; 4 Urban Revolution from 4,500 ybp to the present; 5 Imperial Revolution – great pre-industrial agrarian civilizations from 4,000 ybp (which produced all the great religions still practiced today); 6 Maritime Expansion, Re-invasion of the Americas, European Renaissance from 1450; 7 Scientific Revolution and the European Enlightenment ('science and liberty' – for some) from 1600; 8 Industrial Revolution from 1750 (coal power); 9 Age of Oil from 1920 (The 'Modern Age'); 10 The Nuclear Age, Decolonization and the Origins of Environmentalism from 1950; 11 Digital Age, Globalization and Sustainability from 1980; 12 Sustainability Revolution – Global Re-industrialization and the Feminization of Power (viewed from 2050).

Wednesday was devoted to two discussions: the history of the global economy since the late-1800s and how economic theories changed over time. Four paradigms were explained: liberalism/neoliberalism, Keynesianism/social democracy, socialism and developmental statism. This was followed by an overview of the evolution of development theory, from modernization through basic needs, poverty reduction and now sustainable development.

Thursday focused on theories of change (radical, adaptive, transformative), followed by a session on the power of narratives and myth over the ages. The point of this session was to demonstrate in more detail how we make sense of the world via the narratives we construct. If these narratives are inappropriate for comprehending our context, actions to make a better world are impossible. This was followed by the documentary on Ladakh which illustrates the impact of modernization on a traditional society. This led into Friday for a facilitated discussion at a beautiful place by the river about African conceptions of deep ecology led by the renowned feminist writer Mmatshilo Motsei (Motsei, 2007). By working with what each student believed was their totem, Motsei led them into a discussion of African perspectives on relatedness – in short, what *Ukama* meant to them.

The following Monday began with a morning session on African feminism facilitated by Azille Coetzee from the Philosophy Department. This lecture demonstrated how 'relational subjectivities' lie at the heart of Sub-Saharan African Philosophy. It is not only about *Ubuntu* – meaning 'humanness', *u*hich is the well-known African conception of 'I am because we are'. It is also about *ukama* – the Shona word for 'relatedness' to all beings, species and things (as discussed in Chapters 1 and 2). From this perspective, the classical Cartesian notion that 'I think therefore I am' makes no sense. An *ukamian* response to this claim would be: "Ok, but where and with whom?" This was followed by an afternoon session with Eve Annecke on embodied experiential knowing. The sessions by Motsei, Coetzee and Annecke all complemented each other, subtly connecting feminism, Afro-centric deep ecology and spirituality. These themes, however, were not confined to classroom discussion. Every morning students embodied these themes when they worked with Yoliswa Mahobe, the permaculturalist who cares for the organic gardens. She introduced herself to each group by telling them how she accidentally ended up studying horticulture, which eventually led to her discovering permaculture. Learning to farm by working with rather than against nature brought home in real ways what *Ukama* means in daily life.

Tuesday began with a lecture on long-wave transitions (based on Chapter 4 of this book) as a way of understanding the complex dynamics underway in the world that are shaping the transition to a more sustainable world. This was followed by groups doing land art in the gardens as a way of reconnecting them with nature. This was followed by a deeply emotional exercise that reconnected them to the youngest and oldest relatives they recalled on the first day: I read out a short story by Ben Okri called a Prayer from the Living. In essence, this is a short story about a man looking for his lover during the Rwandan genocide. After this reading, students were asked to write a letter to a child which would be read by that child when s/he grew up. What would you want that grown-up child to read? What will you have done to ensure that that child lives in a better world?

The rest of the course was essentially about processes and case studies of sustainability transitions in practice. This included a lecture on the global renewable energy transition on Tuesday afternoon and many case studies and short documentaries during the course of Wednesday and Thursday. The aim here was to build a sense of hope and a realization that there are literally hundreds of thousands of initiatives all over the world motivated by the vision of a more equitable and sustainable world.

On Friday, they presented their group assignments. Each of the 12 groups had spent their afternoons over the previous two weeks investigating an aspect of the Lynedoch EcoVillage – from waste, water and energy systems to youth education, the primary school, food production and the Montessori educational activities. Overflowing with the energies built up over the previous two weeks, each group playfully and creatively engaged with their topics. Instead of relying on PowerPoint, they combined poetry, music, soulful reflections, insightful technical descriptions, video, graphics, role play, creative criticism and humour to recount their journeys of discovery of each other and their 'home-from-home' for the next two years.

Reflections in hindsight

Students are not required to write an exam. Instead, we assess a range of outputs that make up a kind of pedagogical mosaic. Each element has a distinct character, but it is also interconnected with the other elements in a coherent and holistic manner. As I argued at the start of our curriculum design process back in 2003, Aristotle defined three forms of knowledge: techné, episteme and phronesis. Techné is technical know-how (how to do specific largely practical things), episteme is general wisdom and phronesis means practical judgement appropriate to the context. There is, unfortunately, no appropriate English equivalent for phronesis except possibly prudence, which is too weak a word to be appropriate.

We prepare students with a balance between techné, episteme and phronesis, but with a strong bias towards phronesis. I start every day of every module at the SI by getting everyone to sit in silence. But as I am getting people to settle down I say: "Switch off all electronics, put your feet flat on the ground, sit up with a straight back, and bring yourself into the room because being present in the context is what sustainability is all about". I am subtly drip feeding into them this notion of phronesis. Without phronesis, we have general wisdom (theory) and technical know-how. But without phronesis, this combination has reinforced the disconnect from context, nature and our bodies – a disconnect that originates in the Cartesian notion that "I think therefore I am". Surely, 'I am' is also because 'I feel', as well as because 'I'm connected to others and a place'. An *ukamian* sense of phronesis is the wisdom needed to counteract the European Cartesian disconnect. Inculcating this sensibility is what an evolutionary pedagogy of the present is all about.

Contextual thinking rather than subordinating context to theory is the lasting and essential epistemological contribution of complexity theory and is the essence of *ukama*. Contextual thinking, however, means breaking from reductionism, especially when it comes to research. This, in turn, is reflected in the growing worldwide interest in recent years in transdisciplinary research methodology.

Continuous assessment rather than exam-centred testing is profoundly formative – I like the French word 'formation' (phonetically, 'for-ma-shion') which in my view has the correct emphasis on the process of 'forming the whole person' and not just cognitive development. We ask students to write a journal in the first person so that they can learn to reflect experientially, and this is formally assessed. We require them do a class test of some sort – could be creative (a photographic pic) or literature review, or practical task. The two essays that must be handed in six weeks after the end of a module have a clear purpose. Firstly, in the literature review, we want them to engage with different theories of the world in order to realize that the world is not a given, but is (partially) constructed. Students must engage with theories that challenge their basic assumptions. As they start to see things differently, we make sure that for the entire duration of the degree they are discouraged from 'settling' as they move from one module to the next. Our job is to perturb, disturb, upset, reconstruct and then let go so that they find themselves. This is achieved via the literature review – the first assignment (worth 25% of the mark).

Then we challenge them to apply their theory/conceptual framework to a specific empirical challenge in the second essay (worth 25%). Knowing full-well they have not 'resolved' what they think their worldview is, we invite them to apply their argument developed in the literature review to a particular empirical challenge. What is happening here is that we are asking them to see things through the conceptual lens constructed in the first essay. In other words, we don't want them to just play around with worldviews and arbitrarily select one or the other just because they have to complete the assignment by a deadline. Knowing that they have to apply their argument to a particular empirical context keeps them honest. However, it also makes them realize that reality can be seen from multiple angles. This is absolutely critical when it comes to phronesis, the essence of which is humility. What works in one context will not necessarily work in another. Context, in short, matters.

Unless we train students to really deeply read and appreciate what is specific about a context, they will never overcome the arrogance that higher education tends to create, that is, a sense that "I know, and you don't"/"I know better because I have an education". To be blunt, they know certain things only but cannot know others. When students realize that education only brings them closer to the mystery, they become more humble. And that happens when they engage with the complexities of each empirical context, discovering that unless they listen and observe attentively every time, they will end up making the most common mistake of all, that is, imposing what they assume is right on a context that may not conform in reality to how they assume it works.

In other words, our practice is about effectively inviting the student to traverse the interactive relationship between epistemology and ontology. Unfortunately, the Western Enlightenment tradition is disrespectful of context because of a long history of reducing ontology to epistemology (see Chapter 2). Without making explicit these confusing terms, we are effectively in practice creating a more appropriate balance between the two. This, in turn, is consistent with the break from reductionism that complexity theory has achieved. And it is perfectly consistent with the Sub-Saharan African philosophy of place-based relatedness – *Ukama*. I have not come across many places that have embedded this worldview in practical pedagogical processes.

We are up against a very long tradition that is guilty of preparing people with an ethically abhorrent tolerance of poverty, racism and sexism and almost totally disconnected from nature. Usually done by omission than explicit commission, this is the consequence of teaching theory or methods (in particular models) without reference to context. This is why we have a polycrisis – very educated people created it. We at the SI are part of a small group of institutions around the world experimenting with and pioneering an 'evolutionary pedagogy of the present' that actively strives to achieve a balance between episteme, techné and phronesis. For us, our inspiration stems from this remarkable meeting point between Sub-Saharan African philosophies of *Ukama*, Western post-humanism and Metatheory 2.0 (as discussed in Chapter 2) – a meeting point we initially only grasped intuitively but which has slowly emerged in more explicit documented form in recent years.

Transdisciplinary research practices

Our online Research Methodology course that all our students must complete starts with an essay on 'activist researchers'. We explicitly state that our aim is to develop researchers who become change agents. This does not mean doing research *about* change. It means becoming involved in incremental change processes (see Chapter 5) as researchers who use research methods to generate information and knowledge that is useful to those with whom they work. However, they are also encouraged to critically assess these change processes. What matters is timing: during the research, they do action research to support the change process; later on, while writing up their research, they reflect and constructively criticize what they have witnessed. To be activist researchers, however, means mastering transdisciplinary research methods.

Mono-disciplinary analysis does not help us understand and grapple with emerging complex socio-ecological challenges. The application of single discipline knowledge produces partial solutions but not the much-needed long-term, integrated and sustainable solutions. An emerging body of literature argues that contemporary socio-ecological challenges warrant transdisciplinary responses that embrace what is referred to as knowledge co-production between science and society. This literally refers to a process that requires researchers to engage and collaborate with practitioners to co-generate knowledge to address problems that emerge in real-world situations. The concept of transdisciplinary research (TDR), which has emerged over the last two decades, is not a new science per se but rather a new way of doing science. Instead of doing "science for society", the aim is to do "science with society" (Gibbons, Limoges, Nowotny, Schwartzman, Scott, et al., 1994; Nowotny, Scott and Gibbons, 2001; Hadorn and Pohl, 2008; Scholz, 2011; Becker, 2012; Lang, Wiek, Bergmann, Stauffacher, Martens, et al., 2012; Bergmann, Jahn, Knoblach, Krohn, Pohl, et al., 2013; Seidl, Brand, Stauffacher, Krutli, Le, et al., 2013).

TDR builds on a much longer tradition of 'interdisciplinary' research. More well-known as "Mode 2" research (Gibbons, Limoges, Nowotny, Schwartzman, Scott, et al., 1994; Nowotny, Scott and Gibbons, 2001), this tradition has always been interested in the social contextualization of knowledge production (Rip, 2011). In the African context, Bagele Chilisa from the University of Botswana, has done seminal work on the integration of 'indigenous knowledge systems' into a wider conception of 'post-colonial research' (Chilisa, 2012, 2017) that aligns closely with the ukamian perspective. Our collaborations with her have profoundly influenced our thinking.

With few exceptions (e.g., Chilisa, 2012, 2017), Mode 2 research has emerged largely in the global North (Gibbons, Limoges, Nowotny, Schwartzman, Scott, et al., 1994; Nowotny, Scott and Gibbons, 2001), where well-endowed research institutions can choose to engage with formalized, legitimate and institutionalized stakeholders largely around the challenges of late industrial modernity. However, research contexts in the global South – especially in Africa – are rarely this well-structured. Indeed, informality is often the norm in these contexts. How, then, does one conduct TDR in these contexts?

A key reason for conducting "science with society" is not only explaining and understanding complex societal challenges but also generating implementable solutions (Stauffacher, Walter, Lang, Wiek and Scholz, 2006; Scholz, 2011; Seidl, Brand, Stauffacher, Krutli, Le, et al., 2013; Miller, Wiek, Sarewitz, Robinson, Olsson, et al., 2014; Wiek and Lang, 2016). The TDR approach rests on a distinction between three types of knowledge: systems knowledge (what exists and why), target knowledge (policy proposals for what should/could exist) and transformation knowledge (how systems change over time in a certain direction).

It is not possible to implement transformative TDR approaches in developing world contexts by uncritically replicating and transferring the ideal – typical approaches developed in the developed world. Although transformative in orientation, TDR approaches that have emerged in the global North seem to have in common the fundamental assumption that formal stakeholder engagement is the primary means of engagement.

After nearly a decade and a half of theoretical reflection and on-the-ground experimentation in the South African context, we have developed an approach for conducting TDR in conditions where extensive social informality makes it difficult to assume the existence of legitimated stakeholders. Besides the profound influence of the African literature (Chilisa, 2012, 2017; Coetzee, 2017), this approach has drawn on the following complementary literatures:

- complexity theory (Cilliers, 1998; Juarrero, 2002; Snowden and Boone, 2007; Vester, 2007; Mingers, 2014; Boulton, Allen and Bowman, 2015);
- emergent design theory (Cavallo, 2000; Hasan, 2006; Jonas, 2007; Hesse-Biber, 2010; Hesse-Biber and Leavy, 2010; Sanders and Stappers, 2013);
- assemblage theory (De Landa, 2006; Latour, 2007; Harman, 2008; Farias and Bender, 2010; McFarlane, 2011);
- learning theory (Argyris, 2002; Corcoran and Wals, 2012; Taylor and Cranton, 2012; Tosey, Visser and Saunders, 2012; Kolb, 2014; Medema, Wals and Adamowski, 2014; Wals and Rodela, 2014);
- narrative theory (Czarniawska, 2004; Edelman, 2006; Heinen and Sommer, 2009; Snowden, 2010; Klein, Snowden and Pin, 2011; Kurtz, 2014).

A key insight drawn from the integration of this diverse body of literature for developing a context-sensitive TDR approach is to link the notion of human agency in social actor networks to the broader notion of complex systems change. In our understanding, this means that, when complex systems change, social actors not only make sense of what is happening in order to adapt, but they also act to change their context (Latour, 2007).

In our view, the existing literature on TDR has not yet generated an adequate set of context-relevant guiding logics and principles. Without this, there is no methodology that can be used for navigating TDR processes in social conditions that are highly fluid, relatively unstructured and informalized (for an elaboration of this general problem, see van Breda and Swilling, 2018).

Since 2005, our efforts to bring into conversation European approaches to TDR, African approaches (Chilisa, 2012, 2017) and the richness of our own empirical research conducted by dozens of postgraduate researchers over the years have resulted in a distinctive body of methodological knowledge (see van Breda and Swilling, 2018). The most significant is a set of guiding logics and principles for conducting what we have called Emergent Transdisciplinary Design Research (ETDR). These guiding logics and principles should be seen as cognitive facilitators of imaginative and iterative decision-making processes. These processes are, by definition, incrementalist in that they tend to get driven forward by those who are best placed to ask 'what is the next step' (see Chapter 5) during the unfolding of the applied research processes. Rather than having to predict or know too far in advance exactly what the consequences of embarking on a particular vector or direction of change may be, it is strategically and practically more important to figure out the next step and then see where that may lead to within a rapidly changing context.

Those who ask 'what is the next step?' are effectively asking the group to imagine and create spaces for the 'adjacent possible' (Unger, 2007; Snowden, 2016), 'in-between' or 'third-paces' (Vilsmaier and Lang, 2015) where (radical) experimentation can be explored and promoted. The guiding logics and principles presented later create a cognitive framework for performing acts of 'side-casting' (Snowden, 2012) rather than, by way of contrast, conducting the teleologically orientated 'forecasting' or 'back-casting' activities advocated in the transdisciplinarity literature (Stauffacher, Walter, Lang, Wiek and Scholz, 2006; Scholz, 2011; Wiek and Lang, 2016) and, indeed, in nearly all variations of strategic planning or futuring. Whether planning forwards, towards or backwards from the future, these teleological approaches see the present from the vantage point of an idealized future imaginary. These approaches are often disconnected from the realities and complexities of the current situation.

In the ETDR approach outlined here, however, the role and function of the guiding logics and principles is to nudge the activist/research process towards discovering the evolutionary potential of the present (Snowden, 2015). In this sense, the present is not a burning platform between the past and future but rather where both meet in a 'thick present' where contested uncertainties can be expressed in a multiplicity of experimentations and processes aimed at building broad-based coalitions for change inspired by ever-evolving imaginaries of the future (see Chapter 5).

Overall, the five basic principles for guiding TDR processes are as follows: 1 perturbing the system; 2 innovating through exaptation; 3 multiloop learning; 4 allowing for emergence; 5 absorbing complexity. These principles should be seen as the emergent outcome of an iterative process of critical reflection on specific empirical research experiences that occurred within the South African context since 2005 (for real-world applications of these principles, see van Breda and Swilling, 2018).

Perturbing the system

The principle of "perturbing the system" comes from complex adaptive systems theory, which holds that systems are self-organizing and self-adapting. Small changes in one part of the system can affect bigger changes in other parts of the system, thereby making possible wider systemic change under certain conditions (Chu, Strand and Fjelland, 2003; Wright and Meadows, 2012). Sometimes it is necessary to actively perturb a well-established system so that it tilts into a state of disorder. By consciously using strategic leverage – or what was referred to in Chapter 5 as 'social acupuncture' – change dynamics can be catalysed in key nodes that can have disproportionately large system-wide effects. Indeed, while most attempts at system-wide change in complex systems result in (often unwanted) unintended consequences, it is possible to focus on strategic leverage points that catalyse change processes that evolve and expand over time into system-wide change as an emergent outcome (Wright and Meadows, 2012).

Even state systems with considerable system-wide strategic leverage find it difficult to engineer planned system-wide change. Strategic leverage can instigate multiple, contextual, (relatively) small-scale social experiments that get implemented over a period of time (Snowden and Boone, 2007; Snowden, 2010). Similar to what is described as 'niche innovations' in the multi-lateral perspective (Geels, 2005), these experiments are usually somewhat protected from market forces, regulation and policy regimes so that they can evolve and test out new ways of doing things.

These small-scale or safe-to-fail social experiments are essentially about the co-construction of 'something' (Cavallo, 2000) that acts as a 'boundary object' (Star and Griesemer, 1989; Star, 2010) or 'social attractor' (Snowden, 2010). They are situated at the intersection of particular socio-technical and/or socio-ecological systems in need of broader systemic change.

Innovating through exaptation

There is a growing impatience with innovations that are merely about adaptation ('making do') within the existing systems – sometimes referred to as social 'bricolage'. 'Exaptation', by contrast, means going beyond 'bricolage' (Kincheloe and Berry, 2004). Under certain circumstances, adaptation may well be a major achievement, especially where it can literally mean the difference between life and death (in, for example, informal settlements or refugee camps). The incumbent systems, however, remain largely intact.

When innovations emerge (often serendipitously) that generate solutions to new problems that transcend the boundaries of the current systems, that is exaptation. Exaptation enables the emergence of these solutions by working with what exists *and* the endogenous potentials for radical change (see discussion on postcapitalism in Chapter 1). However, this is achieved without undertaking the traditional TDR practice of first establishing some normative end (normally in the form of a shared vision and values) and then finding the most effective and efficient means with which to achieve these normative ends. Instead of visioning, researchers and community members co-design and implement provisional safe-to-fail experiments relatively quickly (Snowden, 2010; Klein, Snowden and Pin, 2011). The upshot is a set of boundary objects that, in turn, stimulate new narratives about the potential for wider and more radical systemic change.

Multi-loop transformative learning

The basic idea of "multi-loop learning" comes from Bateson (2002), namely that learning is an iterative process whereby people go through many loops of learning. These comprise three distinct levels: "learn" something (level 1), "learn how to learn" about something (level 2), and "learn how to learn how to learn" (level 3). More specifically,

- level 1 signifies the acquisition of new technical knowledge and skills (techné);
- level 2 denotes the learning of learning, figuring out how to share and transfer newly acquired knowledge to others in order to do things more efficiently (techné plus transfer);

- level 3 occurs when a critical awareness of the consequences and direction of the learning process emerges, followed by self-conscious adjustments to the underlying logic and principles of the learning process. Transformative learning (phronesis) happens at this level.

Co-producing systems, target and transformation knowledge (Hirsch-Hadorn, Bradley, Pohl, Rist and Wiesmann, 2006; Hadorn and Pohl, 2008) is fundamental to the ETDR approach. The challenge is how to learn how to co-produce these three different types of knowledge in the fluid, emerging informal environments found in many African cities. The underlying ideas of multi-loop learning are particularly useful (Bateson, 2002; Tosey, Visser and Saunders, 2012; Medema, Wals and Adamowski, 2014) in this regard for making sense of the continuous flow of experiences, reflections, ideas, theorizing and actions that occur in any transformative change process.

While all three levels of learning are necessary, transformative learning (phronesis) occurs at level 3 as the deeper strategic insights and reflection about the learning process itself are generated. Level 3 learning goes beyond traditional cognition. It extends learning into the aesthetic and axiological dimensions of learning. The architecture of the Sustainable Development module described earlier is designed to facilitate level 3 learning.

Anticipating and allowing for emergence

The aim of perturbing the system by implementing multiple safe-to-fail social experiments is to create the conditions necessary for longer term solutions to emerge. It is critical to nudge the TDR research process to avoid premature convergence, thus enabling emergence to occur (Snowden, 2006, 2011; Snowden and Boone, 2007).

During any transformation process, key leverage points emerge that become opportunities for intervention. They are, in reality, bifurcations: as a process matures, it can split and move into different directions and across multiple sites of instability. This is why they are ripe with potential emergent solutions. Transdisciplinary researchers must allow for these serendipitous solutions to emerge. Rather than resisting them because they do not fit into the original plan, they should be allowed to morph into new processes and entities.

The guiding principle of 'allowing space for emergence' has three important aspects. Firstly, there is an expectation that the emergent outcome will be more than the sum total of its parts – in this case, more than the combined results of individual research activities and implementation of small-scale interventions. One such emergent outcome would be a newly established culture of working together (Sennett, 2012).

An idealized version of the future is not a fundamental prerequisite for initiating TDR processes. It may be more desirable to start with practical small-scale projects rather than aim for system-wide changes that may be unattainable. In line with the notion of radical incrementalism in Chapter 5, TDR generates an unfolding set of knowledge platforms in the present that allow for normative imaginaries of the future to emerge from the dynamics of actual processes in the present. Allowing for emergence implies that a culture of working together is not fixed or stable, which also means it cannot be taken for granted. Instead, it gets re-negotiated for each setting and context (Latour, Jensen, Venturini, Grauwin and Boulier, 2012).

Absorbing complexity

When working in complex real-world contexts, it is better to use a research approach that 'absorbs complexity' ('working with complexity', 'making it work for you') rather than attempting to reduce complexity ('simplify') in order to 'increase control' (Snowden, 2011). Attempts to overly structure the TDR process to provide certainty in an uncertain environment are likely to lead to premature convergence and hasty conclusions. This requires researchers to retain some measure of cognitive agility by being open to unanticipated outcomes. In highly fluid social conditions, it is not possible to follow the mainstream TDR guideline to reduce complexity (Hadorn and Pohl, 2008) and create conflict-free zones for conducting dialogues (Scholz, 2011).

A TDR research strategy that aims to perturb particular contexts warrants an approach that "absorbs complexity" by finding ways of working with conflict rather than seeking to contain it. This approach can only succeed if there is trust (Tait and Richardseon, 2010). Trust, however, cannot be structured or even negotiated. It is an emergent outcome of the entangled socio-technical relationships that are painstakingly assembled in and around small-scale safe-to-fail experiments. Trust has to be built both within and outside the TDR team: at the interpersonal level within the team and amongst those involved in the wider project.

Conclusion

After nearly two decades of experimentation with place-based discussion learning and TDR embedded within the complexities of the South African context, a core body of knowledge-in-practice has emerged within the SI-CST partnership. A younger generation of teachers and researchers are now taking this evolutionary pedagogy of the present to a higher level as they grapple with rapidly changing conditions across the global-local spectrum.

In a world that seems to overvalue 'newness', innovation and instantaneous fixes, it is not easy to appreciate the significance of 20 years of everyday routines. And yet these repetitive routines are crucial for sustaining the commitments to change. Consider this reflection:

> As the sun rises over the distant Helderberg mountains, the early mornings at Lynedoch are always heralded by the chorus of birds. As residents awaken, the first arrivals on site are the Montessori pre-schoolers. They chatter away

as they head towards the pre-school building powered by solar panels, often led by a parent. Soon after, the first primary school pupils arrive in their smart uniforms and school bags. Nearly all from black families, they are bright-eyed, energised, self-contained and greet anyone passing by with wide smiles and innocent eyes. Their day starts with a gathering on the tennis court where the whole school dances and exercises to loud contemporary music. That is when the undergraduate and masters students start to arrive. They gather in the amphitheatre under the giant Ficus trees, and as the music sub-sides on the tennis court, the students huddle to listen to someone reading a meaningful poem or passage and then spread out to do their body stretches. My favourite is when Ross van Niekerk leads the exercises with a Thai Chi routine: the same routine she's taught the pre-schoolers for many years now. I'm always amazed as I watch Ross: a fiery personality with a political activist background, she was part of the founding group that established Lynedoch. She has often served as a Trustee of the Home Owners Association and has learned to become a Montessori care giver in the pre-school. She qualified for a government housing subsidy and lives with her large three generation family in one of the first adobe houses built at Lynedoch.

After stretching, students break into their respective groups and head off to their allocated work stations until the start of the first session at 9.30 a.m. To date, their morning group activity has included gardening, tree planting, weeding, litter gath-ering, food preparation and sweeping the buildings. However, this year learning to play music with Tau, art with Fiona or a mindfulness session led by one of the Lynedoch EcoVillage residents were added to the repertoire.

After nearly 20 years of the same basic morning routines, our way of doing things has become a culture – the imprimatur of the so-called SI experience that binds the graduates together in a shared memory of the future. They fan out across South Africa, Africa and the globe as part of a network that assumes we are in the age of sustainability and their job is to ensure that the just transition is realized. As one of them working in a government-created agency to promote a 'green economy' put it recently: "There are the SI graduates in the organization, and the rest – and we see things very differently".

The themes and issues addressed in the chapters of this book have emerged from this evolutionary pedagogy of the present and then fed back into the curriculum. The place-based learning that this unique context makes possible, coupled to the discussion-oriented classroom learning methods and persistent emphasis on the importance of learning to write well, has generated immense creativity. As a result, there have been engagements with vast swathes of diverse literatures over the years and the compilation by excellent students of a wide range of case studies. Inevitably, these all fuse together into a richly textured repertoire of themes and explorations. The preceding chapters are in many ways distillations of this accumulated knowl-edge and wisdom. This chapter elucidates how we facilitate the learning processes that prepare students for living, working and acting in the age of sustainability.

Note

1 This chapters draws on the extensive joint work with John van Breda over the past 15 years – for the basis for this chapter see the synthesis of our collaboration in van Breda and Swilling (2018).

References

Annecke, E. (2013) 'Radical openness and contextualisation: Reflections on a decade of learning for sustainability at the Sustainability Institute', in McIntosh, M. (ed.) *The Necessary Transition: The Journey Towards the Sustainable Enterprise Economy*. Sheffield, UK: Greenleaf Publishing, pp. 41–51.

Argyris, C. (2002) 'Double-loop learning, teaching and research', *Academy of Management Learning & Education*, 1, pp. 206–218.

Bateson, G. (2002) *Mind and Nature: A Necessary Unity*. New York: Hampton Press.

Becker, E. (2012) 'Social-ecological systems as epistemic objects', in Glaser, M., et al. (eds.) *Human-Nature Interactions in the Anthropocene: Potentials of Social-Ecological Systems Analysis*. London: Routledge, pp. 37–59.

Bergmann, M., Jahn, T., Knoblach, T., Krohn, W., Pohl, C., et al. (2013) *Methods for Transdisciplinary Research: A Primer for Practice*. Zurich: Campus Verlag GmbH.

Boulton, J., Allen, P. and Bowman, C. (2015) *Embracing Complexity: Strategic Perspectives for an Age of Turbulence*. Oxford: Oxford University Press.

Cavallo, D. (2000) 'Emergent design and learning environments: Building on indigenous knowledge', *IBM Systems Journal*, 39, pp. 768–781.

Chilisa, B. (2012) *Indigenous Research Methodologies*. Thousand Oaks, CA: Sage.

Chilisa, B. (2017) 'Decolonising transdisciplinary research approaches: An African perspective for enhancing knowledge integration in sustainability science', *Sustainability Science*, 12, pp. 813–827. doi: 10.1007/s11625-017-0461-1.

Chu, D., Strand, R. and Fjelland, R. (2003) 'Theories of complexity', *Complexity*, 8, pp. 19–30.

Cilliers, P. (1998) *Complexity and Postmodernism: Understanding Complex Systems*. London: Routledge.

Coetzee, A. (2017) *African Feminism as a Decolonising Force: A Philosophical Exploration of the Work of Oyeronke Oyewumi*. Stellenbosch: Stellenbosch University.

Corcoran, P. and Wals, E. (2012) *Learning for Sustainability in Times of Accelerating Change*. Wageningen: Wageningen Academic Publishers.

Czarniawska, B. (2004) *Narratives in Social Science Research*. Thousand Oaks, CA: Sage.

De Landa, M. (2006) *A New Philosophy of Society: Assemblage Theory and Social Complexity*. London: A & C Black.

Edelman, G. (2006) *Second Nature: Brain Science and Human Knowledge*. New Haven, CT: Yale University Press.

Farias, I. and Bender, T. (2010) *Urban Assemblages: How Actor-Network Theory Changes Urban Studies*. London and New York: Routledge.

Freeth, R. and Annecke, E. (2016) 'Facilitating social change', in Swilling, M., Musango, J. and Wakeford, J. (eds.) *Greening the South African Economy*. Cape Town: UCT Press, Chapter 21.

Geels, F. W. (2005) *Technological Transitions: A Co-evolutionary and Socio-Technical Analysis*. Cheltenham, UK: Edward Elgar.

Gibbons, M., Limoges, C., Nowotny, H., Schwartzman, S., Scott, P., et al. (1994) *The New Production of Knowledge: The Dynamics of Science and Research in Contemporary Societies*. London: Sage.

Hadorn, G. H. and Pohl, C. (2008) *Handbook of Transdisciplinary Research*. Dordrecht: Springer.

Harari, Y. N. (2011) *Sapiens: A Brief History of Humankind*. London: Vintage.

Harman, G. (2008) 'De Landa's ontology: Assemblage and realism', *Continental Philosophy Review*, 41, pp. 367–383.

Hasan, H. (2006) 'Design as research: Emergent complex activity', in *Proceedings of the 17th Australian Conference of Information Systems (ACIS 2006)*, Adelaide, 6–8 December.

Heinen, S. and Sommer, R. (2009) *Narratology in the Age of Cross-Disciplinary Narrative Research*. Berlin: Walter de Gruyter.

Hesse-Biber, S. (2010) *Mixed Methods Research: Merging Theory with Practice*. New York: Guilford Press.

Hesse-Biber, S. and Leavy, P. (2010) *Handbook of Emergent Methods*. New York: Guilford Press.

Hirsch-Hadorn, G., Bradley, D., Pohl, C., Rist, S. and Wiesmann, U. (2006) 'Implications of transdisciplinarity for sustainability research', *Ecological Economics*, 60, pp. 119–128.

Jonas, W. (2007) 'Design research and its meaning to the methodological development of the discipline', in *Design Research Now, Board of International Resource in Design*. Basel: Birkhäuser.

Juarrero, A. (2002) *Dynamics in Action: Intentional Behavior as a Complex System*. Boston, MA: MIT Press.

Kincheloe, J. and Berry, K. (2004) *Rigour and Complexity in Educational Research: Conceptualizing the Bricolage*. New York: McGraw-Hill.

Klein, G., Snowden, D. and Pin, C. (2011) 'Anticipatory thinking', in *Informed by Knowledge: Expert Performance in Complex Situations*. New York and London: Psychology Press, pp. 235–246.

Kolb, D. (2014) *Experiential Learning: Experience as the Source of Learning and Development*. London: FT Press.

Kurtz, C. (2014) *Working with Stories in Your Community or Organization: Participatory Narrative Inquiry*. Scotts Valley, CA: On Demand Publishing.

Lang, D. J., Wiek, A., Bergmann, M., Stauffacher, M., Martens, P., et al. (2012) 'Transdisciplinary research in sustainability science: Practice, principles, and challenges', *Sustainability Science*, 7(Suppl. 1). doi: 10.1007/s11625–011–0149-x.

Latour, B. (2007) *Reassembling the Social: An Introduction to Actor-Network-Theory*. Oxford: Oxford University Press.

Latour, B., Jensen, P., Venturini, P., Grauwin, S. and Boulier, D. (2012) '"The whole is always smaller than its parts" – A digital test of Garbriel Tardes' monads.' *The British Journal of Sociology*, 63, pp. 590–615. doi: 10.1111/j.1468-4446.2012.01428.x.

McFarlane, C. (2011) *Learning the City: Knowledge and Translocal Assemblage*. Oxford: Wiley Blackwell.

Medema, W., Wals, A. and Adamowski, J. (2014) 'Multi-loop social learning for sustainable land and water governance: A research agenda on the potential of virtual learning platforms', *Journal of Life Sciences*, 69, pp. 23–38.

Miller, T., Wiek, A., Sarewitz, D., Robinson, J., Olsson, L., et al. (2014) 'The future of sustainability science: A solutions-oriented research agenda', *Sustainability Science*, 9, pp. 239–246.

Mingers, J. (2014) *Philosophy and Systems Thinking: A Mutual Synergy*. New York and London: Routledge.

Motsei, M. (2007) *The Kanga and the Kangaroo Court*. Johannesburg: Jacana.

Muhar, A., Visser, J. and van Breda, J. (2013) 'Experiences from establishing structured inter- and transdisciplinary doctoral programs in sustainability: A comparison of two cases in South Africa and Austria', *Journal for Cleaner Production*, 61, pp. 122–129.

Nowotny, H., Scott, P. and Gibbons, M. (2001) *Re-thinking Science: Knowledge and the Public in an Age of Uncertainty*. London: Wiley.

Rip, A. (2011) *The Future of Research Universities*. New York: Prometheus.

Sanders, E. and Stappers, P. (2013) 'Co-creation and the new landscapes of design', *CoDesign*, 4, pp. 5–18.

Scholz, R. W. (2011) *Environmental Literacy in Science and Society: From Knowledge to Decisions*. Cambridge: Cambridge University Press.

Seidl, R., Brand, F., Stauffacher, M., Krutli, P., Le, O., et al. (2013) 'Science with society in the anthropocene', *Ambio*, 42, pp. 5–12. doi: 10.1007/s13280-012-0363-5.

Sennett, R. (2012) *Together: The Rituals, Pleasures and Politics of Cooperation*. London: Penguin.

Snowden, D. (2006) 'Perspectives around emergent connectivity, sensemaking and asymmetric threat management', *Public Money &Management*, 26, pp. 275–277.

Snowden, D. (2010) 'Naturalizing sensemaking', in *Informed by Knowledge: Expert Performance in Complex Situations*. New York and London: Psychology Press, pp. 223–234.

Snowden, D. (2011) 'Good fences make good neighbours', *Information Knowledge Systems Management*, 10, pp. 135–150.

Snowden, D. (2012) *Sidecasting Techniques*. Blog Post. Available at: https://cognitive-edge.com/blog/sidecasting-techniques/ (Accessed: 10 April 2019).

Snowden, D. (2015) *The Evolutionary Potential of the Present*. Blog Post. Available at: https://cognitive-edge.com/blog/the-evolutionary-potential-of-the-present/ (Accessed: 10 April 2019).

Snowden, D. (2016) *The Adjacent Possible*. Blog Post. Available at: https://cognitive-edge.com/blog/the-adjacent-possible/ (Accessed: 10 April 2019).

Snowden, D. and Boone, M. (2007) 'A leader's framework for decision making', *Harvard Business Review*, 85. doi: 10.14236/jhi.v13i1.578.

Star, S. (2010) 'This is not a boundary object: Reflections on the origin of a concept', *Science, Technology, &Human Values*, 35, pp. 601–617.

Star, S. and Griesemer, J. (1989) 'Institutional ecology, translations and boundary objects: Amateurs and professionals in Berkeley's Museum of Vertebrate Zoology', *Social Studies of Science*, 19, pp. 387–420.

Stauffacher, M., Walter, A., Lang, D., Wiek, A. and Scholz, R. (2006) 'Learning to research environmental problems from a functional socio-cultural constructivism perspective: The transdisciplinary case study approach', *International Journal Of Sustainability and Higher Education*, 7, pp. 252–275.

Swilling, M. (2014) 'Rethinking the science-policy interface in South Africa: Experiments in knowledge co-production', *South African Journal of Science*, 10(5/6), p. Art.#2013–0265.7. doi: 10.1590/sajs.2014/2013026.

Swilling, M. and Annecke, E. (2012) *Just Transitions: Explorations of Sustainability in an Unfair World*. Tokyo: United Nations University Press.

Tait, A. and Richardseon, K. (2010) *Complexity and Knowledge Management: Understanding the Role of Knowledge in the Management of Social Networks*. Charlotte, NC: IAP.

Taylor, E. and Cranton, P. (2012) *The Handbook of Transformative Learning: Theory, Research and Practice*. San Francisco: Jossey-Bass.

Tosey, P., Visser, M. and Saunders, M. (2012) 'The origins and conceptualizations of "triple loop" learning: A critical review', *Management Learning*, 43, pp. 291–307.

Unger, R. (2007) *The Self Awakened: Pragmatism Unbound*. Boston, MA: Harvard University Press.

van Breda, J. (2008) 'Overcoming the disciplinary divide: A necessary prerequisite for the establishment of sustainability science: Towards the possibility of a transdisciplinary hermeneutics', in Burns, M. and Weaver, A. (eds.) *Exploring Sustainability Science*. Stellenbosch: Sun Press.

van Breda, J. (2019) *Towards a Transformative Transdisciplinary Research Methodology*. Doctor of Philosophy, Stellenbosch University.

van Breda, J. and Swilling, M. (2018) 'Guiding logics and principles for designing emergent transdisciplinary research processes: Learning lessons and reflections from a South African case study', *Sustainability Science*. doi: 10.1007/s11625-018-0606-x.

Vester, F. (2007) *The Art of Interconnected Thinking: Tools and Concepts for a New Approach to Tackling Complexity*. Munich: MCB Verlag GmbH.

Vilsmaier, U. and Lang, D. (2015) 'Making a difference by marking the difference: Constituting in-between spaces for sustainability learning', *Current Opinion in Environmental Sustainability*, 16, pp. 51–55.

Wals, R. and Rodela, R. (2014) 'Social learning towards sustainability: Problematic, perspectives and promise', *Journal of Life Sciences*, 69, pp. 1–3.

Wiek, A. and Lang, D. (2016) 'Transformational sustainability research methodology', in *Sustainability Science – An Introduction*. Berlin: Springer, pp. 31–41.

Wright, D. and Meadows, D. (2012) *Thinking in Systems: A Primer*. New York and London: Routledge.

11
CONCLUSION
Reflections of an enraged incrementalist

Paris, May 2019: it was a late Friday afternoon and I was in Paris to attend the first meeting of the Scientific Committee of the *Campus de la Transition*[1] (referred to by those involved as 'the Campus'). We met in a small seminar room in the famous Ecole des Hautes Etudes en Sciences Sociales (EHESS) where Algerian-born French philosopher of postmodernism Jacques Derrida used to teach. With various others beamed in via Zoom, we were there to discuss the educational programme of the Campus. The Campus is located in a large French Chateau (with 40 rooms and nine classrooms) located just south of Paris. Donated by the Catholic Church to a mixed group of Jesuits, academics and activists, the vision is to create something similar to the SI in South Africa and the Schumacher College in the United Kingdom. They took over the Chateau in early 2018. I was stunned when the chair of the committee, a professor at EHESS, opened the meeting by saying that their reason for initiating the Campus was to create a space for alternative education programmes that could not be introduced from within the French Universities. "We have tried, and failed", he said. The subsequent conversation was infused by a shared sense that the pent-up demand for inter- and transdisciplinary sustainability-oriented education expressed by young people hungry to understand the state of the world could not be delivered from within some of the most prestigious universities in the world. Without this, I thought to myself, the full potential of the age of sustainability cannot be realized. Inevitably, many more initiatives like the SI, Schumacher College and the Campus de la Transition will emerge (see case studies in Gravata, Piza, Mayumi and Shimhara, 2013).

I left the meeting (and subsequent weekend stay at the Chateau) with renewed appreciation for the support provided by Stellenbosch University that enabled the evolution of the SI (and lately the CST) over the past two decades. These institutions provided the space that a group of us needed to evolve a sustainability-oriented learning and research programme to equip students for a world in transition (see Chapter 10) but from the perspective of a society where it is widely accepted that nothing can remain the same even though there is little clarity about what needs to be done (see Chapter 9). The challenges and failures of the South African transition to democracy justified the creation of a space for experimenting in – and explorations of – sustainable living and learning that were socially inclusive, child-centred and reconnected to the soil. This book was inspired by this milieu. But the story it tells has global resonance. Everyone is implicated in the age of sustainability.

The primary intent of this book is to provide a way of seeing the world that is consistent with a way of acting that can bring about the changes that are needed. For that reason, it is the polar opposite of what Timothy Snyder refers to as the "*politics of inevitability*, a sense that the future is just more of the present, that the laws of progress are known, that there are no alternatives, and therefore nothing really to be done" (2018:7). And like a "ghost from a corpse", the "politics of eternity" follows: the belief that the nation is a victim of history that repeats itself and the leader is its only heroic masculinist defender (Snyder, 2018:15). For Snyder, this is the dominant ideology of those bent on retaining the status quo in the world today – from Trump to Putin, Orban to Duterte, Zuma to Bolsonaro. "To accept this", he argues, "is to deny individual responsibility for seeing history and making change. Life becomes a sleepwalk to a premarked grave in a prepurchased plot" (Snyder, 2018:15). There is no age of sustainability if we are all sleepwalking into premarked graves.

For many who are bitterly opposed to this dystopian *politics of inevitability*, critique is their weapon of choice. But trenchant critique that reveals the contradictions and complexities but fails to also suggest 'what is to be done' is just not good enough – it can unwittingly demoralize us and thus unintentionally reinforce the *politics of inevitability*. And activism without critique can often lack coherent directionality; or else it leaves directionality implicit in the hyper-abstract codes that only the initiated understand which makes it difficult to capture the imaginaries of large swathes of society. Directionality needs to be reconciled with complexity: suppressing the latter to re-create state-centric governance will fail, while ignoring the former by believing in the virtues of the market imperils human civilization as we know it.

If there is one word that captures what this book is about, it would be *Ukama* – the Shona word for relatedness. Relatedness or relationality permeates the arguments developed in all the chapters: the thymotics of the relational self (Chapter 1), an integral complex realism that rebalances epistemology and ontology (Chapter 2), collibratory governance of non-equilibrium economies (Chapters 2 and 7), the emergence of commons-based peer production as a lens for understanding actually existing prototypical ecocultures (Chapters 2 and 6), asynchronous transitional dynamics of actually existing complex adaptive systems (Chapter 4), radical incrementalism (Chapter 5), the potential for commons-based energy democracies (Chapter 8) and an evolutionary place-based relational pedagogy of the present (Chapter 10). Chapter 9, of course, captures the antithesis: the ever-present quest for certainty manifested in the authoritarian populist resistance to the dawn of a new epoch where men, oil and coal will no longer dominate the imaginaries of the future. Today's authoritarian populists have mastered the *politics of inevitability*, including how to harness the power of social media to their own advantage.

Without summarizing the chapters, I want to draw out particularly significant conclusions.

My personal journey as an activist academic described in Chapter 1 has led me to the conclusion that many of the binaries we have inherited are unhelpful. By accepting that the age of sustainability has begun, I am consciously negating the assumption that the only transition that matters is the transition from our current 'structure' to a more sustainable one. My approach is, therefore, profoundly Polanyian – it means surfacing the 'double movement' that shapes the age of sustainability. Take the widely held belief that transition and collapse are opposites. For some, this means concluding that because transition is unlikely ('it's too late'), the alternative is to accept that sustainability is nothing more than retarded collapse. What is not recognized is that transition thinking is about understanding both the dynamics of collapse and the dynamics of regeneration, restoration and transformation without having to assume there is a particular revolutionary moment. Transition is, therefore, about contesting the terms of collapse. This is expressed most clearly in the emergence of ecocultures around the world, including my own two decade experience in the Lynedoch EcoVillage, Stellenbosch.

The metatheoretical synthesis in Chapter 2 synthesizes integral theory, complexity theory and critical realism. This, in turn, provides the conceptual foundation for the integration of three crucial building blocks of the age of sustainability:

- because economies tend towards disequilibrium, it must be accepted that long-term directionality will be impossible without state intervention;
- however, under conditions of increasing complexity, it will not be possible to return to a Weberian golden age of state-centric governance – instead, collibratory modes of governance will be required to strategically manage a wide range of partnerships with a shared sense of directionality; and finally
- given the IPCC's insistence that we have 12 years to turn the ship around, the accelerated open source learning within commons-based peer production systems embedded within collibratory institutional configurations will become an urgent necessity – this being the polar opposite of the anti-innovation intellectual property regimes of late capitalism that will prevent the accelerated learning that is needed within the time frames specified by the IPCC (Standing, 2016).

The research output of the International Resource Panel (IRP) discussed in Chapter 3 has made it very clear that the era of industrial modernity will have to come to an end. Resource depletion and rising resource prices cannot be accommodated by prevailing economic policy frameworks. By documenting the resource limits of the industrial era, the IRP's work is suggestive of the socio-metabolic transition that is now required. This is significant because it is demonstrated that sustainability goes way beyond the mainstream focus on climate change. Even if there was significant decarbonization, the planet would still fall to pieces as ecosystems degrade and resources are overexploited.

As argued in Chapter 4, a deep transition needs to be understood as the emergent outcome of the asynchronous interaction between four long-wave transitions: socio-metabolic transitions (material flows), socio-technical transitions (sectoral change, e.g. energy), techno-industrial transitions (major clusters of energy-mobility-communication technologies) and long-term development cycles (growth rates and prices). However, whether or not the directionality of the coming deep transition will be oriented towards a just transition will depend on the outcome of struggles within the polity between a wide range of organized coalitions that represent divergent interests, specifically

- those who want to replicate the status quo (plus some greening on the side);
- those who believe that radical reforms are needed, but lack the capacity to say how this will be achieved; and
- those searching for a more collective commons-oriented alternative that gives concrete expression to the relational self and relational ways of being.

Futuring and experimentation are two approaches for thinking about futures, with special reference to our urban futures. As argued in Chapter 5, futurists are impatient with the present and derive recommendations for present action from the construction of desired futures (imaginaries). Experimenters are frustrated with 'talk-no-action' and therefore focus on opportunities in the present rather than idealized visions of the future. After contrasting city futures in the global North and South, a synthesis is developed that validates the evolutionary potential of the present. The result is a theory of radical incrementalism. However, following the work of Roberto Unger, the transformative power of radical incrementalism is underrated because of "structure fetishism" – the tendency to think that the only significant change is structural change. This way of seeing tends to non-see the transformative potential of local experimentation, while orienting strategic action exclusively towards the capture of state power (via elections, mass revolutions, insurrection, coups, etc.). This obliterates the rhizomatic dynamics of the Polanyian 'double movement'. Radical incrementalism brings the 'double movement' alive.

Building on the general argument for radical incrementalism developed in Chapter 5, 27 cases of ecocultural commoning from the global South are discussed in Chapter 6. By embodying futures in current ecocultural experiments, radical incrementalists are able to demonstrate in practice what more sustainable and equitable futures could look like. Because many (but by no means all) share a commitment to both social justice and ecological sustainability, they prefigure in practice what a just transition could be on a larger scale. More importantly, they incubate a generation of people who have seen the future.

Developmental states emerged in the late twentieth century to focus on the acceleration of the economic transition from predominantly agricultural to predominantly industrial economies. Their strategic focus was structural transformation. Sustainability transitions are also about structural transformation, but to date the focus has been on social-ecological and socio-technical transitions in the advanced economies. As argued in Chapter 7, these are two very different conceptions of structural transformation: the former is about modernization and the latter is about transcending the ecological limits of modernization. The adoption of the Sustainable Development Goals (SDGs) makes it necessary to synthesize the literature on these two 'directionalities', with special reference to how developmental states can address the need for development within a resource- and carbon-constrained world. This challenge is most clearly defined in the African context. At the centre of this synthesis lies a particular conception of politics, power and the polity that is best captured in Jessop's notion of 'collibration'. As argued in Chapter 2, the 'governance-of-governance' is potentially more productive than a state-centric approach to building sustainability-oriented developmental states.

Whether the emerging 'deep transition' also turns out to be a 'just transition' will depend to a large extent on whether the transition to renewable energy fosters more inclusive democratic modes of socio-economic development. Chapter 8 argues that the global renewable energy revolution is well underway. The decentralized and distributed nature of renewable energy systems that characterize renewable energy provides a unique opportunity for building a new progressive politics of the energy commons. Energy democracies are publicly and/or socially owned renewable energy systems that enhance human well-being, the autonomy of inclusive local economies and the integrity of nature. Energy democracies, in turn, are the most tangible and immediately realizable manifestations of an emergent just transition inspired in part by a sense of the commons. The more extensive energy democracies become, the greater the chances that the emerging deep transition will have a just transition orientation.

Authoritarian populism is emerging around the world. Leaders of this reactionary movement tend to be climate denialists and misogynists, with a penchant for large-scale fossil fuel–based or nuclear-based energy infrastructures. Using South Africa as a case study, Chapter 9 reveals the political dynamics of resistance to the age of sustainability. Drawing on the new literature on 'petro-masculinity' and the well-established literature on neo-patrimonialism, the South African case seems to confirm a new global trend: neo-patrimonial subversion within a toxic masculinist narrative in order to defend elite accumulation strategies based on increasingly costly fossil fuel and nuclear energy systems. The dystopian outcome is what Snyder refers to as the politics of inevitability and eternity (Snyder, 2018).

Finally, Chapter 10 addresses the challenge of teaching and learning to prepare the next generation for the challenges of the age of sustainability. Over the past two decades, a particular approach to 'learning for sustainability' has emerged at the Sustainability Institute, Stellenbosch University, South Africa. Many of the arguments and themes addressed in the chapters of this book have been incorporated into the curricula of the undergraduate and postgraduate degrees. Transdisciplinary research practices have also evolved at the SI's sister institution, the Centre for Complex

Systems in Transition. This experience sheds light on what is referred to as the 'evolutionary pedagogy of the present' – a particular approach to learning and research that has emerged from practice within the South African context in ways that may resonate across many other contexts.

To conclude, I want to return to Rosie Braidotti' sclarion call: "We need a vision of the subject that is 'worthy of the present'" (Braidotti, 2013:51). The relational self – *Ukama* – that has emerged from Western post-humanism and the much longer traditions in Sub-Saharan African worldviews provides the cornerstone of such a present-worthy vision of the subject. On its own, as argued in Chapter 1, *Ukama* is inadequate: yes, relationality is the starting point but the passions for change that are needed to wake people from the slumber of ignorance and the numbness of bewilderment run even deeper. As suggested in Chapter 1, following the first line of the *Iliad*, when the Goddess sings for the rage of Achilles, the thymotics of agency and collective action are unleashed. A rage that aligns with the feminine principle of fertility, care and inclusive justice is what animates the 'double movement'. On the one hand, we have the masculinist politics of inevitability propagated by those who seek to obliterate all memories of the future and degrade the planetary systems needed to sustain that future; on the other hand, we have the fertile politics of hope which is quintessentially about the evolutionary potential of the present. Quoting from the poem by Nigerian poet Ben Okri called *Heraclitus' Golden River*, it is all about "dancing gracefully with change". (Okri, 2012:Location 1175 Kindle). The relational subject engaged in the graceful thymotic dance with change must surely be the subject that is worthy of the present.

Note

1 https://campus-transition.org/en/home/

References

Braidotti, R. (2013) *The Posthuman*. Cambridge: Polity Press.
Gravata, A., Piza, C., Mayumi, C. and Shimhara, E. (2013) *Volta ao Mundo em 13 Escolas*. Sao Paulo: Fundacao Telefonica.
Okri, B. (2012) *Wild*. London: Random House.
Snyder, T. (2018) *The Road To Unfreedom*. London: Penguin Random House.
Standing, G. (2016) *The Corruption of Capitalism: Why Rentiers Thrive and World Does Not Pay*. London: Biteback Publishing.

INDEX

Note: Page numbers in *italics* indicate figures and page numbers in **bold** indicate tables.

greenhouse gas (GHG) emissions 56, 87, 115, 212–213
greening 120
green-tech revolution 122, 126, 130, 133, 230
green urbanism: Bangalore (India) green design 177, 180; ecocultural assemblages and 171, 177–181; Lagos Bus Rapid Transit (LBRT) 177–180; mobility in 178; property development and 177; public-private partnerships in 177–178; resource efficiencies in 178; social justice and 177–178; Songdo (South Korea) 177–180
Groupe Energies Renouvelables, Environnementet Solidarités (GERES) 176
Gupta family 271, 273, 275
Guy, Jeff 17

Haas, W. 122
Haberl, H. 115
Habitat III summit 152
Hajer, M. 58–59, 86–87, 153, 160
Hall, Colin 280
Hartnady, C. 214
Hartwig, M. 36, 39–40, 45–46, 68, 139
Hartwith, M. 43
Hausknost, D. 122
Hedlund, N. 36, 39–40, 43, 45–46, 68, 139
hegemonic masculinity 269, 278–279
Heinberg, R. 268
Henard, E. 146
Himanen, P. 196
Hivre Bazar (India) 174
Hodson, M. 86
Holstenkamp, L. 257
Hood, R. 146
Howard, E. 146
humans: Anthropocene and 8–9, 19–20, 54, 76; complexity and 45; defining 18–22; dominant values of 22; natural resource exploitation and 101–103; othering 19; relational self 19–22; social systems and 54–56, 77; well-being and 122
Hunter, M. 276, 280
hydrocarbons 239–240, *240*
hypermasculinity 269–270, 284

ICLEI 142
ICT *see* information and communication technologies (ICT)
Iliad 24–25

impredicative systems 150–151, 159, 163n2
incrementalism 19, 169, 190, 228; *see also* radical incrementalism
inequality: authoritarianism and 266; commons-based peer production (CBPP) and 67; developing countries and 238; global economic crisis and 103, 267–268; neoliberalism and 103; resource-efficient development pathways and 102; South Africa and 208, 270–272; urban slums and 140–141
information age 120–121, 129
information and communication technologies (ICT): commons-based peer production (CBPP) and 78, 131, 171, 183, 186; economic growth and 125; energy democracy (ED) and 131; mass communication and 65–66; mutual coordination and 66–67; network organization in 65; peer-to-peer (P2P) 64–66; post-capitalism and 132; renewable energy infrastructures and 229, 255; as techno-economic transition 120; as techno-industrial transition 66
infrastructures *see* energy infrastructures; renewable energy infrastructures
innovating through exaptation 309
innovation cycle 14, 112, 119
integral theory 36, 40–41, *41*, **42**
interdisciplinary learning 300–303, 306
Inter-Governmental Agreement (IGA) 273–274
Intergovernmental Panel on Climate Change (IPCC) 75–76, 78–79, 142
Intergovernmental Science-Policy Platform on Biodiversity and Ecosystem Services (IPBES) 78, 142
internal environment 50
International Energy Agency 213–214
International Monetary Fund (IMF) 268
International Resource Panel (IRP): anticipatory science and 78, 103; Cities Working Group 85; decarbonization and 75; decoupling and 79–87, 99, 102, 115; deep transitions and 75, 116; establishment of 76; global resource perspectives of 79–84, 100–102; governance and 79, 99–100, 102; nexus themes of 85–88, 90–92, 101; objectives of 76; on resources 115; scenario-building by 142; socio-metabolic transitions and 78, 83, 100, 115–116; specific resource challenges and 79, 92–99